S0-CAH-398

ELECTRON, SPIN AND MOMENTUM DENSITIES AND CHEMICAL REACTIVITY

Understanding Chemical Reactivity

Volume 21

The titles published in this series are listed at the end of this volume.

Electron, Spin and Momentum Densities and Chemical Reactivity

edited by

Paul G. Mezey
University of Saskatchewan, Saskatoon, Canada

and

Beverly E. Robertson
University of Regina, Regina, Canada

KLUWER ACADEMIC PUBLISHERS
DORDRECHT / BOSTON / LONDON

Library of Congress Cataloging-in-Publication Data is available.

ISBN 0-7923-6085-0

Published by Kluwer Academic Publishers,
P.O. Box 17, 3300 AA Dordrecht, The Netherlands.

Sold and distributed in North, Central and South America
by Kluwer Academic Publishers,
101 Philip Drive, Norwell, MA 02061, U.S.A.

In all other countries, sold and distributed
by Kluwer Academic Publishers, P.O. Box 322,
3300 AH Dordrecht, The Netherlands

Printed on acid-free paper

Printed and bound in Great Britain by Antony Rowe Limited.

Contents

Editorial foreword

The electron density of a non-degenerate ground state system determines essentially all physical properties of the system. This statement of the Hohenberg–Kohn theorem of Density Functional Theory plays an exceptionally important role among all the fundamental relations of Molecular Physics.

In particular, the electron density distribution and the dynamic properties of this density determine both the local and global reactivities of molecules. High resolution experimental electron densities are increasingly becoming available for more and more molecules, including macromolecules such as proteins. Furthermore, many of the early difficulties with the determination of electron densities in the vicinity of light nuclei have been overcome.

These electron densities provide detailed information that gives important insight into the fundamentals of molecular structure and a better understanding of chemical reactions. The results of electron density analysis are used in a variety of applied fields, such as pharmaceutical drug discovery and biotechnology.

If the functional form of a molecular electron density is known, then various molecular properties affecting reactivity can be determined by quantum chemical computational techniques or alternative approximate methods.

Spin densities determine many properties of radical species, and have an important effect on the chemical reactivity within the family of the most reactive substances containing free radicals. Momentum densities represent an alternative description of a microscopic many-particle system with emphasis placed on aspects different from those in the more conventional position space particle density model. In particular, momentum densities provide a description of molecules that, in some sense, turns the usual position space electron density model 'inside out', by reversing the relative emphasis of the peripheral and core regions of atomic neighborhoods.

This book contains a selection of chapter topics based on papers given at the 12th conference of the Commission on Charge, Spin and Momentum Density of the International Union for Crystallography, held in Waskiesiu, Prince Albert National Park, SK, Canada, July 27–August 1, 1997. The choice of topics represents some of the latest advances in the field of electron, spin, and momemtum densities and the analysis of these densities with respect to their roles in determining chemical reactivity.

It is the hope of the editors that this book will provide our readers with an exciting collection of accounts of the latest advances, and also will provide further motivation for new research to address some of the challenging, unsolved problems of the fascinating interrelations between electron, spin, and momemtum densities, and the complex subject of chemical reactivity.

Paul G. Mezey and Beverly Robertson

1

Maximum Entropy charge density studies: Bayesian viewpoint and test applications

PIETRO ROVERSI[1], JOHN J. IRWIN[1] and GÉRARD BRICOGNE[1,2]
[1]*MRC Laboratory of Molecular Biology, Hills Road, Cambridge CB2 2QH, England*
[2]*LURE, Bâtiment 209D, 91405 Orsay, France*

1. Introduction

The Maximum Entropy (abbreviated MaxEnt) method has been used in the field of accurate charge density studies for some time now (see Section 2.2): it has the potential to overcome some of the limitations of traditional multipolar modelling, but great care must be taken not to apply it outside the range of validity of its own foundations.

In this paper, after a brief discussion of the main sources of error affecting the present day implementation of multipolar and MaxEnt charge density studies (Sections 1.1 and 2.2), we present a rationale for the well-known drawbacks of the MaxEnt method as applied to charge density studies. In particular, we will show that the use of a uniform prior-prejudice distribution gives rise to artefacts when the dynamic range of the electron density to be reconstructed is large enough that the exponential modelling of the density requires non-negligible Lagrange multipliers past the resolution limit of the available diffraction data. The artefacts are not due to insufficient numerical precision, but to series termination effects in the Fourier series with Lagrange multipliers as coefficients.

In the last section of the paper, we discuss a Bayesian approach to the treatment of experimental error variances, and its first limited implementation to obtain MaxEnt distributions from a fit to noisy data.

1.1. Model bias in multipolar charge density studies

The main sources of error in charge density studies based on high-resolution X-ray diffraction data are of an experimental nature; when special care is taken to minimise them, charge density studies can achieve an accuracy better than 1% in the values of the structure factor amplitudes of the simplest structures [1, 2]. The errors for small molecular crystals, although more difficult to assess, are reckoned to be of the same order of magnitude.

The challenge is then to achieve the same degree of accuracy in the derived values of the experimental electron density. Recent studies have shown that in some cases this is indeed within the reach of the present-day modelling techniques [3–5]. When the major sources of experimental error have been corrected for the typical root mean square electron density residual can reach values as low as $0.05\,\mathrm{e\,Å^{-3}}$, with maxima below $0.20\,\mathrm{e\,Å^{-3}}$ in absolute value. The observed residuals are usually due to the

Paul G. Mezey and Beverly E. Robertson (eds.), Electron, Spin and Momentum Densities and Chemical Reactivity, 1–26

errors in the experimental data, but high-resolution, high-quality data sets can in some cases bring to light inadequacies of the model.

In practice, the choice of parameters to be refined in the structural models requires a delicate balance between the risk of overfitting and the imposition of unnecessary bias from a rigidly constrained model. When the amount of experimental data is limited, and the model too flexible, high correlations between parameters arise during the least-squares fit, as is often the case with monopole populations and atomic displacement parameters [6], or with exponents for the various radial deformation functions [7].

A main source of model bias lies in the choice of exponents in the single-exponential-type functions $r^n \exp(-\alpha r)$ that are commonly used as the radial parts of the deformation functions: this choice is often 'more of an art than a science' [4]. Very little is known about the optimal values to be used for elements other than those of the first two rows. Selection of the best value for the exponents n is usually carried out by systematically varying exponents and monitoring the effects on the R indices and/or residual densities [8, 9]. The procedure can in some cases be unsatisfactory, as is the case when very diffuse functions centred on one atom are used to model most of the density in the bond, and even some of the density on neighbouring atoms [10].

Extra radial flexibility has been proved necessary in order to model the valence charge density of metal atoms, in minerals [6, 11], and coordination complexes [5], and similar evidence of the inability of single-exponential deformation functions to account for all the information present in the observations have also been found in studies of organic [12, 13] and inorganic [14] molecular crystals.

When atoms occupy highly symmetrical sites, a further limitation of the current multipolar expansions is the limited order of the spherical harmonics employed, that do not usually extend past the hexadecapolar level ($l = 4$). Only two multipolar studies published to date used spherical harmonics to orders higher than $l = 4$: graphite [15] and crystalline beryllium [16]. In the latter work, the most significant contribution to the valence density was indeed shown to be given by a pole of order $l = 6$.

2. MaxEnt charge density studies

Because of the limitation intrinsic to the adoption of an explicit parametrised density model, many crystallographers have been dreaming of disposing of such models altogether. The thermally-smeared charge density in the crystal can of course be obtained without an explicit density model, by Fourier summation of the (phased) structure factor amplitudes, but the resulting map is affected by the experimental noise, and by all 'series-termination' artefacts that are intrinsic to Fourier synthesis of an incomplete, finite-resolution set of coefficients.

A second approach which is not subject to the limitations imposed by the choice of a parametrised model of the density, is the MaxEnt method. The appeal of the method is evident when counting the increasing number of applications to charge density studies that have appeared in the crystallographic literature in the last ten years: see among the most recent ones [17–20], and the works cited in relevant sections of reviews on charge density studies [21] and on MaxEnt methods in crystallography [22]. In principle, MaxEnt maps are not tied to any particular multipolar expansion, or radial deformation function, and can mirror any degree of angular and radial deformation that is present in the observations.

2.1. *The random scatterer model*

All of the studies published so far have been aiming at the reconstruction of the total electron density in the crystal by redistribution of all electrons, under the constraints imposed by the MaxEnt requirement and the experimental data. After the acceptance of this paper, the authors became aware of valence-only MaxEnt reconstructions contained in the doctoral thesis of Garry Smith [58]. The authors usually invoke the MaxEnt principle of Jaynes [23–26], although the underlying connection with the structural model, known under the name of *random scatterer model*, is seldom explicitly mentioned.

According to the latter model, the crystal is described as formed of a number of equal scatterers, all randomly, identically and independently distributed. This simplified picture and the interpretation of the electron density as a probability distribution to generate a statistical ensemble of structures lead to the selection of the map having maximum relative entropy with respect to some *prior-prejudice* distribution $m(\mathbf{x})$ [27, 28].

When it is employed to specify an ensemble of random structures, in the sense mentioned above, the MaxEnt distribution of scatterers is the one which rules out the smallest number of structures, while at the same time reproducing the experimental observations for the structure factor amplitudes as expectation values over the ensemble. Thus, provided that the random scatterer model is adequate, deviations from the prior prejudice (see below) are enforced by the fit to the experimental data, while the MaxEnt principle ensures that no unwarranted detail is introduced.

2.2. *A look at the MaxEnt charge density literature*

Since 1993, a number of studies have been devoted to assessing the limitations of the MaxEnt method when applied to charge density studies, especially in conjunction with uniform prior-prejudice distributions. We summarise here the main points that have arisen from these model studies.

Uneven distributions of residuals. The MaxEnt calculations in presence of an overall chi-square constraint suffer from highly non-uniform distributions of residuals, first reported and discussed by Jauch and Palmer [29, 30]; the error accumulates on a few strong reflexions at low-resolution. The phenomenon is only partially cured by devising an *ad hoc* weighting scheme [20, 31, 32]. Carvalho *et al.* have discussed this topic, and suggested that the recourse to as many constraints as degrees of freedom would cure the problem [33].

Dynamic range of the density and low-density regions in the crystal. In their work cited above, Jauch and Palmer first pointed out the inadequacies of the method in dealing with densities having a large dynamic range. Additional evidence of these inadequacies has come from Papoular *et al.*, who worked on observed and simulated data sets for α-glycine [18]. In the latter study, when all electrons were redistributed with a single-channel approach, the density of the hydrogen atoms was clearly flattened, and features below $2\,e\,\text{Å}^{-3}$ were in general deemed to be scarcely significant, because the large dynamic range of the total density reduced the sensitivity level. A two-channel calculation,[1] fitting structure factors calculated from the deformation density, did not suffer from the same limitations due to the reduced dynamic range of the density to be reconstructed.

Errors in the low-density regions of the crystal were also found in a MaxEnt study on noise-free amplitudes for crystalline silicon by de Vries *et al.* [37]. Data were fitted exactly, by imposing an esd of 5×10^{-4} to the synthetic structure factor amplitudes. The authors demonstrated that artificial detail was created at the mid-point between the silicon atoms when all the electrons were redistributed with a uniform prior prejudice; extension of the resolution from the experimental limit of 0.479 to 0.294 Å could decrease the amount of spurious detail, but did not reproduce the value of the forbidden reflexion $F(222)$, that had been left out of the data set fitted.

Dependence of results from the prior-prejudice distribution. Non-uniform prior-prejudice distributions (NUP for short in what follows) were initially introduced by Jauch and Palmer by centering 3D Gaussian functions at the nuclear positions [29]. They found that the low-density regions of the crystal changed significantly upon introduction of the NUP, but the uneven distribution of errors persisted.

Iversen *et al.*, in their study of crystalline beryllium [32], were the first to make use of NUP distributions calculated by superposition of thermally-smeared spherical atoms. More recently, a superposition of thermally-smeared spherical atoms was used as NUP in model studies on noise-free structure factor amplitudes for crystalline silicon and beryllium by de Vries *et al.* [38]. The artefacts present in the densities computed with a uniform prior-prejudice distributions have been shown to disappear upon introduction of the NUP. No quantitative measure of the residual errors were given.

Finally, recent work of Iversen *et al.* has carefully examined the bias associated to the accumulation of the error on low-order reflexions, and attempted a correction of the MaxEnt density [39]. The study, based on a number of noisy data sets generated with Monte Carlo simulations, has produced less non-uniform distribution of residuals, and has given quantitative estimate of the bias introduced by the uniform prior prejudice. For more details on this work, we refer the reader to the chapter by Iversen that appears in this same book.

---

[1] Two-channel MaxEnt techniques have also been used in the study of magnetization and spin densities [34, 35] and to interpret unpolarised neutron diffraction data [36].

2.3. *The joint use of MaxEnt distributions and structural models*

None of the studies mentioned in Section 2.2 has explicitly addressed the main issue of the redistribution of core electron densities under MaxEnt requirements in the absence of high-resolution observations. This is indeed the key to explaining the unsatisfactory features encountered so far in the applications of the method to charge density studies.

By its very definition, the MaxEnt method is optimally suited to flexibly reconstruct distributions whose main features are well represented in the available data, that is either in the observations or in the prior structural knowledge. When this is the case, the missing structure can be reasonably approximated by a collection of randomly and independently distributed constituents (by 'missing structure' here we mean all those structural details which are not completely defined by the prior knowledge).

If these structural features are not well represented by a mild redistribution of random independent constituents from an initially given *prior prejudice*, and arise instead from some degree of correlation between the scatterers, they cannot be expected to be satisfactorily dealt with by the method. For these reasons, substructures which scatter well beyond the experimental resolution should be left out of the subset of scatterers distributed at random. The data sets commonly collected for charge density studies do not as a rule extend beyond 0.4 Å resolution, but scattering from the atomic core does extend well beyond this limit.[2]

It is therefore clear that MaxEnt redistribution of all electrons, using a uniform prior prejudice and carried out in the absence of very high-resolution diffraction measurements, cannot be expected to reproduce a physically acceptable picture of atomic cores. The reconstruction of total electron densities from limited-resolution diffraction measurements amounts to a misuse of the MaxEnt method, especially when the prior prejudice is uniform.

Within the multichannel Bayesian formalism of structure determination, it is indeed possible to make use of MaxEnt distributions to model systems whose missing structure can be reasonably depicted as made of random independent scatterers. This requires that the structural information absent in the diffraction data be obtained from some other experimental or theoretical source. The known substructure can be described making use of a parametrised model.

2.4. *The MaxEnt equations and density: a brief reminder*

The general computational mechanism of Bayesian crystal structure determination in presence of various sources of partial phase information was first outlined by

[2]When low-temperature studies are performed, the maximum resolution is imposed by data collection geometry and fall-off of the scattered intensities below the noise level, rather than by negligible high-resolution structure factor amplitudes. Use of Ag $K\alpha$ radiation would for example allow measurement of diffracted intensities up to 0.35 Å for amino-acid crystals below 30 K [40]. Similarly, model calculations show that noise-free structure factors computed from atomic core electrons would be still non-zero up to 0.1 Å.

Bricogne [41]; a status report, now somewhat dated, about its actual implementation for a number of crystallographic problems was given by the same author in [42].

In this section, we briefly recall the MaxEnt equations and the functional form of the MaxEnt probability distribution; the formulation is the one obtainable for randomly and independently distributed electrons, in the presence of a subset of electrons whose distribution is assumed to be known. The latter structure will be denoted as 'fragment'.

Let us consider a collection $H = (\mathbf{h}_1, \mathbf{h}_2, \ldots, \mathbf{h}_M)$ of symmetry-unique reflexions. We denote by $\mathbf{F}^*_{\mathbf{h}_j}$ the 'target' phased structure factor amplitude for reflexion $\mathbf{h}_j$, and with $\mathbf{F}^{\text{frag}}_{\mathbf{h}_j}$ the contribution from the known substructure to the structure factor for the same reflexion. We are interested in a distribution of electrons $q(\mathbf{x})$ that reproduces these phased amplitudes, in the sense that, for each structure factor in the set of observations H,

$$\mathbf{F}^*_{\mathbf{h}_j} = \mathbf{F}^{\text{frag}}_{\mathbf{h}_j} + \mathbf{F}^{\text{rand}}_{\mathbf{h}_j}, \tag{1}$$

where the contribution $\mathbf{F}^{\text{rand}}_{\mathbf{h}_j}$ of the random scatterers is related to $q(\mathbf{x})$ by

$$\mathbf{F}^{\text{rand}}_{\mathbf{h}_j} = nf|G| \int_V q(\mathbf{x}) e^{2\pi i \mathbf{h}_j \cdot \mathbf{x}} \, d^3\mathbf{x}. \tag{2}$$

In this expression, $|G|$ is the number of elements of the space group of the crystal, and f and n are the scattering power and number of the point random scatterers in the asymmetric unit, respectively.

Since all the scatterers are identical, their structure factors can be normalised to unitary structure factors, as is always the case for homogeneous structures of normal scatterers [41]:

$$\mathbf{U}^{\text{rand}}_{\mathbf{h}_j} = \mathbf{F}^{\text{rand}}_{\mathbf{h}_j} / (nf|G|) = \left(\mathbf{F}^*_{\mathbf{h}_j} - \mathbf{F}^{\text{frag}}_{\mathbf{h}_j}\right) \Big/ (nf|G|). \tag{3}$$

Now we make use of the invariance of $q(\mathbf{x})$ under symmetry operations of space group G:

$$q(\mathbf{x}) = (1/|G|) \sum_{g \in G} q\left(\mathbf{R}_g \mathbf{x} + \mathbf{t}_g\right) \tag{4}$$

and of the group structure of G, to rewrite Equation (2) as

$$\mathbf{U}^{\text{rand}}_{\mathbf{h}_j} = \int_V q(\mathbf{x}) \left\{ (1/|G|) \sum_{g \in G} \exp\left[2\pi i \mathbf{h}_j \cdot \left(\mathbf{R}_g \mathbf{x} + \mathbf{t}_g\right)\right] \right\} d^3\mathbf{x}. \tag{5}$$

The quantity in curly brackets in Equation (5) is called the *constraint function* $C_j(\mathbf{x})$.

To deal with all the observations $\mathbf{h}_j \in H$ in compact form, the unitary structure factor components can be arranged in a vector $\mathbf{U}^{\text{rand}}$, and the components of the constraint functions collected in a vector $\mathbf{C}(\mathbf{x})$. The MaxEnt distribution of electrons $q^{\text{ME}}(\mathbf{x})$ then takes the form

$$q^{\text{ME}}(\mathbf{x}; \boldsymbol{\lambda}^*) = \left[m(\mathbf{x})/Z(\boldsymbol{\lambda}^*)\right] \exp\left[\boldsymbol{\lambda}^* \cdot \mathbf{C}(\mathbf{x})\right], \tag{6}$$

where $Z(\boldsymbol{\lambda})$ is a normalising factor for $q(\mathbf{x})$,

$$Z(\boldsymbol{\lambda}) = \int_V m(\mathbf{x}) \exp\left[\boldsymbol{\lambda} \cdot \mathbf{C}(\mathbf{x})\right] \, \mathrm{d}^3\mathbf{x} \tag{7}$$

and the saddle point $\boldsymbol{\lambda} = \boldsymbol{\lambda}^*$ is computed by solving the MaxEnt equations

$$\nabla_\lambda \left(\log Z(\boldsymbol{\lambda})\right) = \mathbf{U}^{\mathrm{rand}}. \tag{8}$$

The name of the distribution is due to the fact that the saddle point $\boldsymbol{\lambda}^*$ can also be obtained as the vector of Lagrange multipliers needed to find the distribution $q = q^{\mathrm{ME}}$ for which the relative entropy,

$$\mathcal{S}_m(q) = -\int_V q(\mathbf{x}) \log\left[q(\mathbf{x})/m(\mathbf{x})\right] \mathrm{d}^3\mathbf{x} \tag{9}$$

is at a maximum [27].

2.5. *MaxEnt deformation density maps*

Most of the relevant features of the charge density distribution can be elegantly elucidated by means of the topological analysis of the total electron density [43]; nevertheless, electron density deformation maps are still a very effective tool in charge density studies. This is especially true for all densities that are not specified via a multipole model and whose topological analysis has to be performed from numerical values on a grid.

Conventional implementations of MaxEnt method for charge density studies do not allow easy access to deformation maps; a possible approach involves running a MaxEnt calculation on a set of data computed from a superposition of spherical atoms, and subtracting this map from q^{ME} [44]. Recourse to a two-channel formalism, that redistributes 'positive-' and 'negative-density' scatterers, fitting a set of difference Fourier coefficients, has also been made [18], but there is no consensus on what the definition of entropy should be in a two-channel situation [18, 36, 41]; moreover, the shapes and number of positive and negative scatterers may need to differ in a way which is difficult to specify.

Thanks to the particular choice made for the NUP, taken equal to a superposition of spherical atoms, it is for the first time possible within the present approach to compute MaxEnt deformation maps in a straightforward manner. Once the Lagrange multipliers $\boldsymbol{\lambda}$ have been obtained, the deformation density is simply

$$\Delta q^{\mathrm{ME}}(\mathbf{x}) = m(\mathbf{x}) \left(\frac{\exp\left[\boldsymbol{\lambda}^* \cdot \mathbf{C}(\mathbf{x})\right]}{Z(\boldsymbol{\lambda}^*)} - 1 \right). \tag{10}$$

This map can have negative as well as positive features, and yet its calculation involves only that of the positive map q^{ME}, thus avoiding the issue of extending the MaxEnt method to two-channel problems.

3. The role of the prior-prejudice distribution

It appears from formula (6) that the prior-prejudice distribution $m(\mathbf{x})$ is a fundamental quantity in the calculation of the MaxEnt distribution of electrons, in that the latter is obtained by *modulation* of $m(\mathbf{x})$. In all those regions where the modulating factor required to fit the observations is unity, the final picture is therefore always going to coincide with the prior expectation itself. For this reason, it is of the greatest importance that some of the prior information available about the system under study be conveyed into the calculation by means of a sensible choice for the prior-prejudice distribution.

This is especially true when the observations are not informative enough, as is the case for total charge density reconstruction based on finite resolution X-ray diffraction data. Even when valence electrons only are redistributed at random, the shell structure of the atomic densities might still require high-order components that are past the experimental resolution [2]. The choice of the uniform prior-prejudice distribution amounts to ignoring the presence of atoms in the crystal, so that its property of being 'maximally non-committal' is no longer a virtue but a vice: it is in fact *too* non-committal.

Not only is the choice of a uniform prior-prejudice distribution not sensible; it also exposes the calculation to two main sources of computational errors, both connected with the functional form of the MaxEnt distribution of scatterers, and with its numerical evaluation: namely series termination ripples and aliasing errors in the numerical sampling of the exponential modulation of $m(\mathbf{x})$. The next two paragraphs will illustrate these issues in some detail.

3.1. The spectrum of the exponential modulation of $m(\mathbf{x})$

As already pointed out by Jauch [30], the series appearing in the exponential factor that modulates $m(\mathbf{x})$ in (6) has a finite number of terms, and can therefore give rise to series termination artefacts. In particular, although the exponentiation will ensure positivity of the resulting density, series termination ripples will be present in the reconstructed map whenever the spectrum of the modulation required by the observations extends significantly past the resolution of the series appearing in the exponential. This in turn will depend both on the 'true' density whose Fourier coefficients are being fitted, and on the choice for the prior prejudice.

The phenomenon can be illustrated by considering a model density $q(\mathbf{x})$, from which diffraction data can be computed at arbitrarily high resolution. The (normalised) exponential factor needed to reconstruct $q(\mathbf{x})$ by MaxEnt modulation of a chosen prior-prejudice distribution $m(\mathbf{x})$ can be written as

$$\frac{1}{Z(\boldsymbol{\lambda})} \exp\left[\boldsymbol{\lambda} \cdot \mathbf{C}(\mathbf{x})\right] = \frac{q(\mathbf{x})}{m(\mathbf{x})}. \tag{11}$$

The series in the exponential is called ω: $\omega(\mathbf{x}) = \boldsymbol{\lambda} \cdot \mathbf{C}(\mathbf{x})$.

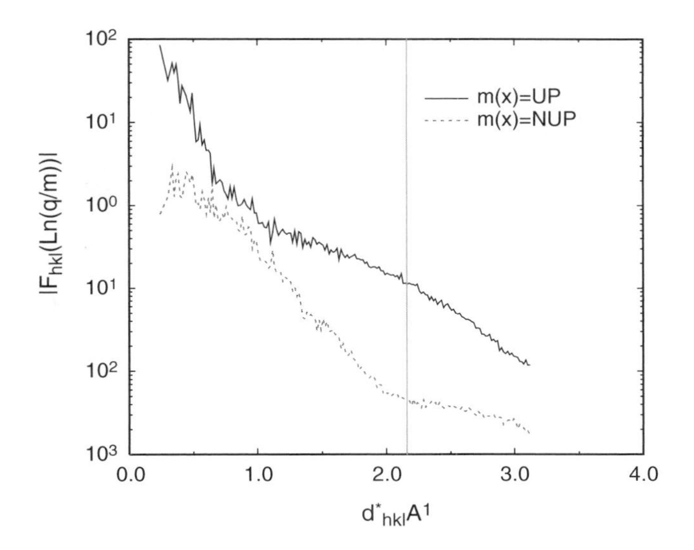

Figure 1. Amplitudes of the Fourier coefficients of $\log(q\mathbf{x})/m(\mathbf{x}))$ in resolution bins for L-alanine at 23 K. $q(\mathbf{x})$: total model density, from a multipolar fit to 23 K diffraction data protect [45]. Continuous line: $m(\mathbf{x})$ = uniform distribution. Dotted line: $m(\mathbf{x})$ = core and valence monopoles. The vertical bar marks the experimental resolution limit 0.463 Å.

Fourier analysis of the logarithm of the ratio $q(\mathbf{x})/m(\mathbf{x})$ can now inform us about the extent to which the finite resolution of the observations fitted is likely to affect the MaxEnt reconstruction, depending on the choice for the prior prejudice. The better guess $m(\mathbf{x})$ is, the smaller the amplitudes of the Lagrange multipliers will be. Finite-resolution effects will be negligible when the use of a good NUP keeps the magnitude of the Lagrange multipliers to a minimum.

Figure 1 shows the average strength of the Fourier coefficients of $\log(q(\mathbf{x})/m(\mathbf{x}))$, with $q(\mathbf{x})$ a multipolar synthetic density for L-alanine at 23 K, and two different prior-prejudice distributions $m(\mathbf{x})$. It is apparent that the exponential needed to modulate the uniform prior still has Fourier coefficients larger than 0.01 past the experimental resolution limit of 0.463 Å. Any attempt at fitting the corresponding experimental structure factor amplitudes by modulation of the *uniform* prior-prejudice distribution will therefore create series termination ripples in the resulting MaxEnt distribution.

The exact amount of error introduced cannot immediately be inferred from the strength of the amplitudes of the neglected Fourier coefficients, because errors will pile up in different points in the crystal depending on the structure factors phases as well; to investigate the errors, a direct comparison can be made in real space between the MaxEnt map, and a map computed from exponentiation of a resolution-truncated 'perfect' ω-map, whose Fourier coefficients are known up to any order by analysing $\log(q(\mathbf{x})/m(\mathbf{x}))$.

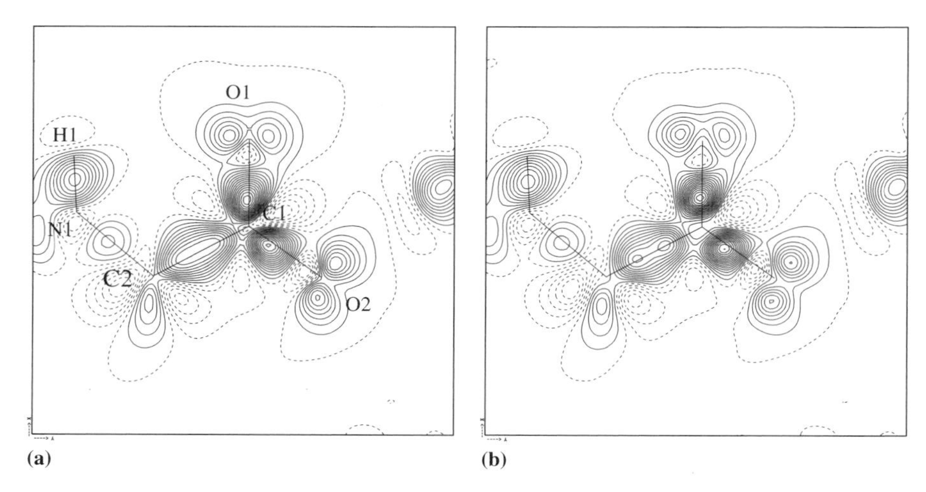

Figure 2. L-alanine. Dynamic deformation density in the COO^- plane. (a) Model dynamic deformation density $\Delta q^{\mathrm{Model}}$. (b) MaxEnt dynamic deformation density $\left(\Delta q^{\mathrm{ME}}_{\mathrm{NUP}}(\mathbf{x})\right)$ map obtained with a non-uniform prior of spherical-valence shells. Map size: 6.0 Å × 6.0 Å. Contour levels: from −1.0 to 1.0 e Å^{-3}, step 0.075 e Å^{-3}.

In particular, if the ω-map suffers from an error $\Delta\omega$ due to its finite resolution

$$\omega^{\mathrm{True}}(\mathbf{x}) = \boldsymbol{\lambda} \cdot \mathbf{C}(\mathbf{x}) + \Delta\omega(\mathbf{x}), \tag{12}$$

$$q^{\mathrm{True}}(\mathbf{x}) = q(\mathbf{x}) + \Delta q(\mathbf{x}) = \frac{m(\mathbf{x})}{Z(\boldsymbol{\lambda})} \exp\left[\omega(\mathbf{x}) + \Delta\omega(\mathbf{x})\right], \tag{13}$$

the error in the final MaxEnt map will be proportional to the density itself:

$$\Delta q(\mathbf{x}) = q(\mathbf{x}) \left[\exp\left(\Delta\omega(\mathbf{x})\right) - 1\right] \approx q(\mathbf{x})\Delta\omega(\mathbf{x}). \tag{14}$$

Errors are therefore enhanced in high-density regions.

3.1.1. L-Ala MaxEnt valence density from noise-free data

To check this prediction, a number of MaxEnt charge density calculations have been performed with the computer program *BUSTER* [42] on a set of synthetic structure factors, obtained from a reference model density for a crystal of L-alanine at 23 K. The set of 1500 synthetic structure factors, complete up to a resolution of 0.555 Å [45], was calculated from a multipolar expansion of the density, with the computer program *VALRAY* [46].

The MaxEnt valence density for L-alanine has been calculated targeting the model structure factor phases as well as the amplitudes (the space group of the structure is acentric, $P2_12_12_1$). The core density has been kept fixed to a superposition of atomic core densities; for those runs which used a NUP distribution $m(\mathbf{x})$, the latter was computed from a superposition of atomic valence-shell monopoles. Both core and valence monopole functions are those of Clementi [47], localised by Stewart [48]; a discussion of the core/valence partitioning of the density, and details about this kind of calculation, may be found elsewhere [49]. The dynamic range of the L-alanine model

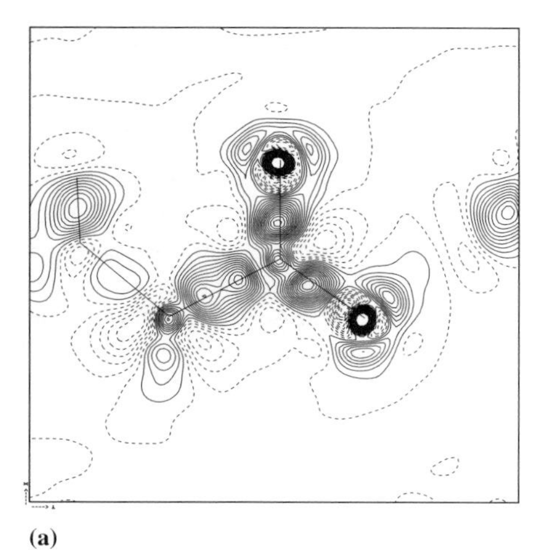

(a)

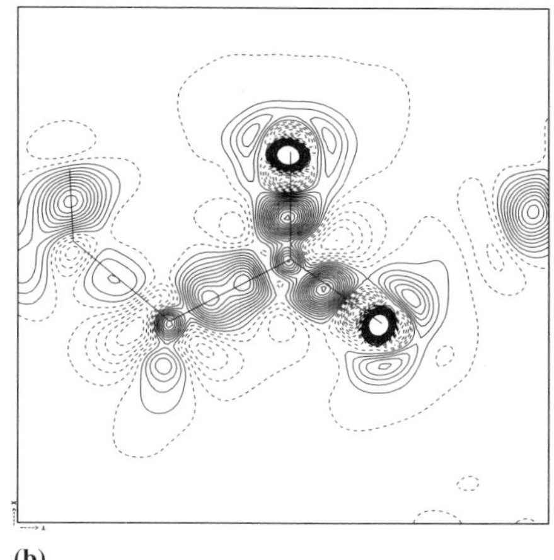

(b)

Figure 3. L-Alanine. Dynamic deformation density in the COO$^-$ plane. (a) $q_{\mathrm{UP}}^{\mathrm{ME}}(\mathbf{x}) - m(\mathbf{x})$. (b) $\exp\left(\omega_{0.555}^{\mathrm{True}}(\mathbf{x})\right) - m(\mathbf{x})$. Map size, orientation and contouring levels as in Figure 2.

valence density at this temperature is $\sim$966; this fairly high value is mainly due to the sharp increase of the valence monopole functions of oxygen atoms at approximately 0.196 Å from the nucleus (see Figure 8).

Uniform prior prejudice. Figure 2(a) shows the model deformation density in the plane of the carboxylate moiety. Figure 3(a) shows the MaxEnt deformation density in the same plane, obtained modulating a uniform prior prejudice for the valence electrons. The valence density is affected by errors up to 22% around the oxygen atoms. Figure 3(b) shows the deformation density computed from exponentiation of the 'perfect' ω-map, truncated at the same resolution used for the MaxEnt calculation. The errors around the oxygen atoms in the two maps of Figure 3 have the same shape; this is a strong indication that these are indeed Fourier-truncation ripples.

We stress here that any low-temperature valence density for a small organic molecule will have a comparably high dynamic range, so that even valence-only MaxEnt studies will always be likely to need a NUP if truncation ripples are to be avoided.

Non-uniform prior prejudice. The dynamic range of the $\exp(\omega)$ map is reduced from 966 to a value of 3.3 when a NUP of spherical valence monopoles is used: as a consequence, the size of the Lagrange multipliers is reduced by between one and two orders of magnitudes, and the error due to series truncation in the ω-map is less than 0.213 e Å^{-3} in absolute value everywhere in the cell, the rms deviation from the model being as low as 0.212 e Å^{-3} (Figure 2(b)).[3]

[3]The value of the rms deviation from the reference density can be deceptively low, due to the fact that in the intermolecular regions the model density is virtually the same as the one made of spherical-valence shells, which was used as a NUP. The agreement between the MaxEnt map and the reference model is very close in those regions.

3.2. Numerical sampling of the exponential modulation of $m(\mathbf{x})$

A second major source of computational difficulties associated with uniform prior-prejudice distributions is connected with the extremely fine sampling grids that are needed to avoid aliasing effects in the numerical Fourier synthesis of the modulating factor in (8). To predict the dependence of aliasing effects upon the prior prejudice, we need to examine more closely the way the MaxEnt distribution of scatterers is actually synthesised from the values of the Lagrange multipliers $\boldsymbol{\lambda}$.

First, we rewrite the constraint functions appearing in the observational equation (5) by taking explicitly into account the phase of the residual target structure factor:

$$\mathbf{U}_{\mathbf{h}_j}^{\text{rand}} = \left|\mathbf{U}_{\mathbf{h}_j}^{\text{rand}}\right| \times \exp(\mathrm{i}\phi_j)\,. \tag{15}$$

Multiplication of the observational equations (5) by a factor $\exp(-\mathrm{i}\phi_j)$, leads to the modified constraint functions,

$$C'_j(\mathbf{x}) = \frac{1}{|G|}\sum_{g\in G}\exp\left[2\pi\mathrm{i}\mathbf{h}_j\cdot(\mathbf{R}_g\mathbf{x}+\mathbf{t}_g) - \mathrm{i}\phi_j\right]. \tag{16}$$

Taking the real and imaginary parts of the left- and right-hand sides of the newly rewritten observational equation, one obtains

$$\int_V q(\mathbf{x})\mathrm{Re}\,C'_j(\mathbf{x})\,\mathrm{d}^3\mathbf{x} = \left|\mathbf{U}_{\mathbf{h}_j}^{\text{rand}}\right|, \tag{17}$$

$$\int_V q(\mathbf{x})\mathrm{Im}\,C'_j(\mathbf{x})\,\mathrm{d}^3\mathbf{x} = 0. \tag{18}$$

Correspondingly, we introduce symbols for the amplitude κ_j and phase θ_j of each complex Lagrange multiplier λ_j: $\lambda_j = \kappa_j\left(\cos\theta_j + \mathrm{i}\sin\theta_j\right)$.

With this choice of constraint functions and Lagrange multipliers, we can rewrite formula (6) and express the MaxEnt distribution of electrons as

$$q^{\text{ME}}(\mathbf{x}) = \frac{m(\mathbf{x})}{Z(\boldsymbol{\kappa},\boldsymbol{\theta})}\exp\left\{\sum_j \kappa_j\left[\cos(\theta_j)\mathrm{Re}\,C'_j + \sin(\theta_j)\mathrm{Im}\,C'_j\right]\right\}. \tag{19}$$

The sum over symmetry operations in formula (16) can be rewritten by considering the effect of multiplying vector $\mathbf{h}_j$ by the rotation matrices $\mathbf{R}_g$. The collection of distinct reciprocal vectors $\mathbf{h}_j\mathbf{R}_g$ is called the *orbit* of reflexion $\mathbf{h}_j$ [27]; Γ_j is the set of symmetry operations in G whose rotation matrices are needed to generate the orbit of $\mathbf{h}_j$; $|\Gamma_j|$ denotes the number of elements in the same orbit [50].

The real part of the constraint function can be written as

$$\begin{aligned}\operatorname{Re} C'_j &= \frac{1}{|G|}\sum_{g\in G}\cos\left[2\pi \mathbf{h}_j\cdot(\mathbf{R}_g\mathbf{x}+\mathbf{t}_g)-\phi_j\right]\\ &= \frac{1}{|\Gamma_j|}\sum_{\gamma\in\Gamma_j}\cos\left[2\pi \mathbf{h}_j\cdot(\mathbf{R}_\gamma\mathbf{x}+\mathbf{t}_\gamma)-\phi_j\right]\end{aligned} \tag{20}$$

and a similar expansion holds for the imaginary part.

Substitution of (20) in (19) gives

$$\begin{aligned}q^{\mathrm{ME}}(\mathbf{x}) = &\frac{m(\mathbf{x})}{Z(\boldsymbol{\kappa},\boldsymbol{\theta})}\\ &\times \exp\left\{\sum_j^M\sum_{\gamma\in\Gamma_j}\frac{\kappa_j}{|\Gamma_j|}\cos\left[2\pi \mathbf{h}_j\cdot(\mathbf{R}_\gamma\mathbf{x}+\mathbf{t}_\gamma)-\psi_j\right]\right\},\end{aligned} \tag{21}$$

where $\psi_j = \phi_j + \theta_j$. This is the actual formula to compute the MaxEnt distribution, by numerical Fourier synthesis followed by exponentiation. As with all Fourier series, aliasing errors can occur when the Fourier coefficients extend very far into reciprocal space, if the grid upon which the density is sampled is not fine enough [50].

To assess the extent to which the *exponentiated* Fourier series has appreciable Fourier amplitudes, and set the sampling grid accordingly, further development of formula (21) is needed. We first rewrite

$$\begin{aligned}q^{\mathrm{ME}}(\mathbf{x}) = &\frac{m(\mathbf{x})}{Z(\boldsymbol{\kappa},\boldsymbol{\theta})}\\ &\times \prod_{j=1}^M\prod_{\gamma\in\Gamma_j}\exp\left\{\frac{\kappa_j}{|\Gamma_j|}\cos\left[2\pi \mathbf{h}_j\cdot(\mathbf{R}_\gamma\mathbf{x}+\mathbf{t}_\gamma)-\psi_j\right]\right\}.\end{aligned} \tag{22}$$

Expanding each of the exponential factors in a series of modified Bessel functions, the MaxEnt distribution can be written:

$$\begin{aligned}q^{\mathrm{ME}}(\mathbf{x}) = &\frac{m(\mathbf{x})}{Z(\boldsymbol{\kappa},\boldsymbol{\theta})}\prod_{j=1}^M\prod_{\gamma\in\Gamma_j}\left\{I_0\left(\frac{\kappa_j}{|\Gamma_j|}\right)\right.\\ &\left.+2\sum_{n=1}^{\infty}I_n\left(\frac{\kappa_j}{|\Gamma_j|}\right)\cos\left\{n\left[2\pi \mathbf{h}_j.(\mathbf{R}_\gamma\mathbf{x}+\mathbf{t}_\gamma)-\psi_j\right]\right\}\right\}.\end{aligned} \tag{23}$$

When the prior prejudice $m(\mathbf{x})$ is uniform, some of the Lagrange multipliers amplitudes are large (of the order of unity or greater). This is especially the case when sharp details are present in the density to be reconstructed and not in the prior prejudice chosen. For a given argument z, the ratio $I_n(z)/I_0(z)$ remains substantial until n exceeds z (see Figure 4), so that large values of the Lagrange multipliers amplitudes $\boldsymbol{\kappa}$ will give rise to appreciable high-resolution coefficients in the Fourier series in (23).

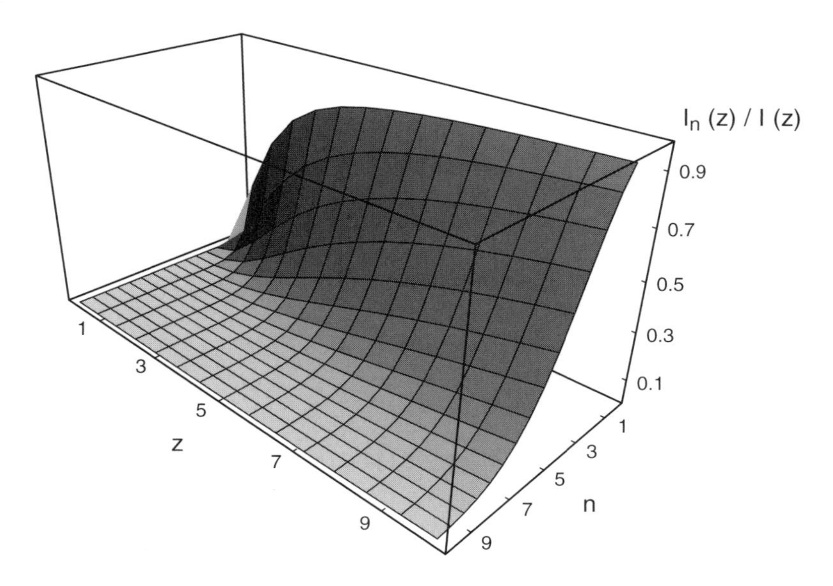

Figure 4. Ratio $I_n(z)/I_0(z)$.

This in turn will require very fine sampling grids along each crystallographic direction, to avoid aliasing effects when the density is synthesised. The size of the arrays needed for the Fourier sampling of $q^{\mathrm{ME}}(\mathbf{x})$ would therefore easily exceed ten million locations for all-electron runs on low-temperature structures. It is clear that MaxEnt distributions of scatterers that contain atomic cores, when obtained by modulation of a uniform prior prejudice, are bound to be spoiled by aliasing effects, unless allowance is made for prohibitively large amounts of memory space.

When the reconstruction of the density is carried out by modulation of a prior prejudice of spherical atoms, only the deformation features have to be accommodated; this can be accomplished relatively easily, and the Lagrange multipliers are usually below 0.01 in modulus, or even smaller for valence-only runs. No aliasing problems occur in the synthesis of $q^{\mathrm{ME}}(\mathbf{x})$.

4. The treatment of the experimental error variances

The calculations discussed in the previous section fit the noise-free amplitudes exactly. When the structure factor amplitudes are noisy, it is necessary to deal with the random error in the observations: we want the probability distribution of random scatterers that is the most probable *a posteriori*, in view of the available observations and of the associated experimental error variances.

In the framework of Bayesian statistics, this can be done by maximising the posterior probability of the Lagrange multipliers defining the distribution [51]; Bayes's theorem gives

$$P^{\text{post}}\big(\boldsymbol{\lambda} \mid |\mathbf{F}|^{\text{obs}}, m(\mathbf{x})\big) = P\big(|\mathbf{F}|^{\text{obs}} \mid \boldsymbol{\lambda}, m(\mathbf{x})\big) P^{\text{prior}}(\boldsymbol{\lambda}, m(\mathbf{x})). \tag{24}$$

In computing the posterior probability, two probability functions are involved:

1. $P^{\text{prior}}(\boldsymbol{\lambda}, m(\mathbf{x}))$: the *a priori* probability is proportional to the exponential of the relative entropy $\mathcal{S}_m$, according to a theorem of Shannon [52]:

   $$P^{\text{prior}}(\boldsymbol{\lambda}, m(\mathbf{x})) \propto \exp\{n\mathcal{S}_m(q(\boldsymbol{\lambda}))\}. \tag{25}$$

   The MaxEnt distribution of scatterers q^{ME}, obtained for $\boldsymbol{\lambda} = \boldsymbol{\lambda}^*$, is also the one that maximises the *a priori* probability in (25):

   $$P^{\text{prior}}(\boldsymbol{\lambda}^*) \simeq \frac{\exp\{n\mathcal{S}_m(q^{\text{ME}}(\boldsymbol{\lambda}^*))\}}{\sqrt{\det(2\pi n \,\text{Hess}(\log Z(\boldsymbol{\lambda}^*)))}}. \tag{26}$$

2. $P(|\mathbf{F}|^{\text{obs}} \mid \boldsymbol{\lambda}, m(\mathbf{x}))$: the conditional probability of the measurements, given a certain set of Lagrange multipliers and a prior-prejudice distribution $m(\mathbf{x})$, can be computed from the likelihood gain Λ_m of the same Lagrange multipliers, given the observed data and the same prior:

   $$P(|\mathbf{F}|^{\text{obs}} \mid \boldsymbol{\lambda}, m(\mathbf{x})) = \Lambda_m(\boldsymbol{\lambda} \mid m(\mathbf{x}), |\mathbf{F}|^{\text{obs}}). \tag{27}$$

   Likelihood has been long proven the optimal criterion to judge whether hypotheses (in this case the values of the Lagrange multipliers) are corroborated by the observations. The recourse to a likelihood gain Λ_m with respect to the prior prejudice $m(\mathbf{x})$ simply reflects the need for a reference point in evaluating the likelihood; the reference chosen is the prior-prejudice distribution $m(\mathbf{x})$, the particular distribution for which all Lagrange multipliers are zero: $q(\mathbf{x}; \boldsymbol{\lambda} = \mathbf{0}) = m(\mathbf{x})$.

   Under the simplifying assumption that the reflexions are independent of each other, Λ_m can be written as a product over reflexions for which experimental structure factor amplitudes are available. For each of the reflexions, the likelihood gain takes different functional forms, depending on the centric or acentric character, and on the assumptions made for the phase probability distribution used in integrating over the phase circle: for a discussion of the crystallographic likelihood functions we refer the reader to the description recently appeared in [51].

Both the *a priori* and the likelihood functions contain exponentials, so that it is convenient to consider the logarithm of the posterior probability, and maximise the *Bayesian score*:

$$B(\boldsymbol{\lambda}) = n\mathcal{S}_m(\boldsymbol{\lambda}) - \tfrac{1}{2}\log\left[\det(2\pi\ \text{Hess}(\log\ Z(\boldsymbol{\lambda})))\right] + \log\ \Lambda_m(\boldsymbol{\lambda}) \tag{28}$$

under the constraint of MaxEnt. $\mathcal{L}_m = \log \Lambda_m$ is called the *log-likelihood gain*. The algorithm implemented to perform this constrained maximisation is an adaptation of the one employed to minimise a χ^2 residual subject to MaxEnt constraints, and described in [27].

4.1. *Likelihood with experimental errors present*

In this section we briefly discuss an approximate formalism that allows incorporation of the experimental error variances in the constrained maximisation of the Bayesian score. The problem addressed here is the derivation of a likelihood function that not only gives the distribution of a structure factor amplitude as computed from the current structural model, but also takes into account the variance due to the experimental error.

Let us assume an experimentally derived distribution $P(R)$ for the amplitude $R = |\mathbf{F}|$ of a reflexion, normalised so as to have: $\int_0^\infty P(R)\,\mathrm{d}R = 1$. The $P(R)$ distribution will be typically Gaussian around the measured $R^{\mathrm{obs}} = |\mathbf{F}|^{\mathrm{obs}}$ with associated variance σ^2. $P(R)$ may take a more involved functional form if the Gaussian has a substantial tail in regions of negative R^{obs}.

The 'error-free' likelihood gain $\Lambda_0(R;\,\Sigma_2)$ gives the probability distribution for the structure factor amplitude as calculated from the random scatterer model (and from the model error estimates for any known substructure). To collect values of the likelihood gain from all values of R around R^{obs}, Λ_0 is weighted with $P(R)$:

$$\Lambda(R^{\mathrm{obs}};\,\sigma^2,\,\Sigma_2) = \int_0^\infty P(R;\,R^{\mathrm{obs}},\,\sigma^2)\Lambda_0(R;\,\Sigma_2)\,\mathrm{d}R. \tag{29}$$

Under general hypotheses, the optimisation of the Bayesian score under the constraints of MaxEnt will require numerical integration of (29), in that no analytical solution exists for the integral. A Taylor expansion of $\Lambda_0(R)$ around the maximum of the $P(R)$ function could be used to compute an analytical expression for Λ and its first and second order derivatives, provided the spread of the Λ_0 distribution is significantly larger than the one of the $P(R)$ function, as measured by σ^2. Unfortunately, for accurate charge density studies this requirement is not always fulfilled: for many reflexions the structure factor variance Σ_2 appearing in Λ_0 is comparable to or even smaller than the experimental error variance σ^2, because the deformation effects and the associated uncertainty are at the level of the noise.

We have for now implemented a drastic simplification, whereby the likelihood function is taken equal to the error-free likelihood, but to the variance parameter Σ_2 appearing in the latter function the experimental error variance is added:

$$\Lambda(R_{\mathrm{obs}}) = \Lambda_0(R_{\mathrm{obs}};\,\Sigma_2'), \qquad \Sigma_2' = \Sigma_2 + \sigma^2 \tag{30}$$

This approximation has already proven very effective in the calculation of likelihood functions for maximum likelihood refinement of parameters of the heavy-atom model, when phasing macromolecular structure factor amplitudes with the computer program *SHARP* [53]. A similar approach was also used in computing the variances to be used in evaluation of a χ^2 criterion in [54].

4.1.1. Effective number of scatterers and variance rescaling
At this stage, two points are worth mentioning:

1. The number of random scatterers appearing in the expression for the Bayesian score does not necessarily correspond to the nominal number of electrons in the system under investigation: the random scatterers bear no physical identity! And yet, the value of n is a key quantity in the optimisation of the Bayesian score, in that it determines the relative weight of the log-likelihood and entropy terms in driving the structure determination process: for example, values of n that are too low will allow a tighter fit of the observations, because of a less stringent entropy requirement, but at the cost of fitting some of the experimental noise.
2. At each stage during the structure determination process, the current structural model gives an estimate of the prediction variance Σ_2 to be associated with the calculated amplitude. The contribution of the random part of the structure to this prediction variance decreases while the structure determination proceeds, and uncertainty is removed by the fit to the observations. Rescaling of Σ_2 would be needed during the optimisation of the Bayesian score.

Both the determination of the effective number of scatterers and the associated rescaling of variances are still in progress within *BUSTER*. The value of n at the moment is fixed by the user at input preparation time; for charge density studies, variances are also kept fixed and set equal to the observational σ^2. An approximate optimal n can be determined empirically by means of several test runs on synthetic data, monitoring the rms deviation of the final density from the reference model density (see below). This is of course only feasible when using synthetic data, for which the perfect answer is known. We plan to overcome this limitation in the future by means of cross-validation methods.

4.2. L-Ala MaxEnt valence density from noisy data

A test of the computational strategy outlined in the previous paragraph has been performed on a set of synthetic noisy structure factor amplitudes. The diffraction data were computed from the same model density for L-alanine at 23 K as the one used for the noise-free calculations described in Section 3.1.

4.2.1. Generation of the noisy data set
Gaussian noise has been added onto the structure factor amplitudes squared as computed from the L-alanine model density; for each datum, the amount of noise added was proportional to the experimental esd for the corresponding intensity measurement:

$$|\mathbf{F}|^2_{\text{Noisy}} = |\mathbf{F}|^2 + \text{Gauss} \times \sigma(|\mathbf{F}|^2_{\text{obs}}), \tag{31}$$

where Gauss is a random deviate of zero mean and unit variance.

From these noisy structure factor amplitudes squared, a sample of 2532 noisy structure factor amplitudes $|\mathbf{F}|_{\text{Noisy}}$ up to 0.463 Å, and the associated standard deviations $\sigma(|\mathbf{F}|_{\text{Noisy}})$, have been computed using the computer program *BAYES* [55]. A number of 2470 of these noisy amplitudes are greater than 2σ, a consequence of the high precision of the experimental data used to calibrate the noise.

BUSTER has been run against the L-alanine noisy data: the structure factor phases and amplitudes for this acentric structure were no longer fitted exactly but only within the limits imposed by the noise. As in the calculations against noise-free data, a fragment of atomic core monopoles was used; the non-uniform prior prejudice was obtained from a superposition of spherical valence monopoles. For each reflexion, the likelihood function was non-zero for a set of structure factor values around this 'procrystal' structure factor; the latter acted therefore as a 'soft' target for the MaxEnt structure factor amplitude and phase.

4.2.2. Initial phase error

The core and valence monopole populations used for the MaxEnt calculation were the ones of the reference density (electrons in the asymmetric unit: $n_{\text{core}} = 12.44$ and $n_{\text{valence}} = 35.56$). The phases and amplitudes for this spherical-atom structure, union of the core fragment and the NUP, are already very close to those of the full multipolar model density: to estimate the initial phase error, we computed the phase statistics recently described in a multipolar charge density study on 0.5 Å noise-free data [56].

For a number of 1907 acentric reflexions up to 0.463 Å resolution, the mean and rms phase angle differences between the noise-free structure factors for the full multipolar model density and the structure factors for the spherical-atom structure (in parentheses we give the figures for 509 acentric reflexions up to 0.700 Å resolution only) were: $\langle\Delta\phi\rangle = 1.012(2.152)^\circ$, $\text{rms}(\Delta\phi) = 2.986(5.432)^\circ$; while the mean and rms arc length errors, normalised so as to have $F_{000} = 100$, were $\langle|\mathbf{F}|\sin\Delta\phi\rangle = 0.034(0.088)$, $\text{rms}(|\mathbf{F}|\sin\Delta\phi) = 0.063(0.116)$.

4.2.3. Computational details

It is of interest to mention some of the details of the valence MaxEnt calculations, performed with the computer program *BUSTER* [42] on an *Alpha Station 500* running at 500 MHz.

BUSTER chooses the minimal grid necessary to avoid aliasing effects, based on the prior prejudice used and on the fall-off of the structure factor amplitudes with resolution: for the 23 K L-alanine valence density reconstruction the grid was (64 144 64). The cell parameters for the crystal are $a = 5.928(1)$ Å; $b = 12.260(2)$ Å; $c = 5.794(1)$ Å [45], so that the grid step was shorter than 0.095 Å along each axis.

The calculation of the thermally-smeared core fragment and the valence monopoles densities was carried out by a Fourier transform of a set of aliased structure factors computed with the program *VALRAY* [46]; details of this calculation have been published elsewhere [49].

The total number of degrees of freedom ($N_{\text{DoF}} = N_{\text{Centric}} + 2N_{\text{Acentric}}$) was 4439; this is also equal to the number of Lagrange multipliers. The constrained maximisation of the Bayesian score converged in less than 40 iterations; sufficient memory and disk

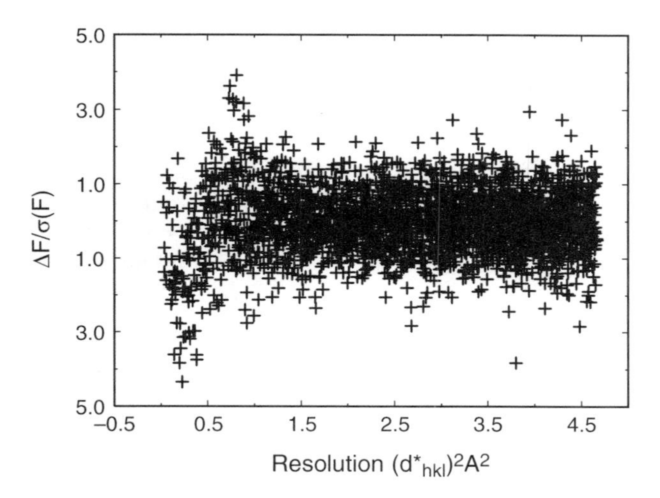

Figure 5. L-Alanine. Fit to noisy data. Calculation A. Distribution of residual structure factor amplitudes at the end of the MaxEnt calculation on 2532 noisy data up to 0.463 Å. Residuals plotted: $\Delta F/\sigma = (|\mathbf{F}|^{\mathrm{ME}} - |\mathbf{F}|^{\mathrm{Noisy}})/\sigma(|\mathbf{F}|^{\mathrm{Noisy}})$.

were available so that the job had about 80% of the CPU, and took about 7 min to complete.

As mentioned in Section 4.1.1, the number of random scatterers n has to be chosen in input. Five *BUSTER* runs used values of n in the series: $n = n_{\mathrm{valence}} \times N^2$, $N = 60, 70, 80, 90, 100$. The rms deviation from the reference map varied between 0.0317 and 0.0293 e Å^{-3}, the latter value pertaining to the run with $N = 90$: this value of n was then used in the calculation described below.

4.2.4. *Quality of the reconstruction*

We briefly discuss in this section the results of the valence MaxEnt calculation on the noisy data set for L-alanine at 23 K: we will denote this calculation with the letter A. The distribution of residuals at the end of the calculation is shown in Figure 5. It is apparent that no gross outliers are present, the calculated structure factor amplitudes being within 5 esd's from the observed values at all resolution ranges.

The same phase statistics mentioned above were computed to obtain an estimate of the phase error for the reconstructed density, for 1907 acentric reflexions up to 0.463 (in parentheses the values for the 509 acentric reflexions up to 0.700 Å): $\langle\Delta\phi\rangle = 0.755(0.854)^\circ$, $\mathrm{rms}(\Delta\phi) = 1.762(1.530)^\circ$; the normalised mean and rms arc lengths are $\langle|\mathbf{F}|_{\mathrm{Noisy}} \sin\Delta\phi\rangle = 0.022(0.040)$ and $\mathrm{rms}(|\mathbf{F}|_{\mathrm{Noisy}} \sin\Delta\phi) = 0.033(0.054)$, respectively. The MaxEnt valence modulation does improve the overall and low-resolution phase error significantly.

The MaxEnt deformation density in the COO^- plane is shown in Figure 6(a). The deformation map shows correct qualitative features; differences between the single C–C bond and the C–O bonds are clearly visible, and so are the lone-pair maxima on the oxygen atoms. If compared to the conventional dynamic deformation density

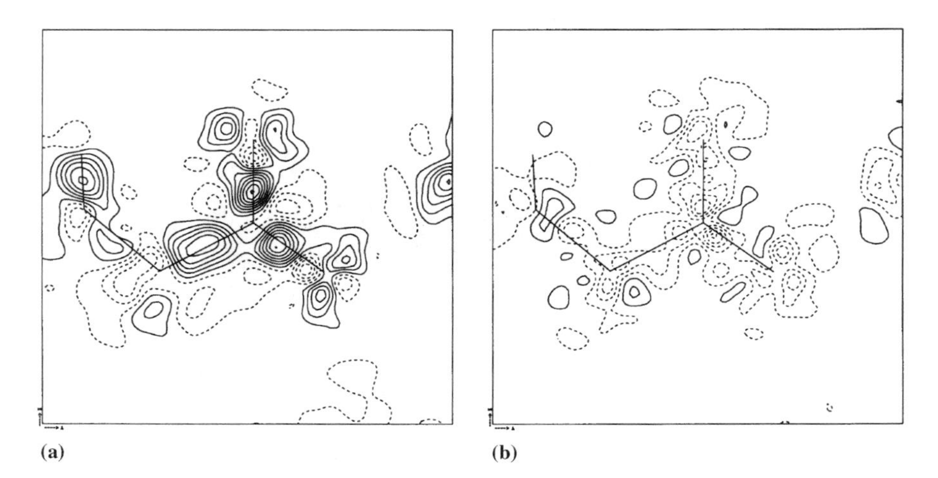

Figure 6. L-Alanine. Fit to noisy data. Calculation *A*. MaxEnt deformation density and error map in the COO^- plane Map size, orientation and contouring levels as in Figure 2. (a) MaxEnt dynamic deformation density Δq^{ME}_{NUP}. (b) Error map: $q^{ME} - q^{Model}$.

maps, usually obtained by Fourier summation, the MaxEnt deformation density is also remarkably clean in intermolecular regions, where the observations do not introduce any modulation in the prior prejudice of atomic valence monopoles.

4.3. The MaxEnt underestimates the deformation features

Figure 6(b) shows the difference between the MaxEnt valence density and the reference density, in the COO^- plane. The error peaks in the bonding and lone-pair regions, where the deformation features are systematically lower than the reference map (negative contours). The deviation from the reference is largest in the region around the C1 atom valence shell, and reaches $-0.406\,\mathrm{e\,Å^{-3}}$.

4.3.1. Intrinsic dispersion of the MaxEnt distribution

The MaxEnt method will always deflate deformation features by the $\langle\delta q\rangle_{\mathrm{rms}}$ corresponding to measurements error [39]. To obtain an empirical estimate of this intrinsic spread allowed by the noise, twenty noisy data sets were generated as in formula (31), and fitted with *BUSTER* using the fragment and NUP already described in the previous paragraph.

The average map and the rms deviation from the average were computed:

$$\langle q^{\mathrm{ME}}(\mathbf{x})\rangle = \frac{1}{N_{\mathrm{maps}}}\sum_{i=1}^{N_{\mathrm{maps}}} q_i^{\mathrm{ME}}(\mathbf{x}), \tag{32}$$

$$\langle\delta q(\mathbf{x})\rangle_{\mathrm{rms}} = \left\{\frac{1}{N_{\mathrm{maps}}-1}\sum_{i=1}^{N_{\mathrm{maps}}}\left[q_i^{\mathrm{ME}}(\mathbf{x}) - \langle q^{\mathrm{ME}}(\mathbf{x})\rangle\right]^2\right\}^{1/2}. \tag{33}$$

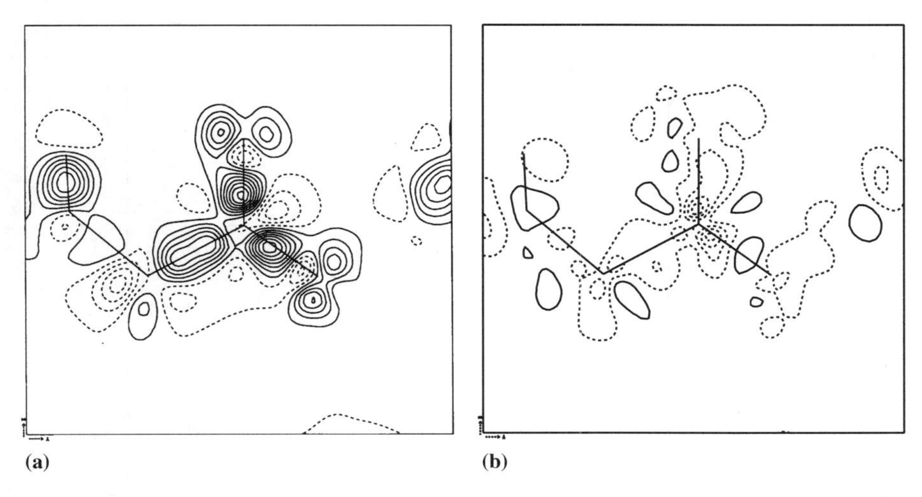

Figure 7. L-Alanine. Fit to noisy data. Calculation B. 10% experimental noise level. MaxEnt deformation density and error map in the COO^- plane. Map size, orientation and contouring levels as in Figure 2. (a) MaxEnt dynamic deformation density $\Delta q^{\mathrm{ME}}_{\mathrm{NUP}}$. (b) Error map: $q^{\mathrm{ME}} - q^{\mathrm{Model}}$.

The $\langle q^{\mathrm{ME}}(\mathbf{x})\rangle$ map is of course less noisy than any of the individual noisy maps; the deviation from the reference model map shows the same systematic underestimation of the deformation features as observed in density A, with a maximum negative error of $-0.362\,\mathrm{e}\,\text{Å}^{-3}$, again in the region of the valence shell of the C1 atom.

The $\langle \delta q(\mathbf{x})\rangle_{\mathrm{rms}}$ map peaks around the two oxygen atoms, where the valence density is highest; the values of $\langle \delta q(\mathbf{x})\rangle_{\mathrm{rms}}$ remain below $0.112\,\mathrm{e}\,\text{Å}^{-3}$. This confirms that the deviations observed in the calculation A are indeed significant with respect to the intrinsic spread brought by the noise in the data.

4.3.2. Dependence of the bias on the noise level

To check for the dependence of this bias on the noise level, a number of 20 noisy data sets were generated with variances lowered to 10% of their experimental values, and MaxEnt calculations run against these low-noise data.

Sections of the density from one of these fits, which we will refer to as calculation B, are shown in Figure 7: the MaxEnt deformation density in the COO^- plane is shown in Figure 7(a); Figure 7(b) is the difference between the MaxEnt valence density and the reference density in the same plane. The lower noise content of the data is clearly visible, when the map is compared with the one for calculation A: in particular, the lone pairs on the oxygen atoms are better defined. The rms deviation from the reference is as low as $0.023\,\mathrm{e}\,\text{Å}^{-3}$.

Still, the deformation features around C1 are systematically underestimated, with a maximum deviation of $-0.0312\,\mathrm{e}\,\text{Å}^{-3}$. As is evident from Figures 6(b) and 7(b), the departure of the MaxEnt distribution from the reference model is most significant

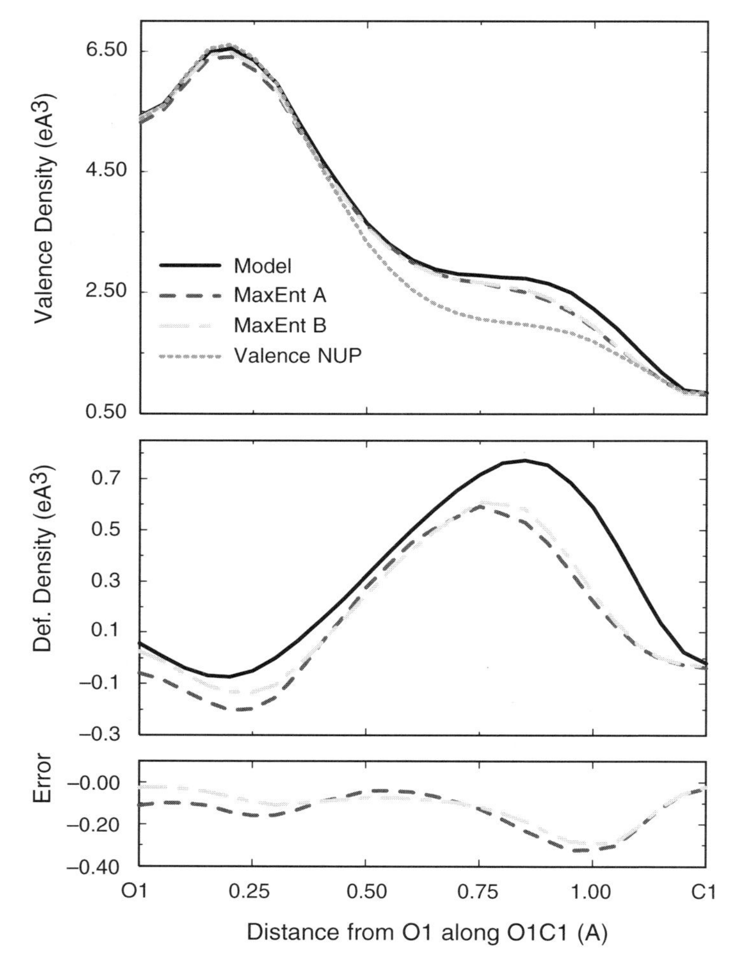

Figure 8. L–Alanine. Fits to noisy data: Calculations *A* (experimental noise) and *B* (10% experimental noise). MaxEnt, deformation and error density profiles along the C1–O1 bond. Solid line: Model valence density. Dashed line: MaxEnt density *A*. Dot-dashed line: MaxEnt density *B*. Dotted line: valence-shells non-uniform prior.

in the regions where the deformation from the prior prejudice of spherical atoms is larger, namely in bonds of order greater than one.

This finding is more evident in the density profiles in Figure 8: both calculations *A* and *B* produce too low a density in the C1–O1 bond. Close to the carbon atom, the profiles depart from the reference density to yield a more 'atom-like' picture of the bond. This bias is milder for low-noise data, because of a tighter constraint from the data.

5. Concluding remarks

The observations presented here are in keeping with the general notion that the MaxEnt method is best understood as a method for testing hypotheses against the experimental data, in the presence of some prior knowledge. In a crystallographic context, the Bayesian viewpoint on crystal structure determination prescribes the use of the MaxEnt method to perform iterative testing of structural hypotheses, and allow model updating [41, 51].

From this viewpoint, it is possible to rationalise the results of the different types of charge density MaxEnt calculations discussed so far. In each case, the calculation provides an answer whose quality is commensurate with the degree of adequacy or inadequacy of the null hypothesis made; these null hypotheses can be ranked in increasing order of information content:

- Many of the MaxEnt calculations described in the literature ignore any knowledge of the atomicity of structures other than that conveyed by the choice made for the target structure factor phases (see Section 2.2): a uniform prior is used, and all electrons are redistributed under the MaxEnt condition. The resulting distribution already contains a clear picture of atoms, with atomic cores and bonding density regions; but the topology of these MaxEnt densities will often be wrong, because the missing structure is not adequately modelled by random independent constituents [57];
- Within the computational scheme described in the course of this work, the available information about the atomic substructure (core+valence) can be taken into account explicitly. In the simplest possible calculation, a fragment of atomic cores is used, and a MaxEnt distribution for valence electrons is computed by modulation of a uniform prior prejudice. As we have shown in the noise-free calculations on L-alanine described in Section 3.1.1, the method will yield a better representation of bonding and non-bonding valence charge concentration regions, but bias will still be present because of Fourier truncation ripples and aliasing errors;
- Full atomicity can be incorporated into the available prior information, using a NUP of spherical-valence shells, together with the atomic cores fragment. The test presented in Section 3.1.1 shows that it is possible to correctly reconstruct the aspherical features in the density, in absence of experimental noise. At this stage, no stereochemical knowledge has yet been used, other than that implicitly conveyed by the geometry of the nuclear framework. The presence of the experimental noise softens the constraints imposed by the observations, so that multiple-order bonds and very sharp non-bonded charge concentration features are deflated (see Section 4.3).
- The next update of the null hypothesis would incorporate a zero-order description of bonding, in terms of a prior prejudice of 'standard' chemical groups. The MaxEnt map then will tell us about the subtle differences induced in formally equivalent chemical bonds by conjugation, stacking, and other intra- and intermolecular interactions. To achieve this degree of accuracy, the refinement of structural parameters

present in the model adopted for the fragment should proceed together with the MaxEnt redistribution of the valence electrons.

We have described in this paper the first implementation of this Bayesian approach to charge density studies, making joint use of structural models for the atomic cores substructure, and MaxEnt distributions of scatterers for the valence part. Used in this way, the MaxEnt method is 'safe' and can usefully complement the traditional modelling based on finite multipolar expansions. This supports our initial proposal that accurate charge density studies should be viewed as the late stages of the structure determination process.

Acknowledgements

This work was partially supported by an International Research Scholarship from the Howard Hughes Medical Institute (to G.B.) and by a collaborative research grant from Pfizer Central Research (to G.B.). *Digital Equipment Corporation* generously loaned the Alpha workstations used for the calculations. Riccardo Destro kindly made the L-alanine diffraction data available to us. Thanks are also due to Mark Spackman, Carlo Gatti and Eric de La Fortelle for useful discussions.

References

1. Larsen, F.K. and Hansen, N.K. (1984) Diffraction studies of the electron density distribution in beryllium metal, *Acta Cryst.*, **B40**, 169–179.
2. Lu, Z.W., Zunger, A. and Deutsch, M. (1993) Electronic charge distribution in crystalline diamond, silicon and germanium, *Phys. Rev.*, **B47**(15), 9385–9410.
3. Destro, R. and Merati, F. (1995) Bond lengths, and beyond, *Acta Cryst.*, **B51**, 559–570.
4. Flensburg, C., Larsen, S. and Stewart, R.F. (1995) Experimental charge density study of methylammonium hydrogen succinate monohydrate. A salt with a very short O–H–O hydrogen bond, *J. Phys. Chem.*, **99**, 10130–10141.
5. Iversen, B.B., Larsen, F.K., Figgis, B.N. and Reynolds, P.A. (1997) X–N study of the electron density distribution in *trans*-tetraammine-dinitronickel(II) at 9 K: transition metal bonding and topological analysis, *J. Chem. Soc., Dalton Trans.* 2227–2240.
6. Nowack, E., Schwarzenbach, D. and Hahn, T. (1991) Charge densities in CoS_2 and NiS_2 (pyrite structure), *Acta Cryst.*, **B47**, 650–659.
7. Spackman, M. (1986) The electron distribution in silicon. A comparison between experiment and theory, *Acta Cryst.*, **A42**, 271–281.
8. Hansen, N.K. and Coppens, P. (1978) Testing aspherical atoms refinements on small-molecule data sets, *Acta Cryst.*, **A34**, 909–921.
9. Espinosa, E., Molins, E. and Lecomte, C. (1997), Electron density study of the one dimensional organic metal bis(thiodimethylene)–tetrathiofulvalene tetracyanoquinodimethane, *Phys. Rev. B*, **56**(4), 1820–1833.
10. Pichon-Pesme, V., Lecomte, C. and Lachekar, H. (1995) On building a data bank of transferable experimental electron density parameters: application to polypeptides, *J. Phys. Chem.*, **99**(16), 6242–6250.
11. Brown, A.S., Spackman, M.A. and Hill, R.J. (1993) The electron distribution in corundum. A study of the utility of merging single-crystal and powder diffraction data, *Acta Cryst.*, **A49**, 513–527.
12. Howard, S., Hursthouse, M.B., Lehmann, C.W. and Poyner, E.A. (1995) Experimental and theoretical determination of electronic properties in L-Dopa, *Acta Cryst.*, **B51**, 328–337.
13. Roversi, P., Barzaghi, M., Merati, F. and Destro, R. (1996) Charge density in crystalline citrinin from X-ray diffraction at 19 K, *Can. J. Chem.*, **74**, 1145–1161.

14. Souhassou, M., Espinosa, E., Lecomte, V. and Blessing, R.H. (1995) Experimental electron density in crystalline H_3PO_4, *Acta Cryst.*, **B51**, 661–668.
15. Chen, R., Trucano, P. and Stewart, R.F. (1977) The valence–charge density of graphite, *Acta Cryst.*, **A33**, 823–828.
16. Stewart, R.F. (1977) A charge-density study of crystalline beryllium, *Acta Cryst.*, **A33**, 33–38.
17. Restori, R. and Schwarzenbach, D. (1995) Maximum-entropy versus least-squares modelling of the electron-density in K_2PtCl_6, *Acta Cryst.*, **B51**, 261–263.
18. Papoular, R.J., Vekhter, Y. and Coppens, P. (1996) The two-channel maximum-entropy method applied to the charge density of a molecular crystal: α-glycine, *Acta Cryst.*, **A52**, 397–407.
19. Takata, M. and Sakata, M. (1996) The influence of completeness of the data set on the charge density obtained with the maximum-entropy method. A re-examination of the electron-density distribution in Si, *Acta Cryst.*, **A52**, 287–290.
20. Yamamoto, K., Takahashi, Y., Ohshima, K., Okamura, F.P. and Yukino, K. (1996) MEM analysis of electron-density distributions for silicon and diamond using short-wavelength X-rays (W$K\alpha_1$), *Acta Cryst.*, **A52**, 606–613.
21. Spackman, M. and Brown, A.S. (1994) Charge densities from X-ray diffraction data, In: *Annual Reports*, Royal Society of Chemistry, Cambridge, pp. 175–212.
22. Gilmore, J.C. (1996) Maximum-entropy and Bayesian statistics in crystallography: a review of practical applications, *Acta Cryst.*, **A52**, 561–589.
23. Jaynes, E. (1957) Information theory and statistical mechanics I, *Phys. Rev.*, **106**, 620–630.
24. Jaynes, E. (1957) Information theory and statistical mechanics II, *Phys. Rev.*, **108**, 171–190.
25. Jaynes, E. (1968) Prior probabilities, In: *IEEE Transactions on Systems Science and Cybernetics*, Vol. SSC-**4**, pp. 227–241.
26. Jaynes, E. (1983) *Papers on Probability, Statistics and Statistical Physics*, R.D. Rosenkrantz, Reidel, Dordrecht, Holland.
27. Bricogne, G. (1984) Maximum entropy and the foundations of direct methods, *Acta Cryst.*, **A40**, 410–445.
28. Bricogne, G. (1991) The phase problem in X-ray crystallography, In: Buck, B. and Macaulay, V.A. (eds.): *Maximum Entropy in Action*, Oxford University Press, pp. 187–216.
29. Jauch, W. and Palmer, A. (1993) The maximum-entropy method in charge-density studies: aspects of reliability, *Acta Cryst.*, **A49**, 590–591.
30. Jauch, W. (1994) The maximum-entropy method in charge-density studies. II. General aspects of reliability, *Acta Cryst.*, **A50**, 650–652.
31. de Vries, R., Briels, W.J. and Feil, D. (1994) Novel treatment of the experimental data in the application of the maximum-entropy method to the determination of the electron-density distribution from X-ray experiments, *Acta Cryst.*, **A50**, 383–391.
32. Iversen, B.B., Larsen, F.K., Souhassou, M. and Takata, M. (1995) Experimental evidence for the existence of non-nuclear maxima in the electron density distribution of metallic beryllium. A comparative study of the maximum entropy method and the multipole method, *Acta Cryst.*, **B51**, 580–591.
33. Carvalho, C.A.M., Hashizume, H., Stevenson, A.W. and Robinson, I.K. (1996) Electron-density maps for the Si(111) 7×7 surface calculated with the maximum-entropy technique using X-ray and electron-diffraction data, *Physica B*, **221**, 469–486.
34. Papoular, R.J. and Gillon, B. (1990) Maximum entropy reconstruction of spin density maps in crystals from polarized neutron diffraction data, *Europhys. Lett.*, **13**(5), 429–434.
35. Zheludev, A., Papoular, V., Ressouche, E. and Schweizer, J. (1995) A non-uniform reference model for maximum-entropy density reconstruction from diffraction data, *Acta Cryst.*, **A51**, 450–455.
36. Sakata, M., Uno, T., Takata, M. and Howard, C. (1993) Maximum-entropy-method analysis of neutron diffraction data, *J. Appl. Cryst.*, **26**, 159–165.
37. de Vries, R., Briels, W.J., Feil, D., te Velde, G. and Baerends, E. (1996) Charge density study with the maximum entropy method on model data of silicon. A search for non-nuclear attractors, *Can. J. Chem.*, **74**, 1054–1058.
38. de Vries, R., Briels, W.J. and Feil, D. (1996) Critical analysis of non-nuclear electron-density maxima and the maximum entropy method, *Phys. Rev. Lett.*, **77**(9), 1719–1722.
39. Iversen, B.B., Jensen, J.L. and Danielsen, J. (1997) Errors in maximum-entropy charge-density distributions obtained from diffraction data, *Acta Cryst.*, **A53**, 376–387.
40. Destro, R. (1996) Personal communication.
41. Bricogne, G. (1988) A Bayesian statistical theory of the phase problem. I. A multichannel maximum-entropy formalism for constructing generalized joint probability distributions of structure factors, *Acta Cryst.*, **A44**, 517–545.

42. Bricogne, G. (1993) Direct phase determination by entropy maximization and likelihood ranking: status report and perspectives, *Acta Cryst.*, **D49**, 37–60.
43. Bader, R.F.W. (1990) *Atoms in Molecules. A Quantum Theory*, The International Series of Monographs in Chemistry. Oxford University Press, Oxford.
44. Lecomte, C. (1995) Personal Communication.
45. Destro, R., Marsh, R.E. and Bianchi, R. (1988) A Low-temperature (23 K) study of L-Alanine, *J. Phys. Chem.*, **92**, 966–973.
46. Stewart, R.F. and Spackman, M. (1983) VALRAY Users Manual. First Edition, Carnegie-Mellon University, Pittsburgh, PA 15213.
47. Clementi, E. (1965) Tables of Atomic Functions. *IBM J. Res. Dev.*, **9**(2), Supplement.
48. Stewart, R.F. (1980) Partitioning of Hartree–Fock atomic form factors into core and valence shells, In *Electron and Magnetization Densities in Molecules and Crystals*, Becker, P. (Ed.), Plenum Press, New York, pp. 427–431.
49. Roversi, P., Irwin, J.J. and Bricogne, G. (1998) Accurate charge density studies as an extension of bayesian crystal structure determination, **A54**(6(2)), 971–996.
50. Bricogne, G. (1993) Fourier transforms in crystallography: theory, algorithms, and applications, In *International Tables for Crystallography*, Vol. B, Reciprocal Space, Shmueli, U. (Ed.), Dordrecht, Kluwer Academic Publishers, Holland, pp. 23–106.
51. Bricogne, G. (1997) The Bayesian statistical viewpoint on structure determination: basic concepts and examples, In *Macromolecular Crystallography*, Vol. 276 of *Methods in Enzymology*, Carter Jr., C.W. and Sweet, R.M. (Eds.), Academic Press, pp. 361–423.
52. Shannon, C.E. and Weaver, W. (1949) *The Mathematical Theory of Communication*, University of Illinois Press, Chicago, Illinois.
53. de La Fortelle, E. and Bricogne, G. (1997) Maximum-likelihood heavy-atom parameter refinement for the MIR and MAD methods, In *Macromolecular Crystallography*, Vol. 276 of *Methods in Enzymology*, Carter Jr., C.W. and Sweet, R.M. (Eds.), Academic Press, pp. 472–494.
54. Xiang, S., Carter Jr., C.W., Bricogne, G. and Gilmore, C.J. (1993) Entropy maximization constrained by solvent flatness: a new method for macromolecular phase extension and map improvement, *Acta Cryst.*, **D49**, 193–212.
55. Blessing, R.H. (1989) '*DREADD* – data reduction and error analysis for single-crystal diffractometer data, *J. Appl. Cryst.*, **22**, 396–397.
56. Spackman, M.A. and Byrom, P.G. (1997) Retrieval of structure factor phases in acentric space groups. Model studies using multipole refinements, *Acta Cryst.*, **B53**, 553–564.
57. Roversi, P., Irwin, J.J. and Bricogne, G. (1996) A Bayesian approach to high-resolution X-ray crystallography: accurate density studies with program *BUSTER*, In *Collected Abstracts of the XVII IUCr Congress and General Assembly, Seattle, Washington, USA*. International Union of Crystallography, p. C343. Also appeared in *Acta Cryst.*, **A52**, C343.
58. Smith, G. (1997) PhD thesis. Durham University, England.

2

Reliability of charge density distributions derived by the maximum entropy method

BO BRUMMERSTEDT IVERSEN

Department of Chemistry, University of Aarhus, DK-8000, Aarhus C, Denmark

1. Introduction

Understanding chemical and physical properties of molecular systems requires knowledge of their charge distributions [1]. Experimentally, electron density distributions (EDDs) can be reconstructed from accurate X-ray diffraction data through a series of elaborate data reduction and data analysis steps [2]. The most widely used method entails least-squares optimization of models containing atom-centered aspherical density functions [3–6]. In the empirical modeling schemes, estimates of errors in the density and in derived properties can be calculated within the framework of the least-squares method. Such estimates rely on several assumptions including the adequacy of the refined model. Several studies [7–9] have shown that even the very sophisticated models currently used in empirical EDD modeling are inadequate to describe very fine density features present in the data and in general, least-squares estimates of EDDs will therefore contain systematic bias due to the model. Nevertheless, the least-squares error estimates allow, to some extent, assessment of the reliability of conclusions drawn from the model densities.

In recent years, a new method, the maximum entropy method (MEM), has been introduced in charge density reconstruction. When X-ray diffraction data are used, the MEM yields the electron density distribution [10, 11], whereas neutron diffraction data allows the direct space nuclear probability density function to be determined [12]. From limited numbers of X-ray diffraction data, EDDs have been reconstructed by the MEM in a number of systems [13–15]. Maps that qualitatively reveal bonding features have been obtained in these and many other studies. Although this is of interest in itself, quantification and detailed analysis of the derived MEM charge densities is highly desirable because chemically important features in molecular electron densities often are very small. It is therefore important that the reliability of MEM densities is scrutinized in order to make the method generally useful. Several authors have pointed out that unphysical features can appear in MEM densities and, depending on the quality and the completeness of the data, fine features in the density may be artifacts of the density reconstruction [16–19]. It has, furthermore, been pointed out that use of an entropy term as a regularizing function in the reconstruction inevitably will introduce systematic bias into the result [20, 21].

Paul G. Mezey and Beverly E. Robertson (eds.), Electron, Spin and Momentum Densities and Chemical Reactivity, 27–36

2. Monte Carlo simulation of errors in MEM densities

In a recent paper we proposed to perform Monte Carlo simulations as basis for the error estimation [22]. The data that are available is a set of observed structure factors with associated standard deviations estimated from the scatter among repeated measurements of equivalent reflections. We assume the error distributions around the true value of each structure factor to be Gaussian and that systematic errors in the data are negligible. From the set of observed structure factors, we can calculate a MEM EDD, ρ^0, using for example the MEED algorithm from Nagoya University [23], but any entropy optimization code may be used. We will not, in this paper, discuss details of the MEM itself but refer readers to the references given in the introduction and to other contributions in this book. The MEM density will have a corresponding set of structure factors F^0. We can construct synthetic data sets by applying random noise to F^0 according to the known error distribution around the true structure factors. The synthetic data sets, F_j^{syn}, can be used as input to a series of Monte Carlo MEM calculations, and the result will be a series of Monte Carlo MEM densities, ρ_j^{MC}. The scatter of these densities can be used to give an estimate of the error in the original MEM density, ρ^0. In Figure 1, a schematic representation of the Monte Carlo calculations is shown.

Once N Monte Carlo densities are available, the estimated standard uncertainty in each pixel of the discretized density can be calculated by

$$\sigma(\rho_x) = \sqrt{\frac{\sum_{i=1}^{N} (\rho_x^i - \bar{\rho}_x)^2}{N-1}}.$$

If no systematic bias is introduced by the MEM algorithm, we will expect that

$$\bar{\rho}_x = \rho_x^{\text{average}} = \frac{1}{N} \sum_{i=1}^{N} \rho_x^i = \rho_x^0.$$

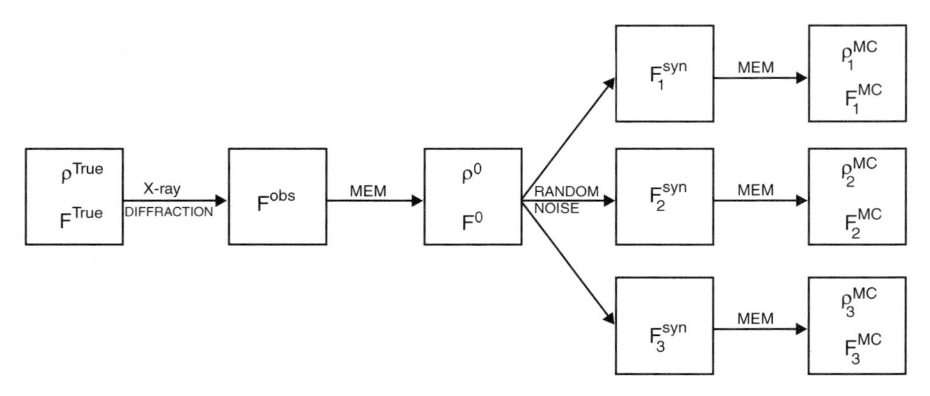

Figure 1. Flow chart of the Monte Carlo calculations to estimate errors in MEM charge densities.

In this formula, ρ_x^0 is the density in pixel number x obtained from the observed structure factors. However, the average value of the Monte Carlo densities turns out to be systematically different from ρ_x^0, as illustrated in Figure 2 where plots of $b(\rho^0) = \rho^{\text{average}} - \rho^0$ are shown for the MEM density of metallic beryllium. The density was calculated based on the very accurate structure factors measured by Larsen and Hansen [24]. In Figure 2(a), the bias in the MEM density calculated with a uniform prior is shown, and in Figure 2(b), the bias obtained with a non-uniform prior is shown. The non-uniform prior corresponds to the EDD of thermally smeared Be atoms placed at their unit cell position (procrystal). It was calculated using wave functions from Clementi and Roetti [25] and neutron diffraction thermal parameters measured by Larsen *et al.* [26]. The maps in Figure 2 indicate where systematic bias, $b(\rho^0)$, is introduced in the density by the MEM algorithm. If a similar systematic bias was introduced in the calculation of ρ^0 from the observed structure factors, then these maps also suggest where ρ^0 may be systematically different from the true EDD, ρ^{true}. We do not know the bias on ρ^{true}, but from the Monte Carlo calculations we know the bias on ρ^0. If ρ^0 is not too different from ρ^{true}, we can assume that the bias in ρ^0 is close to the bias in ρ^{true}. Once we have an estimate of the bias, we can correct the MEM density by subtracting the bias from ρ^0. In Figure 3, bias corrected densities with both uniform and non-uniform priors are shown. The important point to notice is that both types of MEM densities contain considerable systematic bias. However, the calculations show that the MEM bias in the valence regions is smaller when using the non-uniform prior, which indicates that non-uniform priors are preferable to uniform priors. In Figure 4, the random error calculated as the square-root of the variance of the Monte Carlo densities is shown with fine contour intervals of 0.01 e/Å^3. In general, the random error is small in the valence regions.

3. Non-nuclear maxima in hexagonal-close-packed metals

The chemical bonding and the possible existence of non-nuclear maxima (NNM) in the EDDs of simple metals has recently been much debated [13, 27–31]. The question of NNM in simple metals is a diverse topic, and the research on the topic has basically addressed three issues. First, what are the topological features of simple metals? This question is interesting from a purely mathematical point of view because the number and types of critical points in the EDD have to satisfy the constraints of the crystal symmetry [32]. In the case of the hexagonal-close-packed (hcp) structure, a critical point network has not yet been theoretically established [28]. The second topic of interest is that if NNM exist in metals what do they mean, and are they important for the physical properties of the material? The third and most heavily debated issue is about numerical methods used in the experimental determination of EDDs from Bragg X-ray diffraction data. It is in this respect that the presence of NNM in metals has been intimately tied to the reliability of MEM densities.

We originally proposed NNM to be present in metallic beryllium [30] based on analysis of the X-ray diffraction data measured by Larsen and Hansen [24]. Based on Fourier maps and elaborate multipole least-squares modeling, indisputable evidence

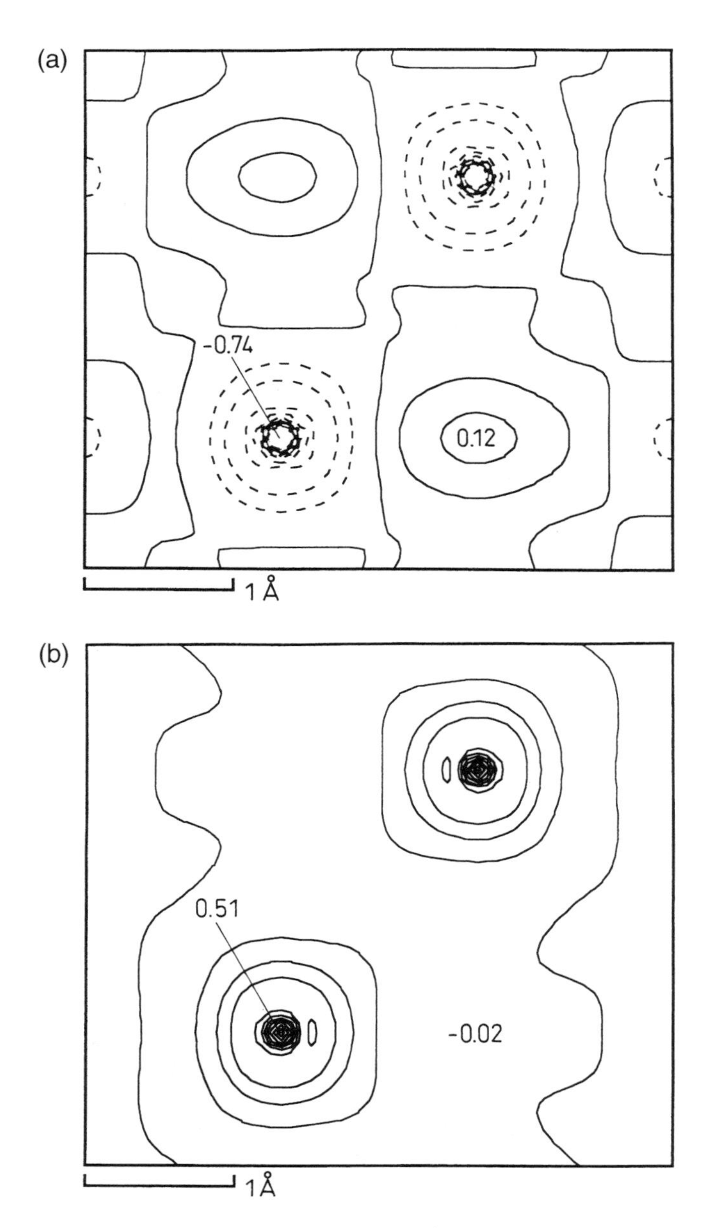

Figure 2. Contour plots of the MEM bias distribution, $b(\rho^0)$, in the (110) plane of the hcp structure of metallic beryllium. The plots are based on 200 Monte Carlo calculations: (a) uniform prior, (b) non-uniform prior. The plots are on a linear scale with 0.05 e/Å^3 intervals. Truncation at -0.5 e/Å^3. Values in e/Å^3 are given for extremum points.

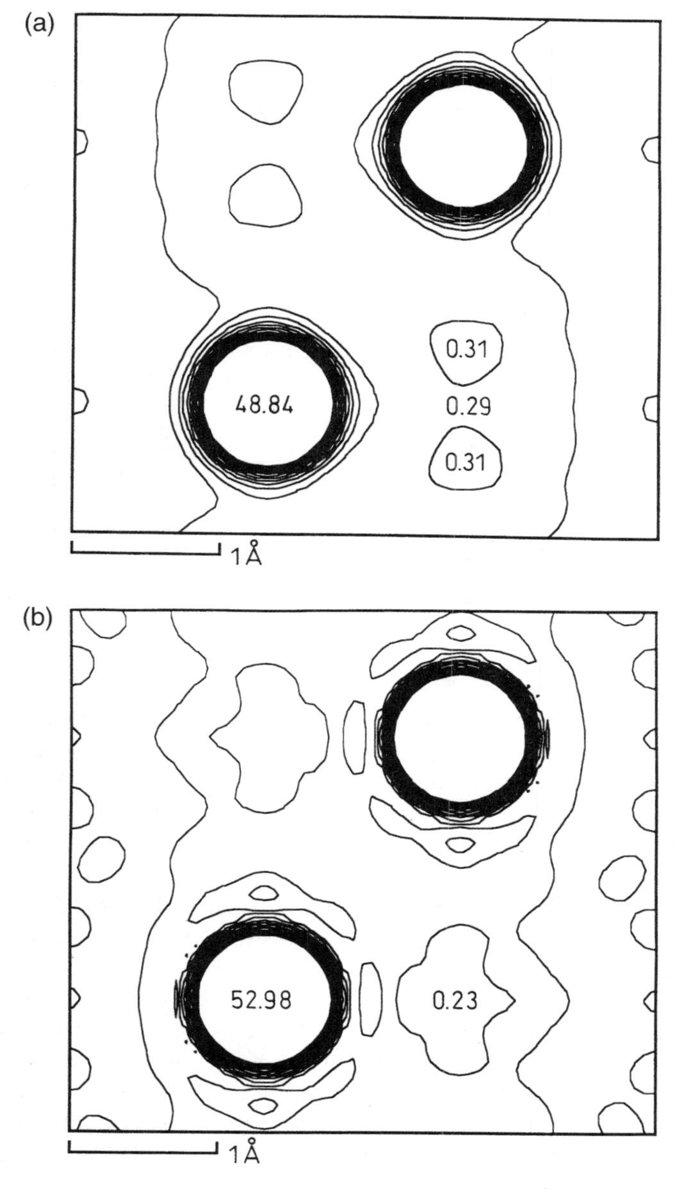

Figure 3. Contour plots of the bias corrected MEM densities in the (110) plane of metallic beryllium: (a) uniform prior, (b) non-uniform prior. The plots are on a linear scale with 0.05 e/Å^3 intervals. Truncation at 1.0 e/Å^3. Maximum values in e/Å^3 are given at the Be position and in the bipyramidal space of the hcp structure.

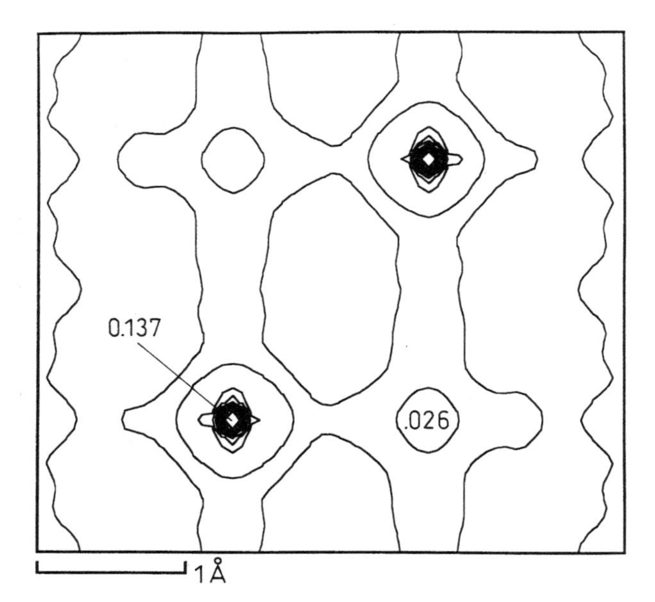

Figure 4. Contour plot in the (110) plane of the estimated random error in the Be MEM densities. The plot is for a uniform prior, but it is essentially identical to the result obtained with a non-uniform prior. The plot is on a linear scale with 0.01 e/Å^3 intervals and 0.1 e/Å^3 truncation. Maximum values in e/Å^3 are given at the Be position and in the bipyramidal space.

revealed that in metallic beryllium, charge is transferred into the bipyramidal region of the structure relative to a model consisting of independent spherical beryllium atoms. Our analysis then addressed whether this redistribution of charge gives rise to NNM in the solid. Topological analysis of the multipole model density showed NNM to be present, but we wanted further confirmation and therefore also employed the MEM. MEM reconstructions were carried out using both uniform and non-uniform prior distributions. All methods pointed to the existence of NNM. The NNM were incorporated into a proposed critical point network for the hcp structure which fulfills the Morse equations [32]. In a recent paper, Vries *et al.* [31] claim that the NNM are artifacts of the MEM used with a uniform prior and conclude that there is no experimental evidence for the existence of NNM in the EDD of metallic Be. In their study, Vries *et al.* neglect to mention that the least-squares multipole model density contains NNM. Furthermore they only cite our results obtained with a uniform prior. Vries *et al.* then show that the use of a procrystal non-uniform prior does not give NNM in the MEM density. Almost exactly the same calculations were already reported in our original beryllium paper [30]. We proposed that the lack of NNM when using a non-uniform prior is due to bias in the prior against moving charge into the valence regions during the MEM optimization. It was shown that if the weight of the low order reflections is increased in the calculations with a non-uniform prior, the NNM reappear. It is in this context we can examine the MEM densities shown in Figures 2 and 3. In the case of a uniform prior, the MEM exaggerates the density in the bipyramidal region of tetrahedral holes of the hcp structure leading to a large NNM at (2/3, 1/3, 1/4). Bias correction diminishes the NNM, but the bipyramidal region still contains a density accumulation relative to the rest of the valence regions. When using a procrystal non-uniform prior, the MEM systematically underestimates the density in the bipyramidal region. After bias correction, a weak maximum reappears in (2/3, 1/3, 1/4). In conclusion, there is a significant, although small, accumulation of density in the bipyramidal space. However, the Be density is very flat in the valence region and when considering the random error in the density it is difficult to be conclusive about possible NNM. The fact that we can reconstruct two MEM densities which differ in the bipyramidal region tells us that the data are not really sensitive about the NNM. Strictly speaking, the MEM neither provides solid evidence for nor against the NNM. We have to include other knowledge/methods to establish the critical point network of beryllium. The Morse equations put strict limitations on the number and types of critical points in a crystal and, to our knowledge, the proposed network is still the only suggestion that fulfills these equations. Furthermore, our network is based on a proper numerical topological analysis of the density and not just drawing sections through the density. In another contribution in the present book, Gatti [33] presents recent theoretical evidence in support of the existence of NNM in Be.

To examine in more detail the questions of NNM and critical point networks we have extended our studies to include metallic magnesium in the hope that comparison with other hcp metals will reveal topology–property relationships. The analysis of the Mg density is based on newly measured single crystal X-ray diffraction data. We have collected a full sphere of very extensive 8(1) K X-ray diffraction data on an almost spherical single crystal of Mg using AgK_α radiation ($\sin\theta_{\max}/\lambda = 1.4\,\mathrm{Å}^{-1}$). Scaled, phased and extinction corrected structure factors suitable for MEM analysis were obtained from multipole modeling with a model similar to the one used for Be [31]. This is necessary because the MEM does not contain a model and therefore cannot filter out systematic errors such as extinction which is quite severe in the present Mg data set ($y_{\max} = 40\%$). A full account of the experimental details as well as the data reduction and the data analysis will appear in a forthcoming paper [34]. In Figure 5 is shown the bias corrected MEM EDD for Mg obtained from 209 unique reflections using a uniform prior. MEM calculations with non-uniform priors as well as theoretical calculations are in progress. Based on the experience with Be, where the bias corrected densities using uniform and non-uniform priors are very similar, we expect the present results to be quite accurate. In Figure 5(a), the bias corrected MEM EDD of Mg is shown, and in Figure 5(b), the corresponding random error estimate in plotted. The density of Mg is much less flat in the valence regions than the EDD of Be. A clear NNM is present in (2/3, 1/3, 1/4) at the center of the bipyramidal space. At a qualitative level, it is clear that the Mg density is more peaked than the Be density. Overall, the topology in the two systems seems to be identical. It should be noted that preliminary topological analysis [34] of a theoretical density calculated with periodic Hartree–Fock and DFT methods [35] also indicates the presence of NNM in Mg. In conclusion the analysis shows that the EDD of metals with the hcp structure probably contain NNM, non-nuclear maxima.

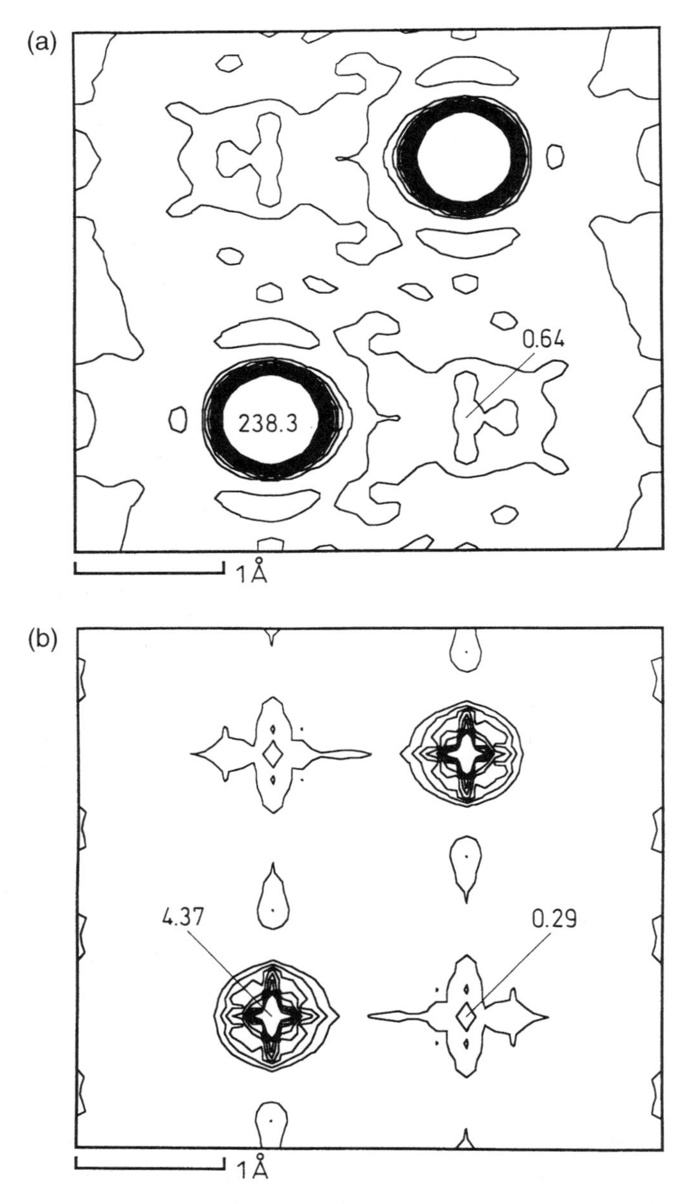

Figure 5. Contour plots in the (110) plane of metallic magnesium: (a) the bias corrected MEM density, (b) the estimated random error in the MEM density. The plots are based on 100 Monte Carlo calculations employing a uniform prior. The plots are on a linear scale, (a) 0.25 e/Å^3 intervals and 5.0 e/Å^3 truncation, (b) 0.1 e/Å^3 intervals and 1.0 e/Å^3 truncation. Maximum values in e/Å^3 are given at the Mg position and in the bipyramidal space.

4. Conclusion

The MEM is a powerful new method which is especially useful in cases with limited data sets (powder diffraction). Monte Carlo simulations have shown that the MEM introduces systematic features into the reconstructed density and caution should be exercised when interpreting fine details of an MEM density. It must be emphasized that because the present MEM algorithms do not contain any models, they cannot filter out inconsistencies in the data stemming from systematic errors. The MEM densities may therefore contain non-physical features not only because of systematic bias in the calculation but also because of systematic errors in the data.

Acknowledgements

Financial support for this work from the Carlsberg Foundation and the Danish Natural Science Research Council is gratefully acknowledged. Lektor Finn Krebs Larsen is thanked for numerous fruitful discussions.

References

1. Bader, R.F.W. (1991) *Atoms in Molecules. A Quantum Theory*, Oxford University Press.
2. Iversen, B.B., Larsen, F.K., Figgis, B.N. and Reynolds, P.A. (1996) *Acta Cryst.*, **B53**, 923–932.
3. Stewart, R.F. (1976) *Acta Cryst.*, **A32**, 565–574.
4. Hirshfeld, F.L. (1977) *Isr. J. Chem.*, **16**, 226–229.
5. Hansen, N.K. and Coppens, P. (1978) *Acta Cryst.*, **A34**, 909–921.
6. Figgis, B.N., Reynolds, P.A. and Williams, G.A. (1980) *J. Chem. Soc. Dalton Trans.*, 2339–2347.
7. Figgis, B.N., Iversen, B.B., Larsen, F.K. and Reynolds, P.A. (1993) *Acta Cryst.*, **B49**, 794–806.
8. Chandler, G.S., Figgis, B.N., Reynolds, P.A. and Wolff, S.K. (1994) *Chem. Phys. Lett.*, **225**, 421–426.
9. Iversen, B.B., Larsen, F.K., Figgis, B.N. and Reynolds, P.A. (1997) *J. Chem. Soc. Dalton Trans.*, 2227–2240.
10. Collins, D.M. (1982) *Nature*, **298**, 49–51.
11. Sakata, M. and Sato, M. (1990) *Acta Cryst.*, **A46**, 263–270.
12. Sakata, M., Uno, T., Takata, M. and Howard, C.J. (1993) *J. Appl. Cryst.*, **26**, 159–165.
13. Takata, M., Kubota, Y. and Sakata, M. (1993) *Z. Naturforsch.*, **48**, 75–80.
14. Takata, M., Umeda, B., Nishibori, E., Sakata, M., Saito, Y., Ohno, M. and Shinoshara, H. (1995) *Nature*, **377**, 46–49.
15. Takata, M., Takayama, T., Sakata, M., Sasaki, S., Kodama, K., Sato, M. (1996) *Physica C*, **263**, 340–343.
16. Jauch, W. and Palmer, A. (1993) *Acta Cryst.*, **A49**, 590–591.
17. Iversen, B.B., Nielsen, S.K. and Larsen, F.K. (1995) *Phil. Mag.*, **A72**, 1357– 1380.
18. Papoular, R., Vechter, Y. and Coppens, P. (1996) *Acta Cryst.*, **A52**, 397–408.
19. Takata, M and Sakata, M. (1996) *Acta Cryst.*, **A52**, 287–291.
20. Jauch, W. (1994) *Acta Cryst.*, **A50**, 650–652.
21. Donoho, D.L., Johnstone, I.M., Hoch, J.C. and Stern, A.S. (1992) *J. R. Statist. Soc.*, **B54**, 41–81.
22. Iversen, B.B., Jensen, J.L. and Danielsen, J. (1997) *Acta Cryst.*, **A53**, 376–387.
23. Kumazawa, S., Kubota, Y., Takata, M., Sakata, M. and Ishibashi, Y. (1993) *J. Appl. Cryst.*, **26**, 453–457.
24. Larsen, F.K. and Hansen, N.K. (1984) *Acta Cryst.*, **B40**, 169–179.
25. Clementi, E. and Roetti, C. (1974) *At. Data Nucl. Tables*, **14**, 177–478.
26. Larsen, F.K., Brown, P.J., Lehman, M.S. and Merisalo, M. (1982) *Philos. Mag.*, **B45**, 31–50.
27. Cao,W.L., Gatti, C., MacDougall, P.J. and Bader, R.F.W. (1987) *Chem. Phys. Lett.*, **141**, 380.
28. Johnson, C. (1992) Am. Crystallogr. Assoc. Ann. Meet., University of Pittsburg, Abstract PA99, 105.
29. Kubota, Y., Takata, M. and Sakata, M. (1993) *J. Phys. Condens. Matter* **H5**, 8245–8254.
30. Iversen, B.B. Larsen, F.K., Souhassou, M. and Takata, M. (1995) *Acta Cryst.*, **B51**, 580–592.
31. Vries, R.Y, Briels, W.J. and Feil, D. (1996) *Phys. Rev. Lett.*, **7**, 1719–1722.

32. Zou, P.F. and Bader, R.F.W. (1994) *Acta Cryst.*, **B50**, 714–725.
33. Gatti, C. (1997) In *Understanding Chemical Reactions*, Mezey, P. and Robertson, B. (Eds.), Kluwer Academic Publishers.
34. Iversen, B.B., Larsen, F.K, Madsen, G.H.K., Cargnoni, F. and Gatti, C. (1997) in preparation.
35. Dovesi, R., Sauders, V.R. and Roetti, C. (1992) CRYSTAL92 User documentations, University of Torino.

3

Maximum entropy reconstruction of spin densities involving non-uniform prior

J. SCHWEIZER[1], R.J. PAPOULAR[2], E. RESSOUCHE[1], F. TASSET[3] and A.I. ZHELUDEV[4]

[1]*DRFMC/SPSMS/MDN CEA-Grenoble, 17 rue des Martyrs, 38054 Grenoble, France*
[2]*Laboratoire Leon Brillouin, CEA-Saclay, BP 2, 91191 Gif sur Yvette, France*
[3]*Institut Laue Langevin, BP 156, 38042 Grenoble, France*
[4]*Physics Department, Brookhaven National Laboratory, Upton, NY 11973, USA*

Diffraction experiments give microscopic information on structures in crystals. Such investigations correspond to Bragg intensity measurements. More and more accurate experiments are performed, which produce accurate maps of the scattering density itself: charge density in the case of X-ray, density of nuclear scattering amplitude in the case of unpolarized neutron and spin (magnetization) gdensity in the case of polarized neutron experiments.

In crystals, the scattering densities are periodic and the Bragg amplitudes are the Fourier components of these periodic distributions. In principle, the scattering density $\rho(\mathbf{r})$ is given by the inverse Fourier series of the experimental structure factors. Such a series implies an infinite sum on the Miller indices h, k, l. Actually, what is performed is a truncated sum, where the indices are limited to those reflections really measured, and where all the structure factors are noisy, as a result of the uncertainty of the measurement. Given these error bars and the limited set of measured reflections, there exist a very large number of maps compatible with the data. Among those, the truncated Fourier inversion procedure selects one of them: the map whose Fourier coefficients are equal to zero for the unmeasured reflections and equal to the exact observed values otherwise. This is certainly an arbitrary choice.

An alternative method, which uses the concept of maximum entropy (MaxEnt), appeared to be a formidable improvement in the treatment of diffraction data. This method is based on a Bayesian approach: among all the maps compatible with the experimental data, it selects that one which has the highest prior (intrinsic) probability. Considering that all the points of the map are equally probable, this probability (flat prior) is expressed via the Boltzman entropy of the distribution, with the entropy defined as

$$S[\rho(\mathbf{r})] = -\sum_i \rho_i \ln(\rho_i).$$

This method has been used for the reconstruction of charge densities from X-ray data [1–3], for maps of nuclear densities from unpolarized neutron data [4–6] as well as for distributions of spin (magnetization) density [7–9]. The density

Paul G. Mezey and Beverly E. Robertson (eds.), Electron, Spin and Momentum Densities and Chemical Reactivity, 37–44

maps obtained by this method, as compared to those resulting from the usual inverse Fourier transformation, are tremendously improved. In particular, any substantial deviation from the background is really contained in the data, as it costs entropy compared to a map that would ignore such features.

However, in most of the cases, before the measurements are performed, some knowledge exists about the distribution which is investigated. It can range from the simple information of the type of scattering electrons (electrons p, d or f) to an elaborate theoretical model. In these cases, the uniform prior which considers all the different pixels as equally likely is too weak a requirement and has to be replaced. In a rigorous Bayesian analysis, Skilling has shown [10] that prior knowledge can be encoded into the MaxEnt formalism through a model $m(\mathbf{r})$, via a new definition for the entropy:

$$S[\rho(\mathbf{r})] = \sum_i \left(\rho_i - m_i - \rho_i \ln \left(\frac{\rho_i}{m_i} \right) \right).$$

In the absence of any data, the maximum of the entropy functional is reached for $\rho(\mathbf{r}) = m(\mathbf{r})$. Any substantial departure from the model, observed in the final map, is really contained in the data as, with the new definition, it costs entropy. This paper presents illustrations of model testing in the case of intermetallic and molecular compounds.

An intermetallic compound: a model for the magnetization density in YCo_5

The magnetic properties of the YCo_5 intermetallic compound have been extensively investigated due to its ferromagnetism with a high Curie point and very high magnetocrystalline anisotropy which makes it a good representative of the RCo_5 permanent magnets. Its crystal structure is represented in Figure 1. It includes one site of Y

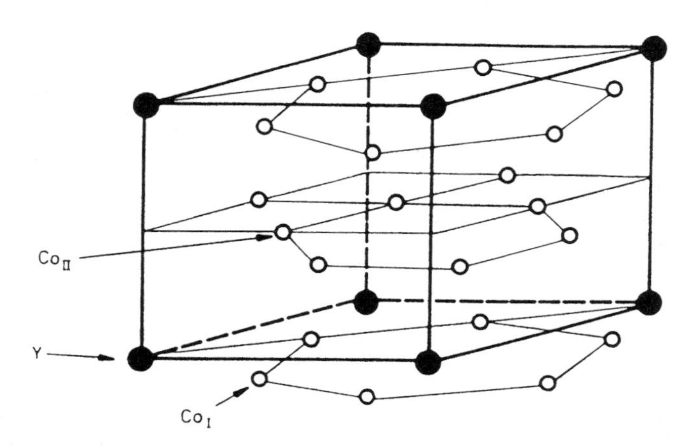

Figure 1. The unit cell of YCo_5.

and two sites of Co: Co_I in the basal plane and Co_{II} in the intermediate plane. Both unpolarized neutron and polarized neutron experiments have been performed at room temperature, in the ferromagnetic state, in order to refine the nuclear structure and to determine the magnetization density.

As at room temperature Bragg reflections contain both nuclear and magnetic structure factors, the nuclear structure was refined from a combination of polarized and unpolarized neutron data. Contrary to the ideal structure where only three atomic sites are present, it has been shown [11, 12] that some Y atoms were substituted by pairs of cobalt. These pairs, parallel to the c-axis are responsible for a structure deformation which shrinks the cobalt hexagons surrounding the substitutions. The amount of these substituted Y was refined to be 0.046 ± 0.008. Furthermore, the thermal vibration parameter of Co_I site appeared to be very anisotropic. The nuclear structure factors F_N were calculated from this refined structure and were introduced in the polarized neutron data to get the magnetic structure factors F_M.

The reconstruction of the magnetization density was done by the MaxEnt method with a uniform prior. The projection on the basal plane is shown in Figure 2. Besides a small contribution at the origin due to the Y substituted by cobalt pairs, the magnetization is well localized on the five atoms of the two cobalt sites.

Therefore, an atomic model, made of a superposition of independent densities centered at the magnetic atoms, was built. The magnetic structure factor can be written as

$$F_M(\mathbf{K}) = \sum_j m_j f_j(K) \exp(\mathrm{i}\mathbf{Kr}) \exp(-W_j),$$

where f_j is the magnetic form factor and m_j the moment of the jth atom. The magnetic form factors are the sum of two contributions: orbital and spin: $f(K) = lf_l(K) + sf_s(K)$. Assuming that the 3d orbital is almost quenched, the orbital form factor was taken as isotropic and equal to $f_l(K) = \langle j_0(K) \rangle + \langle j_2(K) \rangle$. For the spin

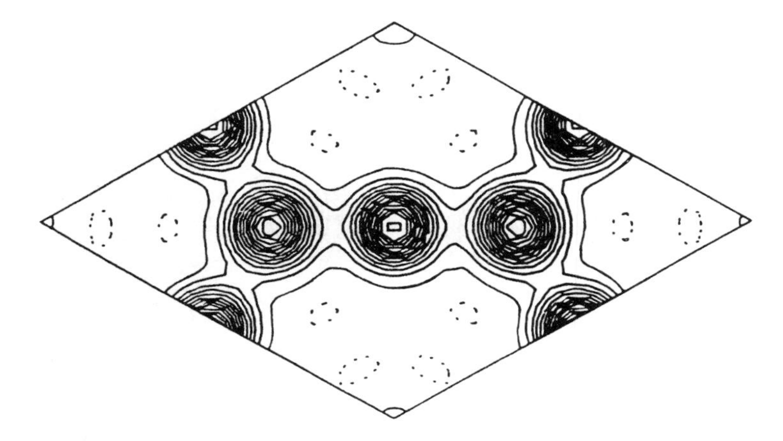

Figure 2. YCo_5: MaxEnt reconstruction with a uniform prior.

part one took into account the anisotropy of the spin density around each magnetic atom: Co_I with one singlet (d_{z^2}) and two doublet (d_{xy} and d_{yz}) and Co_{II} with the five-fold degeneracy completely removed. Altogether 10 parameters were refined from the experimental F_M to determine the atomic magnetic model: the localized moments m_I and m_{II}, the orbital contributions l_I and l_{II} and the occupation numbers: two for the site Co_I and four for the site Co_{II}. The agreement between observed and calculated F_M is very good; the parameters are displayed in Table 1. The magnetization density corresponding to the atomic model and projected on the basal plane is represented in Figure 3. Comparing with the MaxEnt projection (Figure 2) one sees that the distributions are not far from the other, but with more asphericities on the atoms of site Co_I for the refined model.

How to judge the relevance of these asphericities? Are they really compatible with the data or are they simply the biased result of an ill-adapted model? The best way to answer this question is to use this result as a prior probability for a new MaxEnt reconstruction. The map thus obtained, which is given in Figure 4, is striking: the

Table 1. YCo_5: refined parameters for the atomic magnetic model.

Site	Localized moment	Spin proportion	Occupation parameter	
Co_I	1.77 (2)μ_B	0.74 (5)	d_{z^2}	0.23 (3)
			d_{xz}, d_{yz}	0.18 (12)
			$d_{x^2-y^2}, d_{xy}$	0.58
Co_{II}	1.72 (2)μ_B	0.84 (4)	d_{z^2}	0.15 (2)
			d_{xz}	0.24 (4)
			d_{yz}	0.24 (4)
			$d_{x^2-y^2}$	0.22 (3)
			d_{xy}	0.20

Sum of the localized moments in one cell: 8.90 (10)μ_B. Magnetization measured for one cell: 7.99 (2)μ_B.

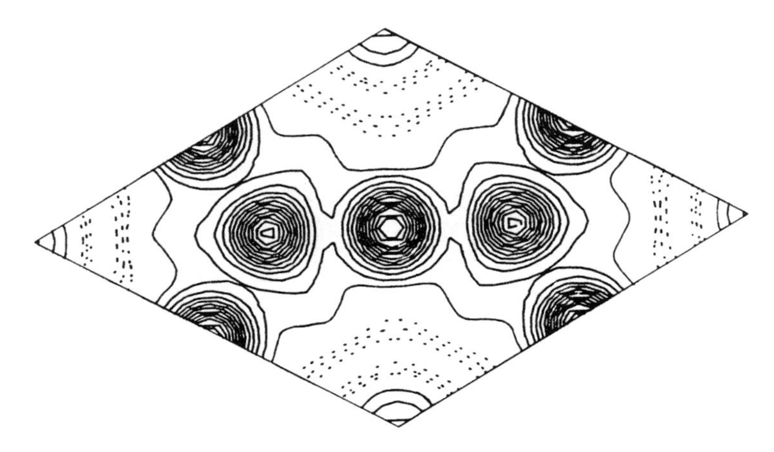

Figure 3. YCo_5: magnetization density of the magnetic atomic model.

Figure 4. YCo_5: MaxEnt reconstruction with a non-uniform prior.

new reconstruction is very similar to that obtained with the uniform prior. All the asphericities which were present in the model have been rubbed out, in spite of the fact that, with the new definition of entropy, it costs entropy. We can conclude that the distribution of the magnetization density which is contained in the data is spherical and that the magnetic model has to be revisited, which is currently being done.

A molecular compound: the antibonding wave function in an imino nitroxide free radical

Conjugated nitroxide free radicals are among the most widely used spin carriers in the design of molecular compounds. As their unpaired electron is delocalized over the different atoms of the molecule, they are convenient building blocks and ideal magnetic bridges between magnetic metals to achieve new compounds with particular magnetic properties. In the case of nitronyl nitroxides, the unpaired electron is supposed to be, in a first approximation, equally shared by the four atoms O, N, N and O, and the single occupied molecular orbital (SOMO) is supposed to exhibit a node on the C atom in between the two NOs (Figure 5(a)). In the case of imino nitroxides, the unpaired electron is mainly carried by the three atoms N, N and O, but, as the symmetry is broken, no node is expected on the central C atom for the SOMO (Figure 5(b)). Several studies of spin densities have been performed on nitronyl nitroxides [13]. We demonstrate here the use of MaxEnt reconstruction with a non-uniform prior for 2-(3-nitrophenyl)-4,4,5,5-tetramethyl-4,5-dihydro-1H-imidazol-1-oxyl (m-NPIN), an imino nitroxide with two non-equivalent molecules in the asymmetric unit cell: molecule A and molecule B.

In order to figure out the F_N's, the nuclear structure was refined from unpolarized neutron data taken at 30 K, in the paramagnetic state, on a 4-circle diffractometer. Furthermore, a set of 248 flipping ratios was measured with polarized neutrons at 1.6 K, with the spin density long range ordered by a 4.65 T applied magnetic field.

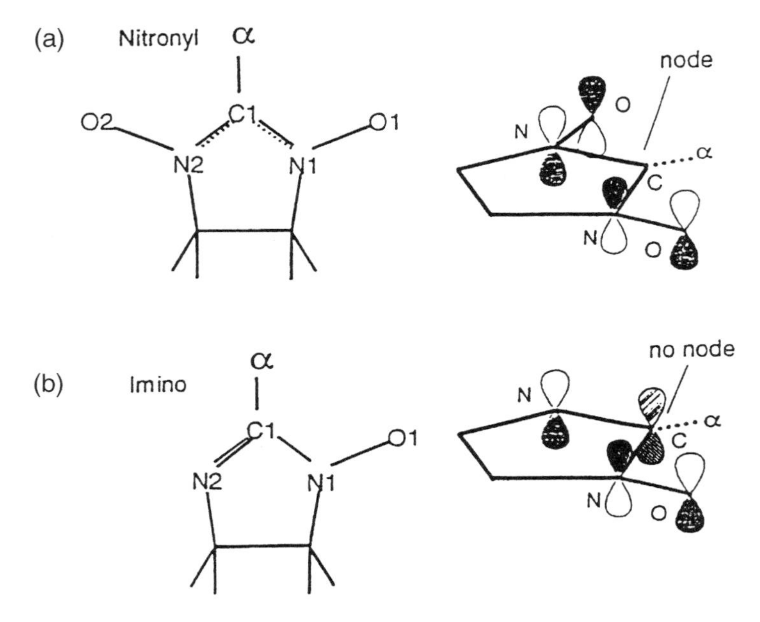

Figure 5. Nitroxide radicals and their SOMO: (a) nitronyl nitroxide and (b) imino nitroxide.

Table 2. m-NPIN: spin populations in the wave function modeling.

Site	Wave function modeling
O1	0.322 (9)
N1	0.258 (9)
C1	−0.042 (7)
N2	0.193 (7)

An approach to solving the inverse Fourier problem is to reconstruct a parametrized spin density based on axially symmetrical p orbitals (p_z orbitals) centered on all the atoms of the molecule (wave function modeling). In the model which was actually used, the spin populations of corresponding atoms of A and B were constrained to be equal. The 'averaged' populations thus refined are displayed in Table 2. Most of the spin density lies on the O1, N1 and N2 atoms. However, the agreement obtained between observed and calculated data ($\chi^2 = 2.1$) indicates that this model is not completely satisfactory.

The spin density reconstructed from MaxEnt with a uniform prior, and projected on the plane of the molecule, is represented with its low contours and with its high contours for molecules A and B in Figure 6. The majority of the spin resides on the N1, N2 and O1 atoms, equally shared between those sites. On the N1 and O1 sites of both molecules the density is not centered on the nuclei but is slightly shifted away from

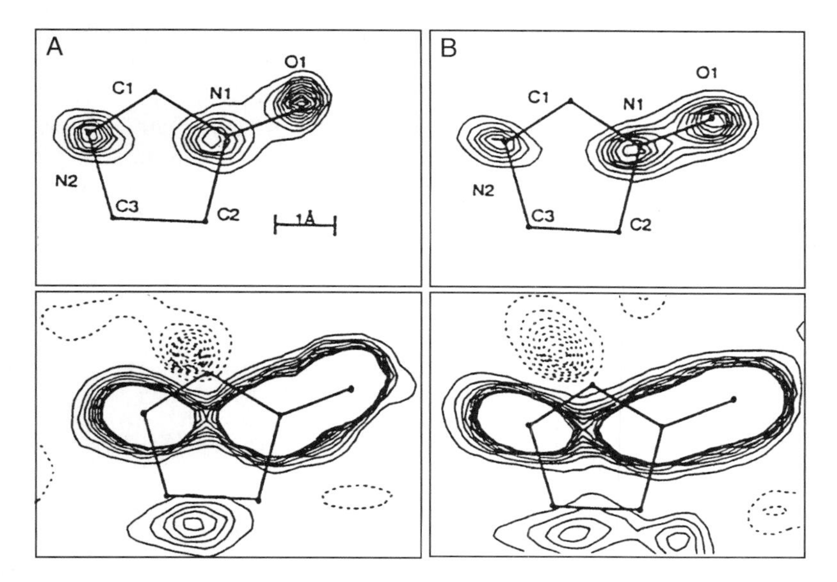

Figure 6. m-NPIN: MaxEnt reconstruction for molecules A and B with a flat prior.

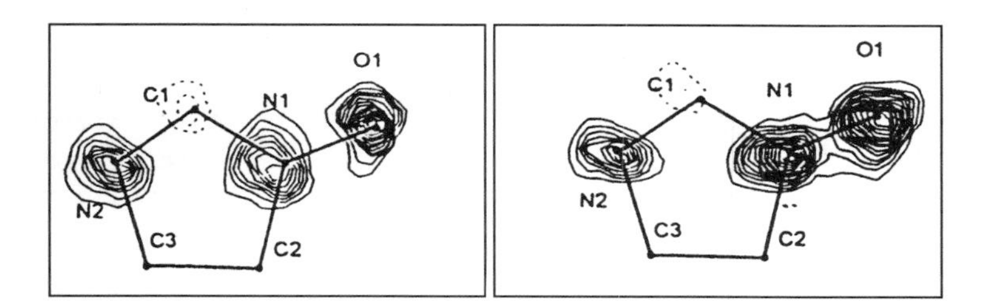

Figure 7. m-NPIN: MaxEnt reconstruction for molecules A and B with a non-uniform prior.

the center of the N1–O1 bond. The effect is more pronounced on the N1 site. On the central C1 carbon atoms, the spin density is negative. Moreover it is off-centered, shifted in the N1–N2 direction.

Are these off-centering real or due to an artifact of the reconstruction? The fact that they occur the same way on two unequivalent molecules is already an indication. The best way to completely answer the question is to reconstruct the spin density with a MaxEnt method and a non-uniform prior, a prior in which the density is centered on the nuclei. We have done this reconstruction, taking as a prior for the two molecules the 'averaged' parametrized spin density refined above. The result is shown in Figure 7. The off-centering of the N1–Ol density and of the negative Cl density is still there, even at the price of a loss of entropy, as it departs from the model.

On the one hand, the antibonding character of the SOMO appears clearly on the N–O bond: the 2p orbitals are slightly bent and pushed away from the center of the bond. On the other hand the observed negative and off-centered density from the carbon nucleus is the result of a competition between spin polarization and spin delocalization. Both are of the same order of magnitude, the spin polarization being slightly larger, providing a negative density and a shift from the central position.

Through these examples we see that we have with the non-uniform prior MaxEnt reconstruction, not only a method which takes advantage of all the knowledge to get the best possible map, but also a very powerful way to tell to what extent a proposed model is compatible with experimental data.

References

1. Sakata, M. and Sato, M. (1990) *Acta Cryst.*, **A46**, 263–270.
2. Papoular, R.J. and Cox, D.E. (1995) *Europhysics Letters*, **32**, 337–342.
3. Papoular, R.J., Vekhter, Y. and Coppens, P. (1996) *Acta Cryst.*, **A52**, 397–407.
4. Sakata, M., Uno, T., Takata, M. and Howard, C.J. (1993) *J. Appl. Cryst.*, **26**, 159–165.
5. Schiebel, P., Wulf, K., Prandl, W., Heger, G., Papoular, R.J. and Paulus, W. (1996) *Acta Cryst.*, **A52**, 176–188.
6. Schiebel, P., Prandl, W., Papoular, R.J. and Paulus, W. (1996) *Acta Cryst.*, **A52**, 189–197.
7. Boucherle, J.X., Henry, J.Y., Papoular, R.J., Rossat-Mignot, J., Schweizer, J., Tasset, F. and Uimin, G. (1993) *Physica B*, **192**, 25–38.
8. Zheludev, A., Ressouche, E., Schweizer, J., Wan, M. and Wang, H. (1994) *J. Mag. Magn. Mat.*, **135**, 147–160.
9. Papoular, R.J., Zheludev, A., Ressouche, E. and Schweizer, J. (1995) *Acta Cryst.*, **A51**, 295–300.
10. Skilling, J. (1988) In *Maximum Entropy and Bayesian Methods in Science and Engineering*, Vol. I, Erickson, G.J. and Smith, C.R. (Eds.), Kluwer Academic Publishers, pp. 173–187.
11. Schweizer, J. and Tasset, F. (1969) *Mater. Res. Bull.*, **4**, 369–376.
12. Schweizer, J. and Tasset, F. (1980) *J. Phys. F: Metal Phys.*, **10**, 2799–2818.
13. Bonnet, M., Luneau, D., Ressouche, E., Rey, P., Schweizer, J., Wan, M., Wang, H. and Zheludev, A. (1995) *Mol. Cryst. Liquid Cryst.*, **271**, 35–53.

4

Transferability, adjustability, and additivity of fuzzy electron density fragments

PAUL G. MEZEY

Mathematical Chemistry Research Unit, Department of Chemistry and Department of Mathematics and Statistics, University of Saskatchewan, 110 Science Place, Saskatoon, Canada, S7N 5C9

1. Introduction

The molecular electron density cloud is a fuzzy object. For large enough distances from the center of the molecule, the value of electron density converges to zero exponentially, and there is no sharp boundary where a molecule 'ends'. Any such sharp boundary would violate the principle of quantum mechanical uncertainty. If our goal is to study local regions of molecules, it is natural to decompose the molecular density cloud into boundaryless, fuzzy fragments exhibiting convergence behavior analogous to those of density clouds of complete molecules.

From the early advances in the quantum-chemical description of molecular electron densities [1–9] to modern approaches to the fundamental connections between experimental electron density analysis, such as crystallography [10–13] and density functional theories of electron densities [14–43], patterns of electron densities based on the theory of catastrophes and related methods [44–52], and to advances in combining theoretical and experimental conditions on electron densities [53–68], local approximations have played an important role. Considering either the formal charges in atomic regions or the representation of local electron densities in the structure refinement process, some degree of approximate transferability of at least some of the local structural features has been assumed.

In more recent years, additional progress and new computational methodologies in macromolecular quantum chemistry have placed further emphasis on studies in transferability. Motivated by studies on molecular similarity [69–115] and electron density representations of molecular shapes [116–130], the transferability, adjustability, and additivity of local density fragments have been analyzed within the framework of an Additive Fuzzy Density Fragmentation (AFDF) approach [114, 131, 132]. This AFDF approach, motivated by the early charge assignment approach of Mulliken [1, 2], is the basis of the first technique for the computation of *ab initio* quality electron densities of macromolecules such as proteins [133–141].

Approximate transferability of fuzzy density fragments is a key feature of the method, where the fuzzy fragments are 'custom-made' in order to reproduce interfragment interactions. By increasing the size of the 'interaction shell' about each fuzzy density fragment, the error of transferred fragment densities can be reduced below any positive threshold. One tool for this purpose is the Adjustable Density

Paul G. Mezey and Beverly E. Robertson (eds.), Electron, Spin and Momentum Densities and Chemical Reactivity, 45–69

Matrix Assembler (ADMA) method, introduced for the generation of *ab initio* quality approximate density matrices for macromolecules [142–146], and for the computation of approximate macromolecular forces [146], among other molecular properties.

Such fragment density matrices must fulfill a set of constraints, in part to ensure a proper representation of the charge conservation condition, and in part to fulfill the technical requirement of mutual compatibility of fuzzy fragment density matrices within an additive framework. Based on properly combined compositions of Löwdin transforms and inverse transforms [147–149] of density matrices, it is possible to combine the relevant idempotency constraints of the assembled density matrices with the adjustability and additivity conditions of fragment density matrices [146]. With respect to experimental electron density representations, a similar method is applied in Quantum Crystallography [67, 68]. The ADMA approach is suitable to describe a series of deformed electron densities occurring during a formal chemical reaction, and to evaluate the similarities within a family of density matrices of related molecules participating in similar chemical reactions in order to find correlations between their reactivities and similarities.

Simple, approximate methods for the readjustment of fragment electron densities based on exact deformations of nuclear arrangements are the Dimension Expansion–Reduction (DER) and the Weighted Affine Transformation (WAT) techniques [113, 114, 130, 150–152]. In addition, an application of the Löwdin transform–inverse Löwdin transform method also serves as a tool for the generation of approximate macromolecular density matrices for slightly distorted nuclear arrangements, if for the original nuclear arrangement a density matrix is available. These methods have also been suggested as tools in the study of the shape and deformability of quantum chemical functional groups [113, 114, 130, 146].

In a certain sense, the differential-topological and algebraic-topological methods of molecular shape characterization [116–130] imitate the natural process of visual comparisons, based on the detection, analysis, and algebraic characterization of various curvature regions of the object, for example, in the simplest case, the locally convex, concave, or saddle type regions of the object. The results of these topological methods are fully reproducible, a claim that cannot be made for visual inspections. These techniques are not restricted to complete molecules. A topological description of the essential properties of local electron densities also has many advantages. Local electron density fragments exhibit a variety of important topological properties which can be used for their characterization.

The description of fuzzy, local density fragments is facilitated by the use of local coordinate systems, however, some compatibility conditions of such local coordinate systems must be fulfilled, reflecting the mutual relations of the fragments within the complete molecule. Manifold theory, topological manifolds, and in particular, differentiable manifolds [153–158], are the branches of mathematics dealing with the general properties of compatible local coordinate systems.

A special technique, the Alexandrov one-point compactification method, often used by topologists within a differential-topological framework, has been applied in the proof of the 'Holographic Electron Density Fragment Theorem' [159–161].

Earlier density extension results were proven only for parts of artificial molecular electron densities, where the complete molecule was assumed to be confined to a finite, bounded region of the three-dimensional space [21], a condition that violates quantum mechanics. However, the new 'Holographic Electron Density Fragment Theorem' quoted here proves the unique extension property of parts of quantum-mechanically correct, boundaryless electron densities of molecules. This new theorem is of special importance with respect to transferability, establishing that for complete, boundaryless molecular electron densities no actual fragment density of sharp boundaries is perfectly transferable. This result has implications on using averaged electron densities for similarity analysis [162].

These topics provide the motivation for a brief topological review in Section 2, followed by some of the details of transferability properties in Section 3. Also in Section 3, some of the consequences of the 'Holographic Electron Density Fragment Theorem' will be discussed, as well as the general proposition that 'No physical system with more than one quantum state is rigorously transferable'. In fact, even atomic nuclei within molecules are not rigorously transferable.

The approximate transferability of fuzzy fragment density matrices, and the associated technical, computational aspects of the idempotency constraints of assembled density matrices, as well as the conditions for adjustability and additivity of fragment density matrices are discussed in Section 4, whereas in Section 5, an algorithm for small deformations of electron densities are reviewed. The Summary in Section 6 is followed by an extensive list of relevant references.

2. Some topological concepts relevant to the shape of molecular electron densities

In some chemical reactions and conformational changes the molecular interactions are often dominated by the local molecular shape properties. Such local properties often show high degrees of similarities within a family of related molecules, and it is natural to expect some, limited transferability of these local moieties. In such cases it is natural to focus on the corresponding local regions of the molecular electron density. Local characterization of a molecular moiety is facilitated by using local coordinate systems. For example, local curvature properties of Molecular Isodensity Surfaces (MIDCOs) $G(K, a)$ of nuclear configuration K and electron density threshold a are often characterized in terms of local Hessian (curvature) matrices expressed as the matrices of second derivatives of local isodensity surfaces interpreted as being defined over various local tangent planes of the MIDCO surfaces $G(K, a)$. Similarly, local coordinate systems are advantageous when using the three-dimensional local curvatures of the four-dimensional representation of molecular density functions, where in addition to the three spatial coordinates, the electron density value is represented along a fourth coordinate axis.

Local coordinate systems can be required to conform with certain mutual compatibility requirements which ensure that the local descriptions are compatible with a global description of the complete system. The branch of topology that deals with such compatible families of local coordinate systems is manifold theory. In the particular case when continuous and differentiable functions are studied within a metric space, such as the three-dimensional electron densities of molecules embedded in the ordinary three-dimensional Euclidean space E^3, the mutual compatibility conditions of local coordinate systems can be formulated in terms of the properties of differentiable manifolds. Such differentiable manifolds provide a framework for a topological analysis of molecular shape in terms of a family of topological similarity measures, based on the very useful concept of topological resolution. In the following paragraphs some of the fundamental concepts of the relevant branches of point set topology and manifold theory are reviewed with special focus on local representations which are relevant to the problem of transferability of subsystems of a system. More details of the fundamentals of topology, as well as some more advanced topological subjects can be found in Refs. [153–158].

Topology is the branch of mathematics that is based on the most general properties of open sets and continuity. Some of the basic concepts can be illustrated using the more familiar setting of a metric space, that is, a space where a distance function, with the intuitively natural properties of distance in the ordinary, three-dimensional space is defined. Within a metric space Y a set A is called an *open set* if around every point y of Y there exists some ball that is also contained within the set A. Open sets of a metric space Y have some fundamental properties that make them very useful, for example, these properties lead to a powerful interpretation (in fact, definition) of continuity of functions: a function f, $f : Y \rightarrow Y'$, assigning points of one metric space Y to points of another metric space Y' is continuous if the inverse image of every open set is also an open set. In a metric space, the definition of openness requires the concept of distance in order to specify the radius of the balls surrounding various points. However, the concept of distance is usually not available if our concern is the topological structure of objects, hence openness, as well as continuity, require an alternative approach in topology. One can, in fact, use some of the very properties of open sets recognized in a metric space as the conditions for openness. These properties themselves may be used to define which sets are to be regarded as open sets. This cannot be done entirely arbitrarily, but there is a surprising degree of freedom in choosing open sets in a mutually consistent way. We say that within a set X a topology $\mathbf{T}$ is defined if a family of subsets of X is specified as the open sets in X, where these sets must fulfill some, not very severe, mutual compatibility conditions.

Specifically, a family $\mathbf{T}$ of subsets of X,

$$\mathbf{T} = \{T_\alpha : X \supset T_\alpha\}, \tag{1}$$

is called a topology on set X, if the following conditions are satisfied:

$$\text{(i)} \qquad X, \varnothing \in \mathbf{T}, \tag{2}$$

where $\varnothing$ is the empty set,

$$\text{(ii)} \qquad \bigcup_\beta T_\beta \in \mathbf{T} \tag{3}$$

for any number of sets in $\mathbf{T}$, and

$$\text{(iii)} \qquad T_\alpha \cap T_\beta \in \mathbf{T} \tag{4}$$

for any two sets $T_\alpha, T_\beta \in \mathbf{T}$.

These three properties, (i)–(iii), are among the properties of open sets in a metric space.

If a set X is provided with a topology $\mathbf{T}$, then the pair $(X, \mathbf{T})$ is called a topological space.

Of course, there are many ways one can select such a family $\mathbf{T}$, and on a given set X one can define many, different topologies. Consequently, when discussing topological properties of a given object (space) X, the actual topology $\mathbf{T}$ must be specified.

Elements of the family $\mathbf{T}$ are called $\mathbf{T}$-open sets. Following some of the natural properties of sets in a metric space, a set C is called a $\mathbf{T}$-closed set if its complement $C^c = X \backslash C$ is a $\mathbf{T}$-open set. A set may be regarded as an open set or a closed set, depending on the topology; if $\mathbf{T}_1$ and $\mathbf{T}_2$ are two different topologies on set X, then a set A in X may be $\mathbf{T}_1$-open but $\mathbf{T}_2$-closed. Note in particular that for each topology $\mathbf{T}$ on X, a topology $\mathbf{T}^c$ can also be defined, where the $\mathbf{T}^c$-open sets are precisely the $\mathbf{T}$-closed sets and vice versa. This topology $\mathbf{T}^c$ on the same set X is called the cotopology of $\mathbf{T}$ of set X.

The comparison of various topologies provides the tools for the introduction of the concept of topological resolution. Assume that for two topologies $\mathbf{T}_1$ and $\mathbf{T}_2$ on set X the following holds: every $\mathbf{T}_1$-open subset of X is also a $\mathbf{T}_2$-open set. Then $\mathbf{T}_1$ is a subfamily of $\mathbf{T}_2$, that is, $\mathbf{T}_2 \supset \mathbf{T}_1$. If this holds, then we say that topology $\mathbf{T}_1$ is coarser (or weaker) than topology $\mathbf{T}_2$, and topology $\mathbf{T}_2$ is finer (or stronger) than topology $\mathbf{T}_1$. Of course, two topologies on the same set X do not need to relate to one another in this manner, and two topologies are called not comparable if neither is weaker than the other. The coarser–finer relation between some of the topologies on a given set X provides a partial ordering of topologies on X.

A set N, $X \supset N$, is called a $\mathbf{T}$-neighborhood of point $\mathbf{r} \in X$ if and only if there exists a $\mathbf{T}$-open set $G \in \mathbf{T}$ such that $\mathbf{r} \in G$, $N \supset G$.

The concepts of base and subbase of topologies are important in the actual construction of a topology that contains a desired family of sets.

A subfamily B, $\mathbf{T} \supset B$, is a base for topology $\mathbf{T}$ if and only if every $\mathbf{T}$-open set $G \in \mathbf{T}$ is a union of some sets in B.

A subfamily S, $\mathbf{T} \supset S$, is a subbase for topology $\mathbf{T}$ if and only if finite intersections of elements of S form a base for $\mathbf{T}$.

Consider a set X. The topological space $(X, \mathbf{T})$ is called a Hausdorff space if for any two distinct points $\mathbf{x}, \mathbf{y} \in X$ there exist disjoint $\mathbf{T}$-open sets T_x, T_y,

$$T_x, T_y \in \mathbf{T}, \tag{5}$$

$$T_x \cap T_y = \varnothing, \tag{6}$$

which contain points $\mathbf{x}$ and $\mathbf{y}$, respectively:

$$\mathbf{x} \in T_x, \tag{7}$$

$$\mathbf{y} \in T_y. \tag{8}$$

If in a topological space $(X, \mathbf{T})$ any two $\mathbf{T}$-closed sets C_α and C_β of X, such that

$$C_\alpha \cap C_\beta = \varnothing, \tag{9}$$

have the property that there exist disjoint $\mathbf{T}$-open sets $T_\alpha, T_\beta \in \mathbf{T}$,

$$T_\alpha \cap T_\beta = \varnothing, \tag{10}$$

such that

$$T_\alpha \supset C_\alpha, \tag{11}$$

and

$$T_\beta \supset C_\beta, \tag{12}$$

then $(X, \mathbf{T})$ is called a normal topological space.

If the topology $\mathbf{T}$ is chosen as the metric topology, that is, if the $\mathbf{T}$-open sets are precisely those which are open in some metric d introduced into the set X, then one obtains the metric topological space $(X, \mathbf{T})$. Note that the metric topological space $(X, \mathbf{T})$ is a Hausdorff space and also a normal space.

Since the specification of topologies implies that all open sets are defined, the concept of continuity can also be generalized to topological spaces, even if distance functions are not given.

Consider two topological spaces, $(X_1, \mathbf{T}_1)$ and $(X_2, \mathbf{T}_2)$, and a function φ from X_1 to X_2. This function φ is continuous if and only if the inverse image of every $\mathbf{T}_2$-open set of X_2 is $\mathbf{T}_1$-open in X_1:

$$\varphi^{-1}(G) \in T_1 \quad \text{if } G \in T_2. \tag{13}$$

A function φ is called one-to-one if it assigns a unique element $\varphi(x) = y \in X_2$ to each element $x \in X_1$.

A function φ is called onto if every element $y \in X_2$ is assigned to some element $x \in X_1$.

A function φ is called bijective if it is both one-to-one and onto.

A function φ is called a homeomorphism if it is bijective and both φ and its inverse φ^{-1} are continuous,

$$\varphi, \varphi^{-1} \in \mathbf{C}, \tag{14}$$

that is, if φ and φ^{-1} are elements of the class $\mathbf{C}$ of continuous functions on X.

For a comparison of objects, one may focus on how well can these objects correspond to each other and possibly replace each other. A correspondence between two objects, specifically, a correspondence between various parts of the two objects, can be described by some functions that assign the points of one object to the points of the other object. Then, properties of these functions can be used to qualify and even quantify the similarity of the two objects. One advantage of topology over straightforward geometrical techniques is the fact that topology allows one to recognize and use less than perfect correspondence between the points of the two objects; by specifying various topologies, and by testing how well correspondences hold up within each topological setting, one can find a detailed and quantifiable description of similarity.

In one extreme case within the topological framework, the two objects can be brought into a perfect correspondence, demonstrating topological equivalence. In a more precise formulation, two topological spaces $(X_1, \mathbf{T}_1)$ and $(X_2, \mathbf{T}_2)$ are called topologically equivalent or homeomorphic if there exists a function

$$f : X_1 \rightarrow X_2, \tag{15}$$

which is bijective and both f and f^{-1} are continuous. Such a function f is called a homeomorphism.

A property is called topological or topological invariant if it is a property of all topological spaces in an equivalence class generated by the equivalence relation 'topologically equivalent'. Many of the familiar concepts often used in a geometrical setting, such as length, boundedness, or being a Cauchy sequence are not topological properties. On the other hand, connectedness and compactness are topological properties; some of the associated elementary results are described below.

If the set X of a topological space $(X, \mathbf{T})$ is a union of two, non-empty, disjoint $\mathbf{T}$-open subsets,

$$X = A \cup B, \quad A, B \neq \varnothing, \quad A \cap B = \varnothing, \quad A, B \in \mathbf{T}, \tag{16}$$

then the topological space $(X, \mathbf{T})$ is disconnected.

Connectedness is defined indirectly as the lack of disconnectedness: a topological space $(X, \mathbf{T})$ is connected if it is not disconnected. A connected open subset is often called a domain.

Consider an n-dimensional set X. Set X is simply connected if and only if every k-dimensional $(k < n)$ topological sphere S^k in set X is contractible to a point.

Take a general set X, a subset A, $X \supset A$. If there exists a class $F = \{F_i\}$ of open subsets of set X such that

$$\bigcup_i F_i \supset A, \tag{17}$$

then F is called an open cover of A.

The family F is called a finite cover if F contains only a finite number of F_i subsets.

If every open cover of a subset A of a topological space X contains a finite subcover, then the subset A of the topological space X is compact. The compactness property is a generalization of the elementary properties of closed and bounded intervals.

Three-dimensional electron densities have no boundaries; they converge to zero exponentially with distance from the nuclei of the peripheral atoms in the molecule. Considering a single, isolated molecule, the exact quantum-mechanical electron density becomes zero in a strict sense only at infinite distance from the center of mass of the molecule. Consequently, the electron density is not a compact set, just as the embedding three-dimensional Euclidean space E^3 is not compact either. However, the three-dimensional Euclidean space E^3, as a subset of a four-dimensional Euclidean space E^4, can be 'slightly' extended (for example, by adding one point) and 'made' compact by various compactification techniques.

One such compactification technique is the Alexandrov one-point compactification method used in the study of the topology of potential energy hypersurfaces and the fundamental group of reaction mechanisms associated with a given stoichiometric family of molecules [118]. The same technique also has been used in the proof of the Holographic Electron Density Fragment Theorem [159–161], establishing for a complete, boundaryless molecular electron density the holographic property of molecular fragments: any non-zero volume fragment density contains the full information about the electron density of the entire, boundaryless molecule.

Some non-compact topological spaces $(X, \mathbf{T})$ can be converted into some compact topological spaces $(X_\infty, \mathbf{T}_\infty)$ by a technique called the Alexandrov one-point compactification. Here

$$X_\infty = X \cup \{\infty\}, \tag{18}$$

that is, a single point, distinct from every other point of X, is added to X. This additional formal point, denoted by ∞, is analogous to the 'ideal point' of infinity in projective geometry.

The topology $\mathbf{T}_\infty$ consists of the following sets:

$$\mathbf{T}_\infty = \mathbf{T} \cup \{A\colon A = X_\infty \backslash B,\ X \backslash B \in \mathbf{T},\ B \text{ compact in} X\}, \tag{19}$$

that is, the family $\mathbf{T}_\infty$ contains the following sets:

(i) each $\mathbf{T}$-open set;
(ii) the complement in X_∞ of each closed and compact subset of X.

Evidently, the topological space $(X, \mathbf{T})$ is embedded in the compact topological space $(X_\infty, \mathbf{T}_\infty)$, since $(X, \mathbf{T})$ is homeomorphic to a subspace of $(X_\infty, \mathbf{T}_\infty)$, as it follows from the definitions given above.

More details of examples of the chemical applications of the Alexandrov one- point compactification method can be found in Refs. [118] and [159].

Sets of local coordinate systems describing certain local features of complicated objects are often advantageous when compared to a single, global coordinate system. Within a topological framework, the general theory of sets of local coordinate systems is called manifold theory. Often, the local coordinate systems are interrelated, and these relations can be expressed by continuous, and in the case of differentiable manifolds, by differentiable mappings, called homeomorphisms (see Equation (15)), and diffeomorphisms, respectively.

A function φ is called a diffeomorphism if φ is a homeomorphism and both the function φ and its inverse φ^{-1} are infinitely differentiable, that is, both φ and φ^{-1} belong to the class $\mathbf{C}^{\infty}$ of functions:

$$\varphi, \varphi^{-1} \in \mathbf{C}^{\infty}. \tag{20}$$

In differentiable manifolds the local coordinate systems must fulfill some compatibility conditions ensuring that in any overlapping region of two local coordinate systems any additional, differentiable functions expressed in either coordinate system are meaningful and differentiable in the other coordinate system as well.

In many applications it is customary to define local coordinate systems indirectly by establishing their connection with the cartesian coordinates in some underlying Euclidean space E^n, if there is one. By labeling the points within each actual space (of local coordinate system) with the coordinate values in the underlying Euclidean space E^n, there is a common reference for all local coordinate systems, and the compatibility conditions can be formulated within the Euclidean space E^n of familiar and intuitively simple properties.

The underlying Euclidean space E^n also simplifies the definition of individual coordinate systems considerably.

An n-dimensional coordinate system $\varphi^{(i)}$ of a $\mathbf{T}$-open set $G^{(i)}$ of a Hausdorff topological space $(X, \mathbf{T})$ is a homeomorphism $\varphi^{(i)}$ between $G^{(i)}$ and an open set $H^{(i)}$ of the Euclidean space E^n.

Informally, a set X is an n-dimensional topological manifold if X is covered by domains of n-dimensional coordinate systems $\varphi^{(i)}$, $i = 1, 2, \ldots$

If differentiability is also ensured, then one obtains a differentiable manifold.

More precisely, a Hausdorff space X covered by countable many $\mathbf{T}$-open sets $G^{(1)}, G^{(2)}, \ldots$, is an n-dimensional differentiable manifold if it satisfies the following conditions:

(i) for each $\mathbf{T}$-open set $G^{(i)}$ of X there exists an n-dimensional coordinate system $\varphi^{(i)}$;

(ii) if the condition of overlap

$$G^{(i)} \cap G^{(j)} \neq \varnothing \tag{21}$$

holds then the function $\varphi^{(ij)}$ defined as

$$\varphi^{(ij)} : \varphi^{(j)}(G^{(i)} \cap G^{(j)}) \to \varphi^{(i)}(G^{(i)} \cap G^{(j)}) \tag{22}$$

is differentiable.

If space X is an n-dimensional differentiable manifold and if Y is a subset of X, then Y is called an m-dimensional submanifold of X if the following additional conditions hold for Y:

(i) Y itself is an m-dimensional differentiable manifold;

(ii) for every point $\mathbf{y} \in Y$ there exists a local coordinate neighborhood G of point $\mathbf{y}$ in X with a local coordinate system having the following properties:

$$\varphi : G \to H, \tag{23}$$

where H, $E^n \supset H$ is a $\mathbf{T}$-open set, and in the subset

$$\varphi(G \cap Y), \quad H \supset \varphi(G \cap Y), \tag{24}$$

the coordinates are constrained to zero:

$$x_{m+1} = x_{m+2} = \cdots = x_n = 0. \tag{25}$$

The function φ, when restricted to the subset $G \cap Y$, is a local coordinate system for Y around point $\mathbf{y}$.

In representations of electron densities, the presence or lack of boundaries plays a crucial role. A quantum mechanically valid electron density distribution of a molecule cannot have boundaries, nevertheless, artificial electron density representations with actual boundaries provide useful tools of analysis. For these reasons, among the manifold representations of molecular electron densities, manifolds with boundaries play a special role.

The role of a boundary in a manifold with boundary can be interpreted with reference to a hyperplane within a Euclidean space E^n, using the concept of half-space, where the hyperplane is in fact the boundary of the half-space. By appropriate reordering of the coordinates, a half-space H^n becomes the subset of a Euclidean space E^n containing all points of E^n with non-negative value for the last coordinate.

A space M where each point $\mathbf{x} \in M$ has an open neighborhood homeomorphic to a set open within a Euclidean half-space H^n, is an n-dimensional manifold with boundary.

3. Limits to transferability

Transferability of subsystems of large systems is an assumption often invoked in the study of physical objects where a direct analysis of the complete system is cumbersome. The study of subsystems, either in isolation or as parts of a smaller object is often simpler than the study of the original large system; yet in many instances, some of the results obtained for the subsystem can be safely extrapolated to the large system. Whereas transferability has proved to be a very useful concept that leads to important and valid results when used with appropriate caution, it is also a concept that is sometimes poorly justified and may lead to erroneous conclusions.

Although transferability of properties associated with local molecular moieties, for example, the transferability of the expected types of reactions and the degree of reactivities of chemical functional groups, are among the most commonly used assumptions of classical chemistry, nevertheless, within a quantum-mechanical framework, transferability has some natural limitations.

One fundamental limitation can be phrased as a formal statement on the interactions between a quantum system and its surroundings:

Theorem No physical system with more than one quantum state is rigorously transferable.

Proof If there are two or more possible quantum states of a system, interactions with the system may change the quantum state of the system, hence interactions may change the system. Consequently, the system is not necessarily rigorously identical to the system obtained by placing this same system into a different environment. Hence, the system is not rigorously transferable.

In fact, in a precise sense, no molecular fragment is rigorously transferable, although approximate transferability is an exceptionally useful and, if used judiciously, a valid approach within the limitations of the approximation. In particular, it is possible to define non-physical entities, such as fuzzy fragment electron densities, which do not exist as separate objects, yet they show much better transferability properties than actual, physically identifiable subsystems of well-defined, separate identity. This aspect of specially designed, 'custom- made', artificial subsystems of nearly exact additivity has been used to generate *ab initio* quality electron densities for proteins and other macromolecules.

The non-transferability of actual subsystems is manifested on all levels, even on the level of atomic nuclei. Although chemists often regard two nuclei of the same isotope as interchangeable, even such nuclei of identical lists of nucleons are not fully transferable, as evidenced, for example, by NMR spectroscopy. Chemical shifts of nuclei of identical lists of nucleons are different, precisely as a consequence of the nuclei being slightly different, caused by their different interactions with their different surroundings. Consequently, even nuclei are not rigorously transferable.

In a rigorous sense, non-transferability of molecular parts has profound implications on chemical conclusions based on electron densities. Since some of the original results on the utility and reliability of transferred electron densities have been derived within the framework of density functional theory, here we shall follow this approach, and describe a recent result on a general, 'holographic' property of electron density fragments of complete, boundaryless molecular electron densities.

These results, as most related results of density functional theory, have direct connections to the fundamental statement of the Hohenberg–Kohn theorem: the non-degenerate ground state electron density $\rho(\mathbf{r})$ of a molecule of n electrons in a local spin-independent external potential V, expressed in a spin-averaged form as

$$\rho(\mathbf{r}) = n \sum_{s_1} \cdots \sum_{s_n} \int \cdots \int \left|\Psi(\mathbf{r}, s_1, \mathbf{r}_2, s_2, \ldots, \mathbf{r}_n, s_n)\right|^2 \mathrm{d}^3\mathbf{r}_2 \cdots \mathrm{d}^3\mathbf{r}_n, \tag{26}$$

fully determines all properties, including the electronic energy E of the molecule.

The Hamiltonian H of a molecule M can be expressed as

$$H = \sum_{i=1}^{n} V(\mathbf{r}_i) + T + V_{\mathrm{ee}} \tag{27}$$

where the usual notations are used for the kinetic energy operator T, the electron–electron repulsion operator V_{ee}, and external potential $V(\mathbf{r})$,

$$V(\mathbf{r}) = \sum_{i=1}^{n} V(\mathbf{r}_i). \tag{28}$$

In the latter expression, $V(\mathbf{r}_i)$ is the electron–nuclear attraction operator describing the interaction of the ith electron of the molecule with the set of nuclei.

As a consequence of the Hohenberg–Kohn theorem [14], a non-degenerate ground state electron density $\rho(\mathbf{r})$ determines the Hamiltonian H of the system within an additive constant, implying that the electron density $\rho(\mathbf{r})$ also determines all ground state and all excited state properties of the system.

The original Hohenberg–Kohn theorem was directly applicable to complete systems [14]. The first adaptation of the Hohenberg–Kohn theorem to a part of a system involved special conditions: the subsystem considered was a part of a finite and bounded entity regarded as a hypothetical system [21]. The boundedness condition, in fact, the presence of a boundary beyond which the hypothetical system did not extend, was a feature not fully compatible with quantum mechanics, where no such boundaries can exist for any system of electron density, such as a molecular electron density. As a consequence of the Heisenberg uncertainty relation, molecular electron densities cannot have boundaries, and in a rigorous sense, no finite volume, however large, can contain a complete molecule.

It is possible, however, to avoid any violation of these fundamental properties, and derive a result on the local electron densities of non-zero volume subsystems of boundaryless electron densities of complete molecules [159–161]. A four-dimensional representation of molecular electron densities is constructed by taking the first three dimensions as those corresponding to the ordinary three-space E^3 and the fourth dimension as that representing the electron density values $\rho(\mathbf{r})$. Using a compactification method, all points of the ordinary three- dimensional space E^3 can be mapped to a manifold S^3 embedded in a four- dimensional Euclidean space E^4, where the addition of a single point leads to a compact manifold representation of the entire, boundaryless molecular electron density.

The actual properties of this transformation combined with the convergence properties of molecular electron densities implies analyticity almost everywhere on the compact manifold. Consequently, this four-dimensional representation of the molecular electron density satisfies the conditions of a theorem of analytic continuation, that establishes the 'holographic properties' of molecular electron densities represented on the compact manifold S^3.

The non-degenerate ground state electron density $\rho'_d(\mathbf{r}')$ over any subset d of manifold S^3, $S^3 \supset d$, where subset d has non-zero volume on S^3, determines uniquely the ground state electron density $\rho'(\mathbf{r}')$ of the complete molecule over the entire manifold S^3.

This result, in turn, implies the following 'holographic properties' of complete, boundaryless molecular electron densities within the ordinary three-dimensional space [159–161].

The non-degenerate ground state electron density $\rho_D(\mathbf{r})$ over any subset D of the ordinary three-dimensional space E^3, where $E^3 \supset D$, and D has non-zero volume, determines uniquely the ground state electron density $\rho(\mathbf{r})$ of the complete molecule over the entire three-dimensional space E^3.

This result, the 'Holographic Electron Density Fragment Theorem', is a negative statement on the transferability of electron density fragments, since the unique extension property implied by the theorem also implies that any given electron density fragment can be transferred only to an environment that is exactly identical to its original environment.

Nevertheless, approximate transferability is a valid concept and in the next section a particular approach will be discussed, based on fuzzy subsystems of molecular electron densities.

4. Approximate transferability of fuzzy fragment density matrices

If the electron density partitioning results in subsystems without boundaries and with convergence properties which closely resemble the convergence properties of the complete system, then it is possible to avoid one of the conditions of the 'Holographic Electron Density Fragment Theorem', by generating fuzzy electron density fragments which do not have boundaries themselves, but then the actual subsystems considered cannot be confined to any finite domain D of the ordinary three-dimensional space E^3.

Transferred electron density fragments obtained by AFDF method can provide excellent approximations. One such approach, formulated in terms of transferability of fragment density matrices within the AFDF framework is a tool that has been suggested as an approach to macromolecular quantum chemistry [114, 115, 130, 142–146] and to a new density fitting algorithm in the crystallographic structure refinement process [161].

The AFDF approach and the ADMA method have been reviewed in detail [142, 146] and here only a shortened version of the main features of these methods will be given.

The fundamental tool for the generation of an approximately transferable fuzzy electron density fragment is the additive fragment density matrix, denoted by $\mathbf{P}^k$ for an AFDF of serial index k. Within the framework of the usual SCF LCAO *ab initio* Hartree–Fock–Roothaan–Hall approach, this matrix $\mathbf{P}^k$ can be derived from a complete molecular density matrix $\mathbf{P}$ as follows.

In order to assign fuzzy, additive electron density fragments

$$F_1, F_2, \ldots, F_k, \ldots, F_m, \tag{29}$$

represented by fragment density functions

$$\rho^1(\mathbf{r}), \rho^2(\mathbf{r}), \ldots, \rho^k(\mathbf{r}), \ldots, \rho^m(\mathbf{r}) \tag{30}$$

to various subgroups of nuclei, the complete family of nuclei of the molecule is subdivided into m mutually exclusive families,

$$f_1, f_2, \ldots, f_k, \ldots, f_m. \tag{31}$$

The molecular electronic density $\rho(\mathbf{r})$ of a fixed nuclear geometry K is expressed in terms of the complete density matrix $\mathbf{P}$ of dimensions $n \times n$, and a set of n atomic orbitals $\varphi_i(\mathbf{r})(i = 1, 2, \ldots, n)$, as

$$\rho(\mathbf{r}) = \sum_{i=1}^{n} \sum_{j=1}^{n} P_{ij}\varphi_i(\mathbf{r})\varphi_j(\mathbf{r}). \tag{32}$$

As proposed in [131, 132], the general AFDF scheme can be given in terms of an atomic orbital membership function $m_k(i)$ defined as

$$m_k(i) = \begin{cases} 1, & \text{if AO } \varphi_i(\mathbf{r}) \text{ is centered on a nucleus of nuclear set } f_k, \\ 0, & \text{otherwise.} \end{cases} \tag{33}$$

Using weighting factors w_{ij}, w_{ji}, constrained by the relations

$$w_{ij} + w_{ji} = 1, \quad w_{ij}, w_{ji} > 0, \tag{34}$$

the elements P_{ij}^k of the $n \times n$ fragment density matrix $\mathbf{P}^k$ of the kth fuzzy density fragment F_k are defined in terms of these membership functions $m_k(i)$,

$$P_{ij}^k = [m_k(i)w_{ij} + m_k(j)w_{ji}]P_{ij}. \tag{35}$$

The simplest choice of weighting factors,

$$w_{ij} = w_{ji} = 0.5, \tag{36}$$

corresponds to the choice

$$P_{ij}^k = 0.5[m_k(i) + m_k(j)]P_{ij}, \tag{37}$$

equivalent to the Mulliken–Mezey fragmentation scheme used in the MEDLA method and in the simplest version of the more advanced macromolecular density matrix method, the ADMA method [142–146].

If the kth density fragment $\rho^k(\mathbf{r})$ is defined as

$$\rho^k(\mathbf{r}) = \sum_{i=1}^{n} \sum_{j=1}^{n} P_{ij}^k\varphi_i(\mathbf{r})\varphi_j(\mathbf{r}), \tag{38}$$

then these fuzzy electron density fragments $\rho^k(\mathbf{r})$ are exactly additive within the given molecule,

$$\rho(\mathbf{r}) = \sum_{k=1}^{m} \rho^k(\mathbf{r}). \tag{39}$$

This follows from the definition (35) of fragment density matrices $\mathbf{P}^k$ that implies exact additivity of these fragment density matrices, i.e., they add up to the density matrix $\mathbf{P}$ of the complete molecule,

$$P_{ij} = \sum_{k=1}^{m} P_{ij}^k. \tag{40}$$

This, in turn, implies the exact additivity of the fuzzy electron density fragments $\rho^k(\mathbf{r})$ as given by Equation (39).

In the following discussions we shall disregard the small changes of the nuclei induced by their surroundings within molecules, and we shall regard two nuclei identical if their lists of nucleons match.

If two electron density fragments are 'anchored' to two identical sets of nuclei of the same nuclear geometry, and if these two fragments come from two molecules in which these nuclei have locally well-matching surroundings, then the two fragment densities are necessarily very similar and are approximately transferable to replace one another. This fact can be used to build approximate electron densities for macromolecules, by generating fragment densities from small 'parent' molecules where the local surroundings of the 'anchor' nuclei are the same as the local surroundings of an identical set of 'anchor' nuclei in the 'target' macromolecule. By combining fuzzy fragment electron densities, each obtained from an appropriately designed formal 'parent' molecule and 'custom-made' to fit within the appropriate local surroundings within the target macromolecule, approximate electron density can be generated for the entire macromolecule. Applying the AFDF approach within this framework [133–146], such computations have led to the first *ab initio* quality electron densities for proteins and other large molecules.

Whereas the first applications of the AFDF approach were based on a numerical combination of fuzzy fragment electron densities, each stored numerically as a set density values specified at a family of points in a three-dimensional grid, a more powerful approach is the generation of approximate macromolecular density matrices within the framework of the ADMA method [142–146]. A brief summary of the main steps in the ADMA method is given below.

We assume that the nuclear families

$$f_1, f_2, \ldots, f_k, \ldots, f_m \tag{41}$$

of the target macromolecule M are identified and a series of parent molecules

$$M_1, M_2, \ldots, M_k, \ldots, M_m \tag{42}$$

are designed, each parent molecule M_k containing a suitably large 'coordination shell' surrounding the set f_k of 'anchor' nuclei of the fuzzy density fragment F_k, where this coordination shell matches that in the target macromolecule M.

With reference to the individual AO basis sets $\varphi(K_k)$ of fragment density matrices $\mathbf{P}^k(\varphi(K_k))$ obtained from parent molecules M_k of nuclear configurations K_k, on the one hand, and the macromolecular AO basis set $\varphi(K)$ of the macromolecular density matrix $\mathbf{P}(\varphi(K))$ associated with the macromolecular nuclear configuration K, on the other hand, the following mutual compatibility conditions are assumed:

(a) For each fragment density matrix $\mathbf{P}^k(\varphi(K_k))$, the AO basis set $\varphi(K_k)$ is defined in a local coordinate system which has axes parallel and of matching orientations with the axes of the reference coordinate system defined for the macromolecule M.
(b) Each parent molecule M_k contains only complete nuclear families from the sets of nuclear families $f_1, f_2, \ldots, f_k, \ldots, f_m$ specified in the target macromolecule M, with the possible exception of additional nuclei formally connected to the 'dangling bonds' at the peripheries of the parent molecules M_k.

In order to fulfill compatibility condition (a), the local coordinate system of each parent molecule M_k can always be reoriented, resulting in a simple similarity transformation of the original fragment density matrix $\mathbf{P}^k(\varphi'(K_k))$ into a compatible fragment density matrix $\mathbf{P}^k(\varphi(K_k))$,

$$\mathbf{P}^k(\varphi(K_k)) = \mathbf{T}^{(k)}\mathbf{P}^k(\varphi'(K_k))\mathbf{T}'^{(k)}, \tag{43}$$

using a suitable orthogonal transformation matrix $\mathbf{T}^{(k)}$ of the original AO basis set $\varphi'(K_k)$ of improper orientation, converting it into a basis set $\varphi(K_k)$ with proper orientation:

$$\varphi(K_k) = \mathbf{T}^{(k)}\varphi'(K_k). \tag{44}$$

The second compatibility condition can also be fulfilled easily by an appropriate choice of the parent molecules M_k with respect to the selection of the nuclear families f_k of the various fragments within the target macromolecule M.

The AFDF approach fulfilling the above two compatibility constraints is referred to as the mutually compatible AFDF method (MC-AFDF approach).

Within the MC-AFDF ADMA method, the management of multiple index assignments of basis orbitals and individual density matrix elements requires a series of index conversion relations. These relations are briefly reviewed below, using the notations of the original reference [143].

Atomic orbital basis functions have several indices, each referring to a different listing of these basis functions. In order to facilitate the correct index assignment in each case, several auxiliary quantities are defined.

For each index pair k, k' of a pair f_k, $f_{k'}$ of nuclear families, a quantity $c_{k'k}$ is defined as follows:

$$c_{k'k} = \begin{cases} 1, & \text{if nuclear family } f_{k'} \text{ is present in parent molcule } M_k, \\ 0, & \text{otherwise} \end{cases} \tag{45}$$

With respect to the local AO basis set

$$\{\varphi_{a,k'}(\mathbf{r})\}_{a=1}^{n_{k'}} \tag{46}$$

of a nuclear family $f_{k'}$, where the number of AOs in this family is denoted by n_k, the AO basis function of serial number b is referred to as $\varphi_{b,k'}(\mathbf{r})$.

With respect to the AO set

$$\{\varphi_i^k(\mathbf{r})\}_{i=1}^{n_{Pk}} \tag{47}$$

of the kth fragment density matrix $P^k(\varphi(K_k))$ of total number of n_{Pk} AO's, where

$$n_{Pk} = \sum_{k'=1}^{m} c_{k'k} n_{k'}, \tag{48}$$

the notation $\varphi_j^k(\mathbf{r})$ is used for the same AO $\varphi(\mathbf{r})$.

With respect to the basis set

$$\{\varphi_x(\mathbf{r})\}_{x=1}^{n} \tag{49}$$

of the density matrix $\mathbf{P}(K)$ of the target macromolecule M, the same AO $\varphi(\mathbf{r})$ of serial index y is denoted by $\varphi_y(\mathbf{r})$, where the index x for each AO

$$\varphi_x(\mathbf{r}) = \varphi_{a,k'}(\mathbf{r}) = \varphi_i^k(\mathbf{r}) \tag{50}$$

is determined from the index a in the basis set of the nuclear family $f_{k'}$ as follows:

$$x = x(k', a, f) = a + \sum_{b=1}^{k'-1} n_b, \tag{51}$$

where the last entry f in $x(k', a, f)$ indicates that k' and a refer to a family of nuclei, in fact, to the family $f_{k'}$ of the nuclei.

For each index k and nuclear family $f_{k''}$ with indices k and k'' for which $c_{k''k} \neq 0$ holds, three additional quantities are defined:

$$a_k'(k'', i) = i - \sum_{b=1}^{k''} n_b c_{bk}, \tag{52}$$

$$k' = k'(i, k) = \min\{k'' : a_k'(k'', i) \leq 0\}, \tag{53}$$

and

$$a_k(i) = a_k'(k', i) + n_{k'}. \tag{54}$$

In terms of the index function $x(k', a, f)$ with reference to the nuclear family $f_{k'}$, the index function $x = x(k, i, P)$ with respect to the kth fragment density matrix $\mathbf{P}^k(\varphi(K_k))$ is given as

$$x = x(k, i, P) = x(k', a_k(i), f), \tag{55}$$

where the last entry P in the index function $x(k, i, P)$ indicates that indices k and i refer to the fragment density matrix $\mathbf{P}^k(\varphi(K_k))$, and the index $x = x(k, i, P)$ itself is the serial index of an AO basis function in the density matrix $\mathbf{P}(K)$ of target molecule M.

The final macromolecular density matrix $\mathbf{P}(K)$ is rather sparse. The index relations described above help to identify the non-zero matrix elements of $\mathbf{P}(K)$, and the actual computations can be restricted to those. Utilizing these restrictions and carrying out a finite number of steps only for the non-zero matrix elements of each fragment density matrix $\mathbf{P}^k(\varphi(K_k))$, an iterative process is used for the assembly of the macromolecular density matrix $\mathbf{P}(K)$:

$$P_{x(k,i,P),y(k,j,P)}(K) \Leftarrow P_{x(k,i,P),y(k,j,P)}(K) + P^k_{ij}(K_k). \tag{56}$$

This iterative procedure depends linearly on the number of fragments and on the size of the target macromolecule M, as long as the parent molecules M_k are confined to some limited size. The storage of the information on the macromolecular basis set has relatively small computer memory requirements. The computation of the macromolecular electron density from this basis set information and the final macromolecular density matrix $\mathbf{P}(K)$ obtained from the finite iterative process (56) can rely on relation (32). As a consequence of the sparsity macromolecular density matrix $\mathbf{P}(K)$, the computational task has linear computer time requirement with respect to the number of fragments, hence, with respect to the size of the target macromolecule M.

5. Small deformations of electron densities, adjustability and additivity conditions for fragment density matrices

In terms of the three-dimensional local coordinate transformations $\mathbf{R}^{(k)}$ leading to the local basis set transformations $\mathbf{T}^{(k)}$, the entire macromolecular system is naturally covered with a family of local coordinate systems. These local coordinate systems are also pairwise compatible, since the actual transformation $\mathbf{V}^{(k,k')}$ between any two such local systems of some serial indices k and k' can be given explicitly as

$$\mathbf{V}^{(k,k')} = (\mathbf{T}^{(k)})^{-1}\mathbf{T}^{(k')} = (\mathbf{T}^{(k)})'\mathbf{T}^{(k')}, \tag{57}$$

where $(\mathbf{T}^{(k)})'$ stands for the transpose of matrix $\mathbf{T}^{(k)}$, and where the fact that matrix $\mathbf{T}^{(k)}$ is an orthogonal matrix is utilized.

Since the individual coordinate transformations $\mathbf{T}^{(k)}$ depend continuously and differentially on some rotation angles specifying these transformations, the same must hold for the combined transformations $\mathbf{V}^{(k,k')}$ as well, since transposition and matrix multiplication do preserve these properties. Consequently, these local coordinate systems of individual fuzzy electron density fragments and their relations with the global, macromolecular coordinate system satisfy the conditions for a differentiable manifold.

The reference to local coordinate systems may be advantageous if one considers local deformations of macromolecules, such as a small local shape change of the pocket region of an enzyme. If the deformation can be considered as being approximately confined to a few molecular fragments, then within such an approximation it appears justified to retain the local density matrix representations of all other fragments making up the rest of the macromolecule and modify only those fragment density matrices which are assumed to be affected by the deformation.

We shall assume that the fragment density matrix $\mathbf{P}^k(\varphi(K_k))$ is available for the local fragment nuclear geometry K_k, expressed at the corresponding nuclear locations, and with reference to the local basis set $\varphi(K_k)$. If a distorted local nuclear geometry K'_k does not deviate much from the original local nuclear geometry K_k, then a fairly simple matrix transformation of the original fragment density matrix $\mathbf{P}^k(\varphi(K_k))$ can be used to generate an approximate fragment density matrix at the new location K'_k.

In fact, for a simple, but still remarkably useful first approximation of the electronic density of the new nuclear arrangement K'_k, one may use the same density matrix $\mathbf{P}^k(\varphi(K_k))$, but in combination with a new basis set $\varphi(K'_k)$ obtained by simply moving the centers of the old AO basis functions to the new nuclear locations,

$$\rho^k_{\text{apprx}}(\mathbf{r}, K'_k) = \sum_{i=1}^{n}\sum_{j=1}^{n} \mathbf{P}^k_{ij}(\varphi(K_k))\varphi_i(\mathbf{r}, K'_k)\varphi_j(\mathbf{r}, K'_k), \tag{58}$$

where the components of this new local basis set are denoted by $\varphi_i(\mathbf{r}, K'_k)$.

The macromolecular density matrix built from such displaced local fragment density matrices does not necessarily fulfill the idempotency condition that is one condition involved in charge conservation. It is possible, however, to ensure idempotency for a macromolecular density matrix subject to small deformations of the nuclear arrangements by a relatively simple algorithm, based on the Löwdin transform–inverse Löwdin transform technique.

The formal vector $\varphi(K)$ denotes the set of atomic orbital basis functions with centers at the original nuclear locations of the macromolecular nuclear configuration K, where the components $\varphi_i(\mathbf{r}, K)$ of vector $\varphi(K)$ are the individual AO basis functions. The macromolecular overlap matrix corresponding to this set $\varphi(K)$ of AO's is denoted by $\mathbf{S}(K)$. The new macromolecular basis set obtained by moving the appropriate local basis functions to be centered at the new nuclear locations is denoted by $\varphi(K')$, where the notation $\varphi_i(\mathbf{r}, K')$ is used for the individual components of this new basis set $\varphi(K')$. The corresponding new macromolecular overlap matrix is denoted by $\mathbf{S}(K')$.

Pre- and postmultiplication by the matrix $\mathbf{S}(K)^{1/2}$ generates the Löwdin transform of the macromolecular density matrix $\mathbf{P}(K) = \mathbf{P}(\varphi(K), K)$, expressed in terms of the AO basis set $\varphi(K)$:

$$\mathbf{S}(K)^{1/2}\mathbf{P}(\varphi(K), K)\mathbf{S}(K)^{1/2}. \tag{59}$$

For a correct density matrix

$$\mathbf{P}(\varphi(K), K)\mathbf{S}(K)\mathbf{P}(\varphi(K), K) = \mathbf{P}(\varphi(K), K) \tag{60}$$

must hold, consequently, the Löwdin transform $S(K)^{1/2}\mathbf{P}(\varphi(K), K)\mathbf{S}(K)^{1/2}$ of density matrix $\mathbf{P}(\varphi(K), K)$ is idempotent:

$$\begin{aligned}&\mathbf{S}(K)^{1/2}\mathbf{P}(\varphi(K), K)\mathbf{S}(K)^{1/2}\mathbf{S}(K)^{1/2}\mathbf{P}(\varphi(K), K)\mathbf{S}(K)^{1/2}\\ &\quad = \mathbf{S}(K)^{1/2}\mathbf{P}(\varphi(K), K)\mathbf{S}(K)^{1/2}.\end{aligned} \tag{61}$$

The inverse Löwdin transform constructed for the above idempotent matrix $\mathbf{S}(K)^{1/2}\mathbf{P}(\varphi(K), K)\mathbf{S}(K)^{1/2}$, given with respect to the actual new, macromolecular overlap matrix $\mathbf{S}(K')$, is expressed as

$$\mathbf{P}(\varphi(K'), K', [K]) = \mathbf{S}(K')^{-1/2}\mathbf{S}(K)^{1/2}\mathbf{P}(\varphi(K), K)\mathbf{S}(K)^{1/2}\mathbf{S}(K')^{-1/2}. \tag{62}$$

This new, approximate macromolecular density matrix $(\varphi(K'), K', [K])$ for the new, slightly distorted nuclear geometry K' is also idempotent with respect to multiplication involving the actual new overlap matrix $\mathbf{S}(K')$,

$$\mathbf{P}(\varphi(K'), K', [K])\mathbf{S}(K')\mathbf{P}(\varphi(K'), K', [K]) = \mathbf{P}(\varphi(K'), K', [K]). \tag{63}$$

This can be shown as follows. A series of simple substitutions give

$$\begin{aligned}&\mathbf{P}(\varphi(K'), K', [K])\mathbf{S}(K')\mathbf{P}(\varphi(K'), K', [K])\\ &\quad = \mathbf{S}(K')^{-1/2}\mathbf{S}(K)^{1/2}\mathbf{P}(\varphi(K), K)\mathbf{S}(K)^{1/2}\mathbf{S}(K')^{-1/2}\mathbf{S}(K')\mathbf{S}(K')^{-1/2}\\ &\qquad \times \mathbf{S}(K)^{1/2}\mathbf{P}(\varphi(K), K)\mathbf{S}(K)^{1/2}\mathbf{S}(K')^{-1/2}\\ &\quad = \mathbf{S}(K')^{-1/2}\mathbf{S}(K)^{1/2}\mathbf{P}(\varphi(K), K)\mathbf{S}(K)^{1/2}\mathbf{S}(K)^{1/2}\\ &\qquad \times \mathbf{P}(\varphi(K), K)\mathbf{S}(K)^{1/2}\mathbf{S}(K')^{-1/2}\\ &\quad = \mathbf{S}(K')^{-1/2}\mathbf{S}(K)^{1/2}\mathbf{P}(\varphi(K), K)\mathbf{S}(K)\mathbf{P}(\varphi(K), K)\mathbf{S}(K)^{1/2}\mathbf{S}(K')^{-1/2}\\ &\quad = \mathbf{S}(K')^{-1/2}\mathbf{S}(K)^{1/2}\mathbf{P}(\varphi(K), K)\mathbf{S}(K)^{1/2}\mathbf{S}(K')^{-1/2}\\ &\quad = \mathbf{P}(\varphi(K'), K', [K]),\end{aligned} \tag{64}$$

that is, idempotency condition (63) holds.

For the new, slightly distorted macromolecular nuclear geometry K', the electronic density can be expressed as the improved approximation

$$\rho_{\text{apprx}}(\mathbf{r}, K, [K']) = \sum_{i=1}^{n}\sum_{j=1}^{n}\mathbf{P}_{ij}(\varphi(K'), K', [K])\varphi_i(\mathbf{r}, K')\varphi_j(\mathbf{r}, K'). \tag{65}$$

If the original macromolecular density matrix is already available, then such approximate macromolecular electron densities for slightly distorted nuclear geometries are simpler to calculate than the full recalculation of an ADMA macromolecular density matrix that involves a new fragmentation procedure.

Note that for large nuclear displacements, for example, distortions exceeding about 0.3–0.4 a.u., the method based on the Löwdin transform–inverse Löwdin transform technique is not recommended. However, for smaller distortions the method discussed above appears to provide a useful approximation.

6. Summary

Approximate transferability of molecular components is a concept that lies at the foundation of the classification of chemical reactions and molecular families according to functional groups and reaction types. The very definition and choice of molecular components, however, involves questions reaching to the foundations of quantum chemistry, the topological characterization of local and global shape of molecules, the roles of local and global coordinate systems that can be treated within a unified framework using manifold theory, and the limitations on true transferability, as manifested, for example, by the 'holographic electron density fragment theorem', reviewed in this contribution. Approximate transferability, however, remains a useful concept that also serves as the motivation for simple computational algorithms which can utilize common features of slightly distorted macromolecular conformations. These approaches effectively utilize approximate transferability, while maintaining some of the constraints, such as density matrix idempotency, required for consistent electron density representations. After discussions on the theoretical concepts and constraints, some of the relevant computational methods are also reviewed.

References

1. Mulliken, R.S. (1955) *J. Chem. Phys.*, **23**, 1833, 1841, 2338, 2343.
2. Mulliken, R.S. (1962) *J. Chem. Phys.*, **36**, 3428.
3. Hartree, D.R. (1928) *Proc. Cambridge Phil. Soc.*, **24**, 111, 426; *ibid.* (1929) **25**, 225, 310.
4. Fock, V. (1930) *Z. Physik*, **61** 126.
5. Roothaan, C.C. (1951) *Rev. Mod. Phys.*, **23**, 69; *ibid.* (1960) **32**, 179.
6. Hall, G.G. (1951) *Proc. Roy. Soc. London*, **A205**, 541.
7. Löwdin, P.-O. (1955) *Phys. Rev.*, **97**, 1474.
8. McWeeny, R. (1960) *Rev. Mod. Phys.*, **32**, 335.
9. Coleman A.J. (1963) *Rev. Mod. Phys.*, **35**, 668.
10. Sands, D.E. (1969) *Introduction to Crystallography*, Benjamin, New York.
11. Buerger, M.J. (1970) *Contemporary Crystallography*, McGraw-Hill, New York.
12. Woolfson, M.M. (1970) *An Introduction to X-Ray Crystallography*, Cambridge University Press, Cambridge.
13. Glusker, J.P. and Trueblood, K.N. (1972) *Crystal Structure Analysis*, Oxford University Press, New York.
14. Hohenberg, P. and Kohn, W. (1964) *Phys. Rev.*, **136**, B864.
15. Kohn, W. and Sham, L.J. (1965) *Phys. Rev.*, **140**, A1133.
16. Parr, R.G. (1975) *Proc. Natl. Acad. Sci. USA*, **72**, 763.
17. Levy, M. (1979) *Proc. Natl. Acad. Sci. USA*, **76**, 6062.
18. Levy, M. (1979) *Bull. Amer. Phys. Soc.*, **24**, 626.

19. Levy, M. (1982) *Phys. Rev. A*, **26**, 1200.
20. Levy, M. (1990) *Adv. Quant. Chem.*, **21**, 69.
21. Riess, J. and Münch, W. (1981) *Theor. Chim. Acta*, **58**, 295.
22. Coppens, P. and Hall, M.B. (Eds.) (1982) *Electron Distribution and the Chemical Bond* Plenum, New York and London.
23. Ludena, E.V. (1983) *J. Chem. Phys.*, **79**, 6174.
24. Perdew, J.P. (1986) *Phys. Rev. B*, **33**, 8822.
25. Becke, A. (1986) *Phys. Rev. A*, **33**, 2786.
26. Becke, A. (1986) *J. Chem. Phys.*, **84**, 4524.
27. Politzer, P. (1987) *J. Chem. Phys.*, **86**, 1072.
28. Salahub, D.R. (1987) *Adv. Chem. Phys.*, **69**, 447.
29. Becke, A. (1988) *J. Chem. Phys.*, **88**, 1053.
30. Becke, A. (1988) *Phys. Rev. A*, **38**, 3098.
31. Parr, R.G. (1988) *J. Phys. Chem.*, **92**, 3060.
32. Tachibana, A. (1988) *Int. J. Quantum Chem.*, **34**, 309.
33. Tachibana, A. Density functional theory for hidden high-T_c superconductivity. In *High Temperature Superconducting Materials*, Hatfield W.E. and Miller, Jr., J.H. (Eds.), Dekker, New York.
34. March, N.H. (1989) *Electron Density Theory of Atoms and Molecules*, Academic, New York.
35. Parr, R.G. and Yang, W. (1989) *Density Functional Theory of Atoms and Molecules*, Clarendon Press, Oxford.
36. Kryachko, E.S. and Ludena, E.V. (1989) *Density Functional Theory of Many-Electron Systems*, Kluwer, Dordrecht.
37. Ziegler, T. (1991) *Chem. Rev.*, **91**, 651.
38. Pápai, I., Goursot, A., St.-Amant, A. and Salahub, D.R. (1992) *Theor. Chim. Acta*, **84**, 217.
39. Labanowski, J.K. and Andzelm, J. (Eds.) (1991) *Density Functional Methods in Chemistry*, Springer-Verlag, New York.
40. Andzelm, J. and Wimmer, E. (1992) *J. Chem. Phys.*, **96**, 1280.
41. Seminario, J.M. and Politzer, P. (1992) *Int. J. Quantum Chem. Symp.*, **26**, 497.
42. Pichon-Pesme, V., Lecomte, C., Wiest, R. and Benard, M. (1992) *J. Am. Chem. Soc.*, **114**, 2713.
43. Wiest, R., Pichon-Pesme, V., Benard, M. and Lecomte, C. (1994) *J. Phys. Chem.*, **98**, 1351.
44. Collard, K. and Hall, G.G. (1977) *Int. J. Quantum Chem.*, **12**, 623.
45. Tal, Y., Bader, R.F.W., Nguyen-Dang, T.T., Ojha, M. and Anderson, S.G. (1981) *J. Chem. Phys.*, **74**, 5162.
46. Bader, R.F.W. and Nguyen-Dang, T.T. (1981) *Adv. Quantum Chem.*, **14**, 63.
47. Cioslowski, J. (1990) *J. Phys. Chem.*, **94**, 5496.
48. Cioslowski, J., Mixon, S.T. and Edwards, W.D. (1991) *J. Amer. Chem. Soc.*, **113**, 1083.
49. Cioslowski, J. and Fleischmann, E.D. (1991) *J. Chem. Phys.*, **94**, 3730.
50. Cioslowski, J., O'Connor, P.B. and Fleischmann, E.D. (1991) *J. Amer. Chem. Soc.*, **113**, 1086.
51. Cioslowski, J., Mixon, S.T. and Fleischmann, E.D. (1991) *J. Amer. Chem. Soc.*, **113**, 4751.
52. Cioslowski, J. and Mixon, S.T. (1992) *Can. J. Chem.*, **70**, 443.
53. Clinton, W.L., Galli, A.J. and Massa, L.J. (1969) *Phys. Rev.*, **177**, 7.
54. Clinton, W.L., Galli, A.J., Henderson, G.A., Lamers, G.B., Massa, L.J. and Zarur, J. (1969) *Phys. Rev.*, **177**, 27.
55. Clinton, W.L. and Massa, L.J. (1972) *Int. J. Quantum Chem.*, **6**, 519.
56. Clinton, W.L. and Massa, L.J. (1972) *Phys. Rev. Lett.*, **29**, 1363.
57. Clinton, W.L., Frishberg, C., Massa, L.J. and Oldfield, P.A. (1973) *Int. J. Quantum Chem. Quantum Chem. Symp.*, **7**, 505.
58. Henderson, G.A. and Zimmermann, R.K. (1976) *J. Chem. Phys.*, **65**, 619.
59. Tsirel'son, V.G., Zavodnik, V.E., Fonichev, E.B., Ozerov, R.P. and Kuznetsolirez, I.S. (1980) *Kristallogr.*, **25**, 735.
60. Frishberg, C. and Massa, L.J. (1981) *Phys. Rev. B*, **24**, 7018.
61. Frishberg, C. and Massa, L.J. (1982) *Acta Cryst.*, **A38**, 93.
62. Massa, L.J., Goldberg, M., Frishberg, C., Boehme, R.F. and LaPlaca, S.J. (1985) *Phys. Rev. Lett.*, **55**, 622.
63. Frishberg, C. (1986) *Int. J. Quantum Chem.*, **30**, 1.
64. Cohn, L., Frishberg, C., Lee, C. and Massa, L.J. (1986) *Int. J. Quantum Chem., Quantum Chem. Symp.*, **19**, 525.
65. Massa, L.J. (1986) *Chemica Scripta*, **26**, 469.

66. Karle, J. (1991) *Proc. Natl. Acad. Sci. USA*, **88**, 10099.
67. Massa, L., Huang, L. and Karle, J. (1995) *Int. J. Quantum Chem., Quant. Chem. Symp.*, **29**, 371
68. Huang, L., Massa, L. and Karle, J. (1996) *Int. J. Quantum Chem., Quant. Chem. Symp.*, **30**, 1691.
69. Carbó, R., Leyda, L. and Arnau, M. (1980) *Int. J. Quantum Chem.*, **17**, 1185.
70. Carbó, R. and Arnau, M. (1981) Molecular engineering: a general approach to QSAR. In *Medicinal Chemistry Advances*, de las Heras, F.G. and Vega, S. (Eds.), Pergamon Press, Oxford.
71. Carbó, R., Sune, E., Lapena, F. and Perez, B.J. (1986) *J. Biol. Phys.*, **14**, 21.
72. Carbó, R. and Domingo, Ll. (1987) *Int. J. Quantum Chem.*, **32**, 517.
73. Carbó, R. and Calabuig, B. (1989) *Comput. Phys. Commun.*, **55**, 117.
74. Carbó, R. and Calabuig, B. (1992) *Int. J. Quantum Chem.*, **42**, 1681.
75. Carbó, R. and Calabuig, B. (1992) *Int. J. Quantum Chem.*, **42**, 1695.
76. Carbó, R., Calabuig, B., Vera, L. and Besalu, E. (1994) Molecular quantum similarity: theoretical framework, ordering principles, and visualization techniques. In *Advances in Quantum Chemistry*, Vol. 25, Löwdin, P.-O., Sabin, J.R. and Zerner, M.C. (Eds.), Academic Press, New York.
77. Carbó, R. (Ed.) (1995) *Molecular Similarity and Reactivity: From Quantum Chemical to Phenomenological Approaches*, Kluwer Academic Publ., Dordrecht.
78. Hodgkin, E.E. and Richards, W.G. (1987) *Int. J. Quantum Chem., Quant. Biol. Symp.*, **14**, 105.
79. Dean, P.M. (1987) *Molecular Foundations of Drug–Receptor Interaction*, Cambridge University Press, New York.
80. Leicester, S.E., Finney, J.L. and Bywater, R.P. (1998) *J. Mol. Graph.*, **6**, 104.
81. Bywater, R. Communication at the VII Annual Meeting of the Molecular Graphics Society, San Francisco, 1988.
82. Mezey, P.G. (1987) *Int. J. Quantum Chem., Quant. Biol. Symp.*, **14**, 127.
83. Arteca, G.A., Jammal, V.B. and Mezey, P.G. (1988) *J. Comput. Chem.*, **9**, 608.
84. Arteca, G.A., Jammal, V.B., Mezey, P.G., Yadav, J.S., Hermsmeier, M.A. and Gund, T.M. (1988) *J. Molec. Graphics*, **6**, 45.
85. Arteca, G.A. and Mezey, P.G. (1989) *J. Phys. Chem.*, **93**, 4746.
86. Arteca, G.A. and Mezey, P.G. (1989) IEEE Eng. In *Med. Bio. Soc. 11th Annual Int. Conf.*, **11**, 1907.
87. Johnson, M.A. (1989) *J. Math. Chem.*, **3**, 117.
88. Johnson, M.A. and Maggiora, G.M. (Eds.) (1990) *Concepts and Applications of Molecular Similarity*, Wiley, New York.
89. Burt, C., Richards, W.G. and Huxley, P. (1990) *J. Comput. Chem.*, **11**, 1139.
90. Arteca, G.A. and Mezey, P.G. (1990) *Int. J. Quantum Chem. Symp.*, **24**, 1.
91. Mezey, P.G. (1991) *J. Math. Chem.*, **7**, 39.
92. Mezey, P.G. (1991) New symmetry theorems and similarity rules for transition structures. In *Theoretical and Computational Models for Organic Chemistry*, Formosinho, S.J. Csizmadia, I.G. and Arnaut, L.G. (Eds.), Kluwer Academic Publishers, Dordrecht.
93. Harary, F. and Mezey, P.G. (1991) *Theor. Chim. Acta*, **79**, 379.
94. Good, A. and Richards, W.G. (1992) *J. Chem. Inf. Comp. Sci.*, **32**, 112.
95. Zachman, C.-D., Heiden, M., Schlenkrich, M. and Brickmann, J. (1992) *J. Comp. Chem.*, **13**, 76.
96. Allan, N.L. and Cooper, D.L. (1992) *J. Chem. Inf. Comp. Sci.*, **32**, 587.
97. Ponec, R. and Strnad, M. (1992) *J. Chem. Inf. Comp. Sci.*, **32**, 693.
98. Luo, X. and Mezey, P.G. (1992) *Int. J. Quantum Chem.*, **41**, 557.
99. Mezey, P.G. (1992) *J. Math. Chem.*, **11**, 27.
100. Luo, X., Arteca, G.A. and Mezey, P.G. (1992) *Int. J. Quantum Chem.*, **42**, 459.
101. Zabrodsky, H., Peleg, S. and Avnir, D. (1992) *J. Amer. Chem. Soc.*, **114**, 7843.
102. Zabrodsky, H., Peleg, S. and Avnir, D. (1993) *J. Amer. Chem. Soc.*, **115**, 8278.
103. Zachman, C.-D., Kast, S.M., Sariban, A. and Brickmann, J. (1993) *J. Comp. Chem.*, **14**, 1290.
104. Chau, P.-L. and Dean, P.M. (1994) *J. Computer-Aided Molecular Design*, **8**, 545.
105. Mezey, P.G. (1994) *J. Chem. Inf. Comp. Sci.*, **34**, 244.
106. Mezey, P.G. (1994) *Int. J. Quantum Chem.*, **51**, 255.
107. Bywater, R.P. (1995) In *Molecular Similarity and Reactivity: from Quantum Chemical to Phenomenological Approaches*, R. Carbó, (Ed.) Kluwer Academic Publ., Dordrecht, The Netherlands.
108. Cooper, D.L. and Allan, N.L. (1995) In *Molecular Similarity and Reactivity: From Quantum Chemical to Phenomenological Approaches*, Carbó, R. (Ed.), Kluwer Academic Publ., Dordrecht, The Netherlands.
109. Stefanov, B.B. and Cioslowski, J. (1995) *J. Comput. Chem.*, **16**, 1394.
110. Zabrodsky, H. and Avnir, D. (1995) *Adv. Mol. Struct. Res.*, **1**, 1.

111. Zabrodsky, H. and Avnir, D. (1995) *J. Amer. Chem. Soc.*, **117**, 462.
112. Anzali, S., Barnickel, G., Krug, M., Sadowski, J., Wagener, M. and Gasteiger, J. (1996) Evaluation of molecular surface properties using a Kohonen neural network. In *Neural Networks in QSAR and Design*, Devillers, J. (Ed.), Academic Press, London.
113. Mezey, P.G. (1996) Molecular similarity measures of conformational changes and electron density deformations. In *Advances in Molecular Similarity*, Vol. 1, p. 89.
114. Mezey, P.G. (1996) Functional groups in quantum chemistry. In *Advances in Quantum Chemistry*, Vol. 27, p. 163.
115. Mezey, P.G. (1998) Chemical bonding in proteins and other macromolecules. In *Pauling's Legacy: Modern Modelling of Chemical Bonding*, Maksic, Z. and Orville-Thomas, J. (Eds.), Elsevier Science Publ., Amsterdam, The Netherlands.
116. Mezey, P.G. (1986) *Int. J. Quantum Chem., Quant. Biol. Symp.*, **12**, 13.
117. Mezey, P.G. (1987) *J. Comput. Chem.*, **8**, 462.
118. Mezey, P.G. (1987) *Potential Energy Hypersurfaces*, Elsevier, Amsterdam.
119. Mezey, P.G. (1988) *J. Math. Chem.*, **2**, 325.
120. Mezey, G. (1988) *J. Math. Chem.*, **2**, 299.
121. Pipek, J. and Mezey, P.G. (1988) *Internat. J. Quantum Chem. Symp.*, **22**, 1.
122. Pipek, J. and Mezey, P.G. (1989) *J. Chem. Phys.*, **90**, 4916.
123. Mezey, P.G. (1990) Three-dimensional topological aspects of molecular similarity. In *Concepts and Applications of Molecular Similarity*, Johnson, M.A. and Maggiora, G.M. (Eds.), Wiley, New York.
124. Mezey, P.G. (1990) Topological quantum chemistry. In *Reports in Molecular Theory*, Weinstein, H. and Náray-Szabó, G. (Eds.), CRC Press, Boca Raton.
125. Mezey, P.G. (1993) *Shape in Chemistry: An Introduction to Molecular Shape and Topology*, VCH Publishers, New York.
126. Mezey, P.G. (1993) *J. Math. Chem.*, **12**, 365.
127. Zimpel, Z. and Mezey, P.G. (1996) *Int. J. Quant. Chem.*, **59**, 379.
128. Zimpel, Z. and Mezey, P.G. (1997) *Int. J. Quantum Chem.*, **64**, 669.
129. Mezey, P.G. (1997) Fuzzy measures of molecular shape and size. In *Fuzzy Logic in Chemistry*, Rouvray, D.H. (Ed.), Academic Press, San Diego.
130. Mezey, P.G. (1997) Quantum chemistry of macromolecular shape, *Internat. Rev. Phys. Chem.*, **16**, 361.
131. Mezey, P.G. (1995) *Density domain bonding topology and molecular similarity measures.*, In *Topics in Current Chemistry, Vol. 173, Molecular Similarity*, Sen, K. (Ed.), Springer-Verlag, Heidelberg.
132. Mezey, P.G. (1995) Methods of molecular shape-similarity analysis and topological shape design. In *Molecular Similarity in Drug Design*, Dean, P.M. (Ed.), Chapman & Hall, Blackie Publishers, Glasgow, UK.
133. Walker, P.D. and Mezey, P.G. (1993) Program MEDLA 93, Mathematical Chemistry Research Unit, University of Saskatchewan, Saskatoon, Canada.
134. Walker, P.D. and Mezey, P.G. (1993) *J. Amer. Chem. Soc.*, **115**, 12423.
135. Walker, P.D. and Mezey, P.G. (1994) *J. Amer. Chem. Soc.*, **116**, 12022.
136. Walker, P.D. and Mezey, P.G. (1994) *Canad. J. Chem.*, **72**, 2531.
137. Walker, P.D. and Mezey, P.G. (1995) *J. Math. Chem.*, **17**, 203.
138. Walker, P.D. and Mezey, P.G. (1995) *J. Comput. Chem.*, **16**, 1238.
139. Mezey, P.G. (1997) Computational microscopy: pictures of proteins, *Pharmaceutical News*, **4**, 29.
140. Borman, S. (1995) MEDLA Technique Calculates Electron Densities, *Chem. Eng. News*, **73**, 29.
141. Mezey, P.G. (1998) Shape analysis. In *Encyclopedia of Computational Chemistry*, Schleyer, P.V.R. (Ed.), Wiley, Chichester, UK.
142. Mezey, P.G. (1996) Local shape analysis of macromolecular electron densities. In *Computational Chemistry: Reviews and Current Trends*, Leszczynski, J. (Ed.), World Scientific Publ., Singapore.
143. Mezey, P.G. (1995) *J. Math. Chem.*, **18**, 221.
144. Mezey, P.G. (1995) Program ALDA 95, Mathematical Chemistry Research Unit, University of Saskatchewan, Saskatoon, Canada.
145. Mezey, P.G. (1995) Program ADMA 95, Mathematical Chemistry Research Unit, University of Saskatchewan, Saskatoon, Canada.
146. Mezey, P.G. (1997) *Int. J. Quantum Chem.*, **63**, 39.
147. Löwdin, P.-O. (1950) *J. Chem. Phys.*, **18**, 365.
148. Löwdin, P.-O. (1956) *Adv. Phys.*, **5**, 1.
149. Löwdin, P.-O. (1970) *Adv. Quantum. Chem.*, **5**, 185.

150. Mezey, P.G. (1995) Program DER95, Mathematical Chemistry Research Unit, University of Saskatchewan, Saskatoon, Canada.
151. Mezey, P.G. (1995) Program WAT95, Mathematical Chemistry Research Unit, University of Saskatchewan, Saskatoon, Canada.
152. Mezey, P.G. (1996) Program SWAT 96, Mathematical Chemistry Research Unit, University of Saskatchewan, Saskatoon, Canada.
153. Simmons, G.F. (1963) *Introduction to Topology and Modern Analysis*, McGraw-Hill, New York,
154. Gamelin, T.W. and Greene, R.E. (1963) *Introduction to Topology*, Saunders College Publishing, New York.
155. Singer, I.M. and Thorpe, J.A. (1976) *Lecture Notes on Elementary Topology and Geometry*, Springer-Verlag, New York.
156. Morse, M. and Cairns, S.S. (1969) *Critical Point Theory in Global Analysis and Differential Topology: An Introduction*, Academic Press, New York, London.
157. Guillemin, V. and Pollack, A. (1974) *Differential Topology*, Prentice Hall, Englewood Cliffs.
158. Bishop, R.L. and Crittenden, R.J. (1964) *Geometry of Manifolds*, Academic Press, New York.
159. Mezey, P.G. (1999) *Molec. Phys.*, **96**, 169.
160. Mezey, P.G. (1998) *J. Math. Chem.*, **23**, 65.
161. Mezey, P.G. (1998) A crystallographic structure refinement approach using *ab initio* quality additive fuzzy density fragments. In *Advances in Molecular Structure Research*, Hargittai, M. and Hargittai, I. (Eds.), JAI Press, New York.
162. Mezey, P.G. (1998) Averaged Electron Densities for averaged conformations, *J. Comput. Chem.*, **19** 1337.

5

Beyond the local-density approximation in calculations of Compton profiles

YASUNORI KUBO

Department of Physics, College of Humanities and Sciences, Nihon University, 3-chome sakurajosui setagaya-ku, Tokyo 156, Japan

1. Introduction

Generally, all band theoretical calculations of momentum densities are based on the local-density approximation (LDA) [1] of density functional theory (DFT) [2]. The LDA-based band theory can explain qualitatively the characteristics of overall shape and fine structures of the observed Compton profiles (CPs). However, the LDA calculation yields CPs which are higher than the experimental CPs at small momenta and lower at large momenta. Furthermore, the LDA computation always produces more pronounced fine structures which originate in the Fermi surface geometry and higher momentum components than those found in the experiments [3–5].

One obvious drawback of the LDA-based band theory is that the self-interaction term in the Coulomb interaction is not completely canceled out by the approximate self-exchange term, particularly in the case of a tightly bound electron system. Next, the discrepancy is believed to be due to the DFT which is a ground-state theory, because we have to treat quasi-particle states in the calculation of CPs. To correct these drawbacks the so-called self-interaction correction (SIC) [6] and GW-approximation (GWA) [7] are introduced in the calculations of CPs and the full-potential linearized APW (FLAPW) method [8] is employed to find out the effects. No established formula is known to take into account the SIC.

In the present calculation the SIC potential is introduced for each angular momentum in a way similar to the SIC one for atoms [9]. The effects of the SIC are examined on the CPs of three materials, diamond, Si and Cu compared with high resolution CP experiments except diamond [10, 11]. In order to examine the quasi-particle nature of the electron system, the occupation number densities of Li and Na are evaluated from the GWA calculation and the CPs are computed by using them [12, 13].

The purpose of this paper is as follows. Section 2 outlines why we have to go beyond the LDA in the calculations of CPs. The first approach, SIC, beyond the LDA is presented in Section 3, the other approach, GWA, is given in Section 4, and the results are discussed compared with experimental ones in Sections 3 and 4. Section 5 contains the summary and conclusions.

Paul G. Mezey and Beverly E. Robertson (eds.), Electron, Spin and Momentum Densities and Chemical Reactivity, 71–91

2. Why are trials beyond the LDA necessary?

In a typical Compton scattering experiment with unpolarized radiation, the cross section is expressed as

$$\frac{\mathrm{d}^2\sigma}{\mathrm{d}\Omega\,\mathrm{d}\omega} = \left(\frac{\mathrm{d}\sigma}{\mathrm{d}\omega}\right)_0 \left(\frac{\omega_f}{\omega_i}\right) S(\mathbf{q},\omega), \tag{1}$$

where $(\mathrm{d}\sigma/\mathrm{d}\omega)_0$ is the well-known Thomson scattering cross section, and $\hbar\omega = \hbar(\omega_i - \omega_f)$ is the transferred energy. The dynamical structure factor $S(\mathbf{q},\omega)$ is expressed as

$$S(\mathbf{q},\omega) = \sum_f \left| \left\langle f \left| \sum_j \mathrm{e}^{\mathrm{i}\mathbf{q}\mathbf{r}_j} \right| i \right\rangle \right|^2 \delta(E_f - E_i + \omega). \tag{2}$$

In the ideal case being performed at X-ray energy transfers much higher than the characteristic energies of the scattering system, the impulse approximation [14] is applicable. In this case, the dynamical structure factor is directly connected with the electron momentum density $\rho(\mathbf{p})$:

$$S(\mathbf{q},\omega) = \int \rho(\mathbf{p})\,\mathrm{d}\mathbf{r}\,\delta\left(\omega - \frac{\mathbf{q}^2}{2m} - \frac{\mathbf{q}\cdot\mathbf{p}}{m}\right). \tag{3}$$

Taking the photon scattering vector $\mathbf{q}$ in z-direction, the dynamical structure factor is related to the Compton profile $J(p_z)$ by

$$S(\mathbf{q},\omega) = \frac{m}{|\mathbf{q}|} \int \rho(\mathbf{p})\,\mathrm{d}p_x\,\mathrm{d}p_y = \frac{m}{|\mathbf{q}|} J(p_z), \tag{4}$$

$$p_z = \frac{m\omega}{|\mathbf{q}|} - \frac{|\mathbf{q}|}{2}. \tag{5}$$

Here, using electron field operator, momentum density is expressed as

$$\rho(\mathbf{p}) = (2\pi)^{-3} \int \mathrm{d}\mathbf{r} \int \mathrm{d}\mathbf{r}' \exp(\mathrm{i}\mathbf{p}\cdot(\mathbf{r}-\mathbf{r}'))\langle \Psi^*(\mathbf{r},0)\Psi(\mathbf{r}',0)\rangle. \tag{6}$$

Furthermore, the field operator is expanded in the Bloch waves with wave vector $\mathbf{k}$ in the band denoted by b as

$$\Psi(\mathbf{r},t) = \sum_{b,\mathbf{k}} a_{b,\mathbf{k}}(t)\psi_{b,\mathbf{k}}(\mathbf{r}). \tag{7}$$

The momentum density is given by the momentum wave functions and occupation number densities

$$\rho(\mathbf{p}) = \sum_{b,b',\mathbf{k}} \chi^*_{b',\mathbf{k}}(\mathbf{p})\chi_{b,\mathbf{k}}(\mathbf{p})N_{b,b'}(\mathbf{k}). \tag{8}$$

The momentum wave functions are given by

$$\chi_{b,\mathbf{k}}(\mathbf{p}) = \delta(\mathbf{K} + \mathbf{k} - \mathbf{p}) \int \psi_{b,\mathbf{k}}(\mathbf{r}) \exp(-\mathrm{i}\mathbf{p} \cdot \mathbf{r})\, \mathrm{d}\mathbf{r} \tag{9}$$

and the occupation number densities are expressed by annihilation and creation operator, and show translational symmetry with respect to the reciprocal lattice vector **K**.

$$N_{b,b'}(\mathbf{k}) = \langle a^*_{b',\mathbf{k}}(0) a_{b,\mathbf{k}}(0) \rangle, \tag{10}$$

$$N_{b,b'}(\mathbf{k}) = N_{b,b'}(\mathbf{k} + \mathbf{K}). \tag{11}$$

As noticed from this expression, the CP calculation has to be basically carried out on the quasi-particle picture. Formally, quasi-particle energies and wave functions have to be evaluated by solving

$$H_0 \psi_{b,\mathbf{k}}(\mathbf{r}) + \int \mathrm{d}\mathbf{r}' \Sigma(\mathbf{r}, \mathbf{r}'; E_{b,\mathbf{k}}) \psi_{b,\mathbf{k}}(\mathbf{r}') = E_{b,\mathbf{k}} \psi_{b,\mathbf{k}}(\mathbf{r}). \tag{12}$$

Here, H_0 is a Hartree local Hamiltonian that includes the Coulomb effects of both nuclei and average electronic charge distributions,

$$\begin{aligned} H_0 &= -\frac{1}{2}\nabla^2 - \sum_n Z|\mathbf{r} - \mathbf{R}_n|^{-1} + \int \rho(\mathbf{r}')|\mathbf{r} - \mathbf{r}'|^{-1}\, \mathrm{d}\mathbf{r}' \\ &= \frac{\partial}{\partial \rho}\{T[\rho] + U_{\mathrm{ext}}[\rho] + U_{\mathrm{c}}[\rho]\}, \end{aligned} \tag{13}$$

where

$$\rho(\mathbf{r}) = \sum_{b,\mathbf{k}}^{\mathrm{occ}} |\psi_{b,\mathbf{k}}(\mathbf{r})|^2. \tag{14}$$

In Equation (12), the self-energy operator $\Sigma(\mathbf{r}, \mathbf{r}'; E_{b,\mathbf{k}})$ is, in general, non-local and depends on energy. Therefore, to solve the Schrödinger equation, a series of approximations have to be introduced.

First, the self-energy operator is replaced by a local exchange-correlation potential, which is given by the functional derivative of the exchange-correlation energy with respect to the electron density:

$$\int \mathrm{d}\mathbf{r}' \Sigma(\mathbf{r}, \mathbf{r}'; E_{b,\mathbf{k}}) \Rightarrow V_{\mathrm{XC}}(\mathbf{r}) = \frac{\delta E_{\mathrm{XC}}[\rho]}{\delta \rho}, \tag{15}$$

$$E_{\mathrm{XC}}[\rho] = \int \mathrm{d}\mathbf{r}\, \rho(\mathbf{r}) \varepsilon_{\mathrm{XC}}[\rho]. \tag{16}$$

The replacement of Equation (15) corresponds to the density functional method. But the exchange-correlation energy is generally unknown. Therefore, the unknown exchange-correlation energy is replaced by the known form of homogeneous electron gas, which corresponds to the LDA. The replacement is expressed by

$$E_{\mathrm{XC}}^{\mathrm{LDA}}[\rho] = \int \mathrm{d}\mathbf{r}\, \rho(\mathbf{r})\varepsilon_{\mathrm{XC}}[\rho^{\mathrm{h}}], \tag{17}$$

$$V_{\mathrm{XC}}^{\mathrm{LDA}}[\rho] = \frac{\delta E_{\mathrm{XC}}^{\mathrm{LDA}}[\rho]}{\delta\rho}. \tag{18}$$

Thus, the Schrödinger equation (12) is expressed as follows and becomes soluble:

$$\left\{H_0[\rho] + V_{\mathrm{XC}}^{\mathrm{LDA}}[\rho]\right\}\psi_{b,\mathbf{k}}^{\mathrm{LDA}}(\mathbf{r}) = E_{b,\mathbf{k}}^{\mathrm{LDA}}\psi_{b,\mathbf{k}}^{\mathrm{LDA}}(\mathbf{r}). \tag{19}$$

Equation (15) is solved self-consistently employing the FLAPW method. Using the solutions, wave functions and energies, momentum densities in Equation (8) are calculated. In this step, one more drastic approximation we are going to make is that the occupation number in Equation (10) is replaced by the step function

$$N_{b,b'}(\mathbf{k}) \Rightarrow \Theta\left(E_{\mathrm{F}} - E_{b,\mathbf{k}}^{\mathrm{LDA}}\right)\begin{cases}1 & E_{b,\mathbf{k}}^{\mathrm{LDA}} \leq E_{\mathrm{F}},\\ 0 & E_{b,\mathbf{k}}^{\mathrm{LDA}} > E_{\mathrm{F}},\end{cases} \tag{20}$$

where E_{F} is the Fermi energy. From these processes the CP is calculated as follows:

$$\rho^{\mathrm{LDA}}(\mathbf{p}) = \sum_{b,\mathbf{k}} \left|\chi_{b,\mathbf{k}}^{\mathrm{LDA}}(\mathbf{p})\right|^2 \Theta\left(E_{\mathrm{F}} - E_{b,\mathbf{k}}^{\mathrm{LDA}}\right), \tag{21}$$

$$\chi_{b,\mathbf{k}}^{\mathrm{LDA}}(\mathbf{p}) = \delta(\mathbf{K} + \mathbf{k} - \mathbf{p}) \int \psi_{b,\mathbf{k}}^{\mathrm{LDA}}(\mathbf{r}) \exp(-\mathrm{i}\mathbf{p}\cdot\mathbf{r})\, \mathrm{d}\mathbf{r}, \tag{22}$$

$$J^{\mathrm{LDA}}(p_z) = \int \rho^{\mathrm{LDA}}(\mathbf{p})\, \mathrm{d}p_x\, \mathrm{d}p_y. \tag{23}$$

Thus, the obtained CP of Equation (23) corresponds to the so-called conventional band calculation CPs.

Typical CPs calculated by the FLAPW-LDA are shown compared with experiments measured by Sakurai [14] in Figures 1 and 2, for Li and Cu, respectively. As seen in both figures, there are serious discrepancies between the experiments and the calculations. That is, the calculated profiles are higher than the experimental profiles at small momenta and lower at large momenta, as observed consistently in studies of other solids. Therefore, I take this as an indication that we have to go beyond the LDA.

3. Self-interaction correction on CPs

One obvious drawback of the LDA is that, when we replace unknown exchange-correlation energy by the known form of the exchange-correlation for a homogeneous electron gas in Equation (17), we have a problem in that cancelation of self-Coulomb

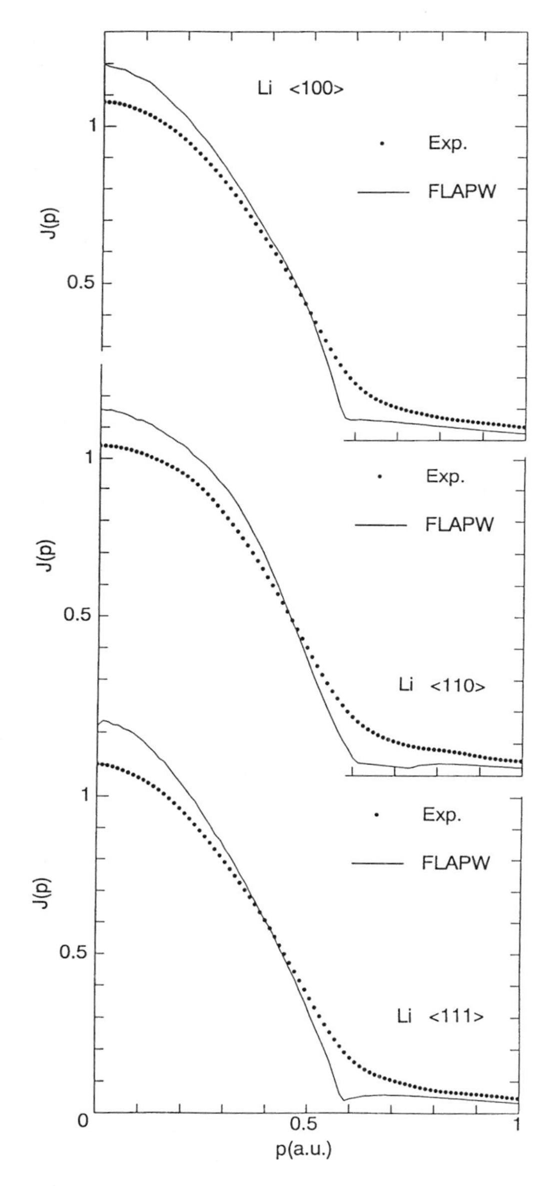

Figure 1. The valence-electron CPs of Li along the three principal symmetry directions. The solid curves represent the FLAPW-LDA calculations. The dots represent the experimental results measured by Sakurai *et al.* [33].

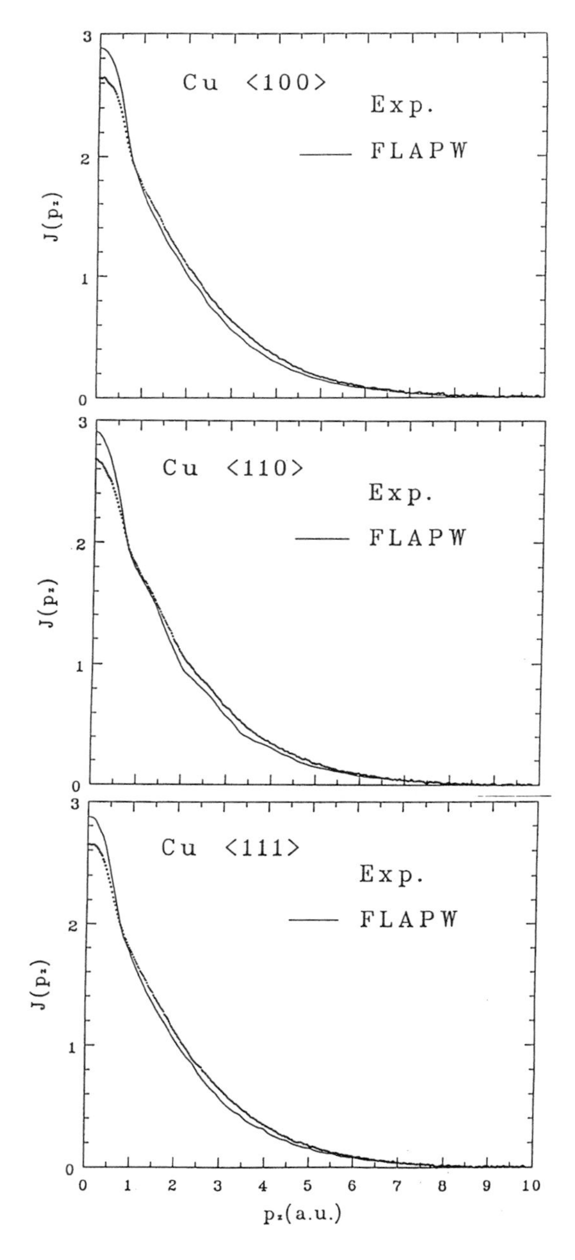

Figure 2. The valence-electron CPs of Cu along the three principal symmetry directions. The solid curves represent the FLAPW-LDA calculations. The dots represent the experimental results measured by Sakurai *et al.* [24].

energy and self-exchange-correlation energy is not generally guaranteed [15], shown as follows:

$$U_{\mathrm{c}}[\rho_{i,\sigma}] = \frac{1}{2}\int \rho_{i,\sigma}(\mathbf{r})\rho_{i,\sigma}(\mathbf{r}')|\mathbf{r}-\mathbf{r}'|^{-1}\,\mathrm{d}\mathbf{r}\,\mathrm{d}\mathbf{r}', \tag{24}$$

$$E_{\mathrm{XC}}^{\mathrm{LDA}}\big[\rho_{i,\uparrow},\rho_{i,\downarrow}\big] = \int \rho_i(\mathbf{r})\varepsilon_{\mathrm{XC}}\big[\rho_{\uparrow}^{\mathrm{h}},\rho_{\downarrow}^{\mathrm{h}}\big]\,\mathrm{d}\mathbf{r}, \tag{25}$$

$$\sum_{i,\sigma}\Big(U_{\mathrm{c}}[\rho_{i,\sigma}] + E_{\mathrm{XC}}^{\mathrm{LDA}}[\rho_{i,\sigma},0]\Big) \neq 0. \tag{26}$$

In these equations, (24)–(26), orthonormal orbits are denoted by indices i's. Equation (26) means that the orbiting electron interacting with itself, that is self-interaction, exists. This is unphysical. In order to remove this unphysical term, the SIC is taken into account by the following procedure. The SIC for the LDA in the density functional method has been treated for free atoms and insulators [16], and found an important role in determining the energy levels of electrons. However, no established formula is known to take into account the SIC for semiconductors and metals. As a way of trial, in the present calculation, the atomic SIC potential is introduced for each angular momentum in a way similar to the SIC potential for atoms [17] as follows:

$$V_{\mathrm{SIC}}^{l} \Rightarrow \begin{cases} 0 & \text{in the region of interstitial,} \\ V_{\mathrm{SIC}}^{l}(\rho_l) & \text{in the region inscribed sphere,} \end{cases} \tag{27}$$

$$V_{\mathrm{SIC}}^{l}(\rho_l) = -w_l\bigg\{\int \mathrm{d}\mathbf{r}'\rho_l(\mathbf{r}')|\mathbf{r}-\mathbf{r}'|^{-1} + V_{\mathrm{XC}}\big[\rho_l,0\big]\bigg\}, \tag{28}$$

$$\big\{H_0[\rho_l] + V_{\mathrm{XC}}^{\mathrm{LDA}}[\rho_l] + V_{\mathrm{SIC}}^{l}[\rho_l]\big\}\phi_l(r) = E_l\phi_l(r), \tag{29}$$

$$\rho_l(r) = \frac{|\phi_l(r)|^2}{4\pi}, \tag{30}$$

$$w_l = \int_{-\infty}^{E_{\mathrm{F}}} D_l(E)\,\mathrm{d}E. \tag{31}$$

That is, the SIC potential is set to be zero in the interstitial region, and inside of the inscribed sphere the SIC potential is calculated in the same way as in the free atom case except that a non-integer occupation number at each angular momentum orbital state denoted by l is allowed. Thus, the SIC potential in the inscribed sphere is given in Equation (28). Here, the effective weight is obtained from the corresponding partial density of states in Equation (31). This angular averaged orbital density in Equation (30) is calculated from the radial Schrödinger equation with the spherical part of the LDA potential plus its SIC potential in Equation (29). This procedure is incorporated in the whole self-consistent scheme of the FLAPW-LDA calculation.

This FLAPW-SIC scheme has been applied to the CP calculations of Cu, Si and diamond. The semiconductor Si and the insulator diamond have energy gaps and the most upper valence electrons are regarded as being a slightly bound state. The noble metal Cu has tightly bound d-electrons.

Effects of the introduction of the SIC on the band structures of Si and diamond are summarized as follows. With the introduction of the SIC, energy gaps of diamond and Si become larger by 20% and 23%, respectively, than those obtained without the SIC [18], which is in better agreement with experiments as shown in Table 1. The bandwidths of diamond and Si become narrower by 17% and 6%, respectively. The CPs calculated by the FLAPW and FLAPW-SIC scheme are plotted for Si in Figure 3 together with the experimental profiles by Sakurai *et al.* [19]. The contributions from the core electrons to the CPs are evaluated from FLAPW calculations with and without the SIC. The difference between the core CPs with and without the SIC is negligibly small for both materials. In the case of Si, the theoretical profiles are convoluted with the experimental overall resolution of 0.13 a.u. The theoretical profiles computed with and without the SIC provide a reasonable overall description of the measured profiles. However, as found in other solids, both theoretical profiles are higher than the measured profiles at small momenta, and there is a crossover around 0.8 a.u. with the situation reversing itself at large momenta. It is seen that introduction of the SIC affects the shape of the profiles in a way that brings the theory into better agreement with the experiment. Although the reduction of the discrepancy is small in the total profile, the effect of the SIC on the valence-electron profiles is better seen in Figure 4, where the characteristic features of each profile are better displayed by the first derivatives, because the contribution from the core to the first derivatives is slowly and monotonously varying. In the case of diamond, introduction of the SIC makes a definite change in the overall shape of the theoretical profiles as seen in Figure 5. Unlike the case of Si, diamond has a large band gap and the wave functions of the valence electrons are more localized. By nature, the SIC acts to enhance this feature as seen in Figure 5 compared to the case of Si. No high resolution experimental profile of diamond is available. We show here an earlier measurement by Reed and Eisenberger [20]. Their profiles are deconvoluted and the process often produces a spurious structure. Therefore, we are not able to make a rigorous comparison between calculation and experiment.

In the case of Cu, the effects of the SIC on the band structure are summarized as follows [21]. The width of the s-type band is not affected. The relative position of the d-bands with respect to the Fermi energy is lowered by 2 eV, and the width of the d-band is reduced by 15%. As a result, the electrons in the d-bands are more localized. The s–d hybridization near the Fermi energy is reduced. Consequently, I have got somewhat controversial results on the geometry of the Fermi surface. As reference,

Table 1. Energy band gaps of diamond and silicon calculated by FLAPW-LDA and FLAPW-SIC schemes. The experimental values [34] are also shown. Units are in eV.

	LDA	SIC-LDA	Experiment
Diamond	4.07	5.17	5.48
Si	0.46	0.73	1.17

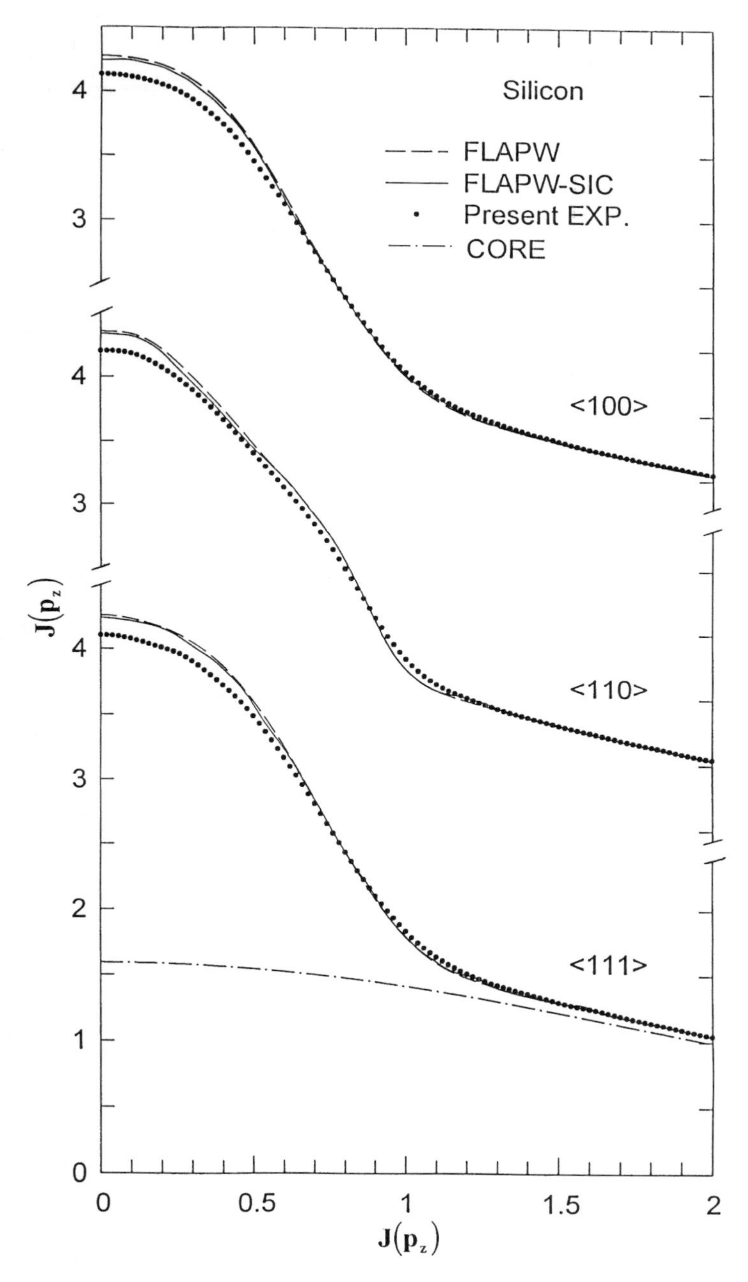

Figure 3. Measured (dotted) and calculated CPs of Si by the FLAPW-LDA (dashed) and the FLAPW-SIC (solid) schemes. The theoretical core profile is represented by a dash-dotted curve (after Kubo *et al.* [10]).

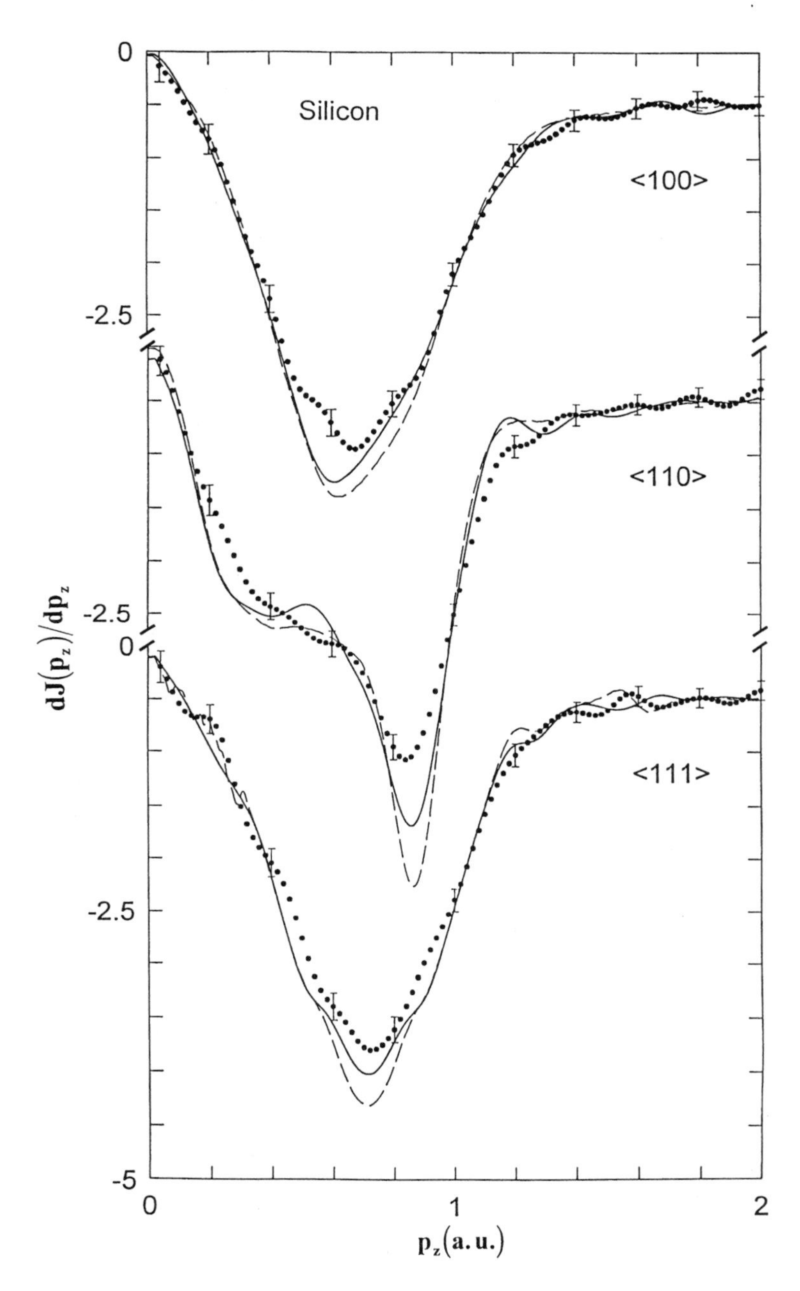

Figure 4. First derivatives of the measured and computed CPs of Si. Explanations are the same as those in Figure 3 (after Kubo *et al.* [10]).

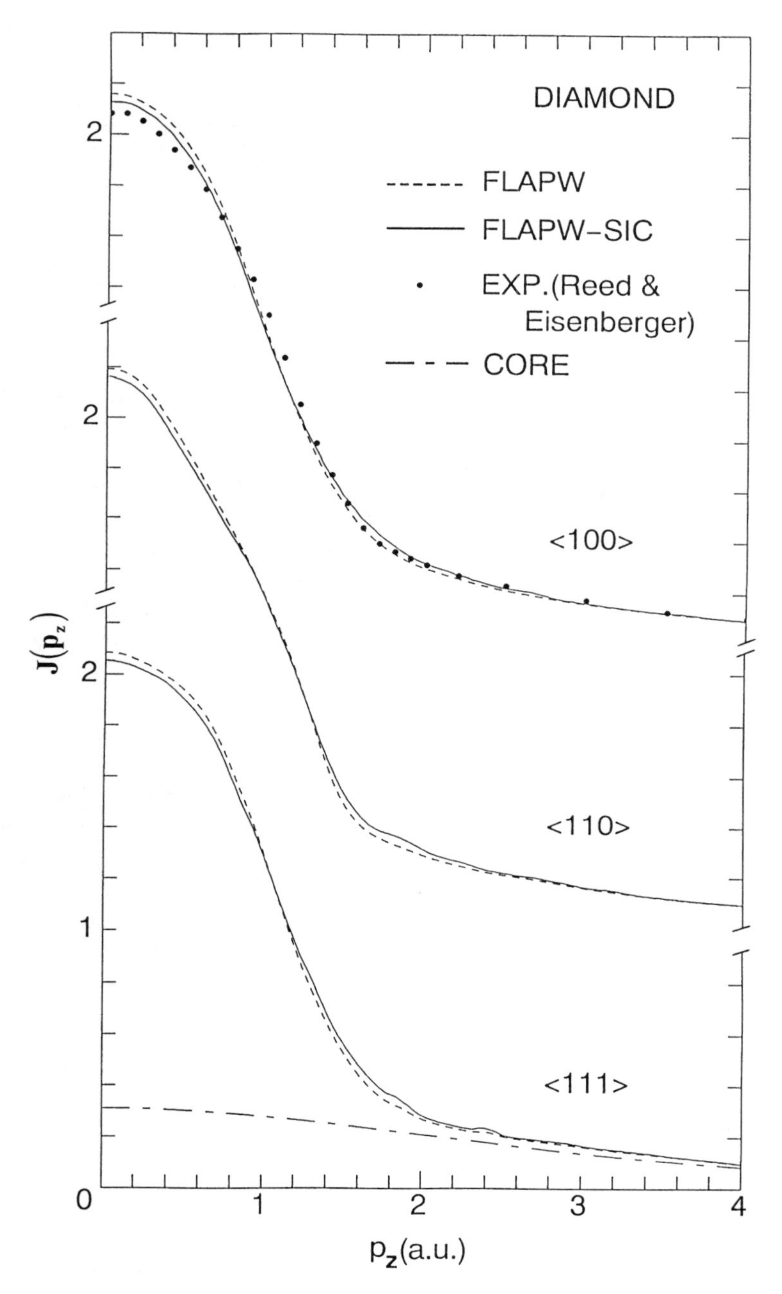

Figure 5. CPs of diamond along the three principal directions calculated by the FLAPW-LDA (dashed) and the FLAPW-SIC (solid) schemes. The dots represent the experimental profile measured by Reed and Eisenberger [20]. The theoretical core profile (dash-dotted) is also shown (after Kubo *et al.* [10]).

Table 2. The dHvA frequencies of some symmetry orbits in Cu calculated by the FLAPW-LDA (denoted as LDA) and FLAPW-SIC (SIC), respectively. The experimental values measured by Shoenberg [22], and Coleridge and Templeton [23], respectively.

Orbit	Notation	Exp.	LDA	SIC
Berry	B_{100}	5.998	6.463	6.343
Berry	B_{111}	5.814	6.167	6.300
Neck	N_{111}	0.218	0.296	0.090
Dog-bone	D_{110}	2.514	2.427	2.637
Rosette	R_{100}	2.462	2.427	2.491

the cross-sectioned area of the Fermi surface obtained by de Haas–van Alphen (dHvA) experiments [22, 23] and computed ones are shown in Table 2. The valence-electron CPs calculated by the FLAPW and FLAPW-SIC schemes are shown in Figure 6 together with the experimental profiles by Sakurai *et al.* [24]. The calculated profiles are convoluted with the experimental overall resolution 0.12 a.u. As shown in Figure 6, the overall shapes of the profiles calculated with the SIC is always lower in small momenta (0–1 a.u.) and higher in the middle momenta (1–4 a.u.) than those calculated without the SIC. Beyond 4 a.u., although they are indistinguishable in the figure, the profiles calculated with the SIC are always slightly higher than those calculated without the SIC.

4. Electron-correlation effects on CPs

As mentioned in Section 2, the CPs of solids have to be calculated on the quasi-particle scheme. In order to calculate the quasi-particle states, non-local and energy-dependent self-energy in Equation (13) must be evaluated in a real system. In practice, the exact self-energy for real systems are impossible to compute, and we always resort to approximate forms. A more realistic but relatively simple approximation to the self-energy is the GWA proposed by Hedin [7]. In the GWA, the self-energy operator in Equation (12) is

$$\Sigma(\mathbf{r}, \mathbf{r}'; E) = \left(\frac{\mathrm{i}}{2\pi}\right) \int G(\mathbf{r}, \mathbf{r}'; E + \omega) W(\mathbf{r}, \mathbf{r}'; \omega) \mathrm{e}^{\mathrm{i}\delta\omega} \, \mathrm{d}\omega. \tag{32}$$

In Equation (32), $G(\mathbf{r}, \mathbf{r}'; \omega)$ is, in principle, the dressed Green's function given as

$$G(\mathbf{r}, \mathbf{r}'; \omega) = 2 \sum \frac{\psi^*_{b,\mathbf{k}}(\mathbf{r}) \psi_{b,\mathbf{k}}(\mathbf{r}')}{\omega - E_{b,\mathbf{k}} \pm \mathrm{i}\delta}. \tag{33}$$

We can properly approximate the dressed Green's function by its LDA counterpart,

$$G_{\mathrm{LDA}}(\mathbf{r}, \mathbf{r}'; \omega) = 2 \sum \frac{\psi^{\mathrm{LDA}*}_{b,\mathbf{k}}(\mathbf{r}) \psi^{\mathrm{LDA}}_{b,\mathbf{k}}(\mathbf{r}')}{\omega - E^{\mathrm{LDA}}_{b,\mathbf{k}} \pm \mathrm{i}\delta}. \tag{34}$$

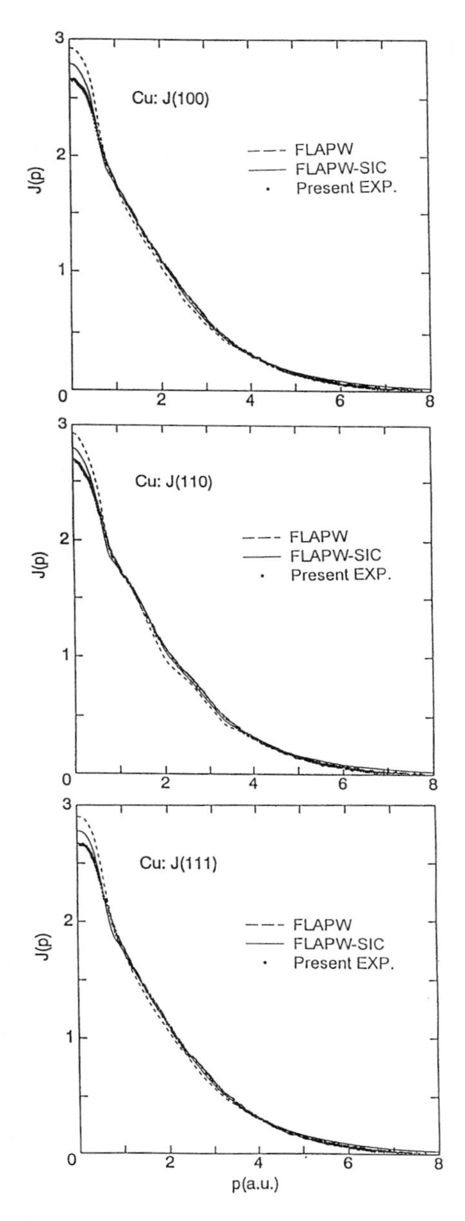

Figure 6. The valence-electron CPs of Cu calculated by the FLAPW-LDA (dashed) and the FLAPW-SIC (solid) schemes. The dots represent the experimental profiles measured by Sakurai *et al.* [24].

The $W(\mathbf{r}, \mathbf{r}'; \omega)$ in Equation (32) is a dynamically screened interaction, and is given as

$$W(\mathbf{r}, \mathbf{r}'; \omega) = \sum_{\mathbf{K},\mathbf{K}',\mathbf{q}} \mathrm{e}^{\mathrm{i}(\mathbf{q}+\mathbf{K})\cdot\mathbf{r}} W_{\mathbf{K},\mathbf{K}'}(\mathbf{q}, \omega) \mathrm{e}^{\mathrm{i}(\mathbf{q}+\mathbf{K}')\cdot\mathbf{r}'}, \tag{35}$$

$$W_{\mathbf{K},\mathbf{K}'}(\mathbf{q}, \omega) = \varepsilon^{-1}_{\mathbf{K},\mathbf{K}'}(\mathbf{q}, \omega) V(|\mathbf{q}+\mathbf{K}|), \tag{36}$$

$$V(|\mathbf{q}+\mathbf{K}|) = \frac{4\pi}{|\mathbf{q}+\mathbf{K}|^2}. \tag{37}$$

Inverse dielectric functions $\varepsilon^{-1}_{\mathbf{K},\mathbf{K}'}(\mathbf{q}, \omega)$ in Equation (36) are calculated within the random phase approximation [25]. Thus, the self-energy operator in Equation (32) is properly expressed by

$$\Sigma(\mathbf{r}, \mathbf{r}'; E) \Rightarrow \Sigma^{\mathrm{LDA}}_{\mathrm{GWA}}(\mathbf{r}, \mathbf{r}'; E) = \int G_{\mathrm{LDA}} W. \tag{38}$$

From the self-energy operator in Equation (38), the self-energy value in GWA is calculated as

$$\Sigma^{\mathrm{LDA}}_{b,b'}(GWA) = \langle b, \mathbf{k} | \Sigma(\mathbf{r}, \mathbf{r}'; E) | b', \mathbf{k} \rangle. \tag{39}$$

It has been suggested that quasi-particle wave functions do not deviate much from LDA wave functions [26]. Furthermore, in the evaluation of momentum densities shown in Figure 9, the characteristics of the quasi-particle states dominantly reflect on the occupation number densities which should be evaluated by using the general quasi-particle Green's function. In GWA, however, the corresponding occupation number densities are

$$N^{\mathrm{GWA}}_{b,b'}(\mathbf{k}) = \pi^{-1} \int_{-\infty}^{\mu} \mathrm{Im}\, G^{\mathrm{GWA}}_{b,b'}(\mathbf{k}, E)\, \mathrm{d}E, \tag{40}$$

$$G^{\mathrm{GWA}}_{b,b'}(\mathbf{k}, E) = \Big[E - E(\mathrm{LDA}) - \Sigma^{\mathrm{LDA}}(\mathrm{GWA}) \Big]^{-1}_{b,b'}. \tag{41}$$

Using $N^{\mathrm{GWA}}_{b,b'}(\mathbf{k})$ in Equation (40), the CP by the GWA is calculated as follows:

$$\rho^{\mathrm{GWA}}(\mathbf{p}) = \sum_{b,b',\mathbf{k}} \chi^{\mathrm{LDA}^*}_{b,\mathbf{k}} \chi^{\mathrm{LDA}}_{b',\mathbf{k}} N^{\mathrm{GWA}}_{b,b'}(\mathbf{k}), \tag{42}$$

$$J^{\mathrm{GWA}}(p_z) = \int \rho^{\mathrm{GWA}}(\mathbf{p})\, \mathrm{d}p_x\, \mathrm{d}p_y. \tag{43}$$

This quasi-particle approach for CPs has been performed on Li and Na [12, 13]. In these materials, only diagonal terms of the occupation number densities are evaluated in a reasonable justification [27]. The GWA occupation number densities (denoted as $N(\mathrm{GWA})$) thus obtained are shown for the three principal directions in Figures 7 and 8 for Na and Li, respectively. For reference, the occupation number densities obtained

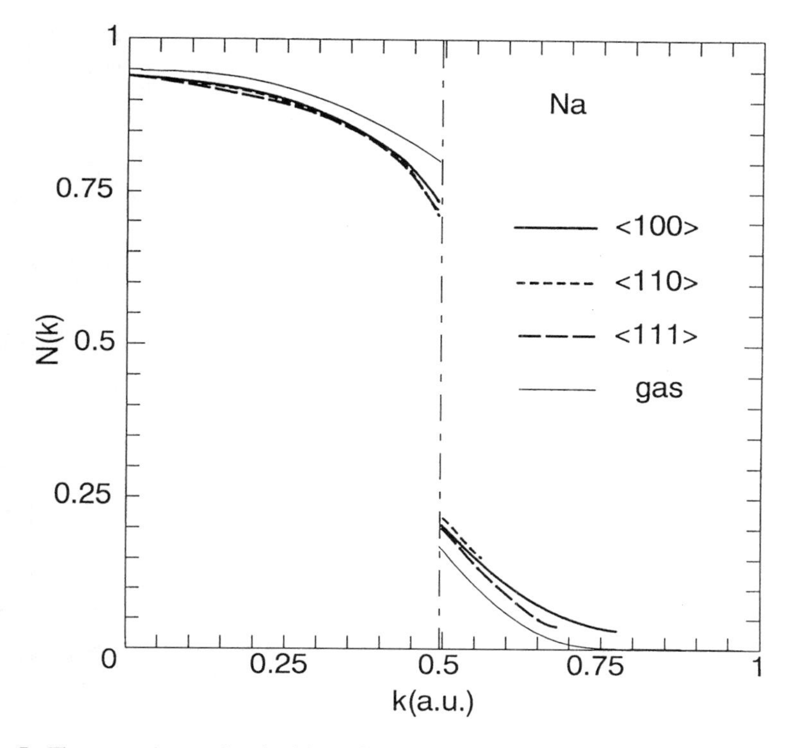

Figure 7. The occupation number densities as functions of wave vector for Na. The thick curves labeled ⟨100⟩, ⟨110⟩ and ⟨111⟩ represent the three principal directions within the first Brillouin zone, obtained by the FLAPW-GWA. The thin solid curve is obtained from an interacting electron-gas model [27]. The dash-dotted line represents the Fermi momentum.

from an interacting electron gas model (denoted as N(gas)) are also displayed in the same figures. In the case of Na, the N(GWA) are similar to its N(gas). On the other hand, N(GWA) of Li shows a remarkable k-dependence, and very different features compared to its N(gas), particularly in the ⟨110⟩ direction.

Using these occupation number densities, CPs of Na are calculated along three principal directions. Since the anisotropy in the CPs is very small and high resolution Compton experiments have been performed only for a polycrystal sample, the averaged GWA CPs are shown compared with the high resolution experiment by Sakurai *et al.* [28] in Figure 9. For comparison, the LDA and free-electron calculations are also shown in the same figure. The calculated profiles are convoluted with the overall momentum resolution of experiment 0.12 a.u. In this figure, the difference between the free-electron and the LDA CPs is regarded as dominantly due to the core-orthogonalization effect, since the lattice potential has a very weak effect on the conduction electrons. The discrepancy between the LDA and the experiment is considerably reduced by introduction of electron-correlation effects by the electron-gas model. Furthermore, the introduction of the electron-correlation effects by the

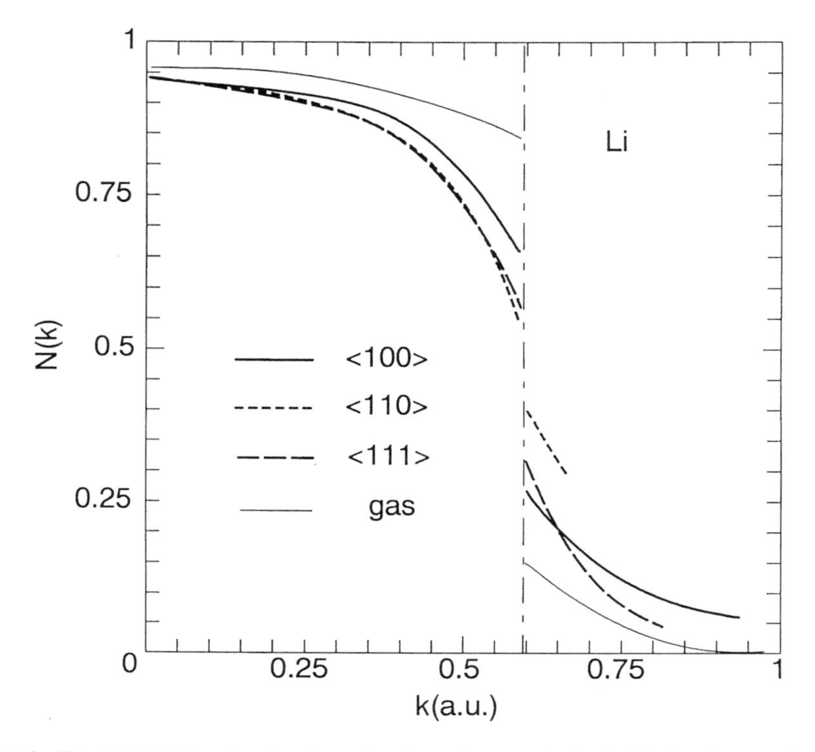

Figure 8. The occupation number densities as functions of wave vector for Li. Explanations are the same as those in Figure 7.

GWA leads to a good agreement between the theory and experiment. This finding can be interpreted as being mainly due to a different behavior between $N(\text{GWA})$ and $N(\text{gas})$ around the Fermi momentum as seen in Figure 7.

In the case of Li, the effect of its lattice potential to the electron states lead to a large anisotropy of the Fermi surface [29], as well known. As a typical phenomenon due to this effect, electron states around N-point in the lowest conduction band lie just above the Fermi level compared with corresponding electron states of Na. The contributions of these features to the self-energy evaluation are remarkably different compared to the case of its electron-gas model, and produce large difference between $N(\text{GWA})$ and $N(\text{gas})$ near the Fermi momentum seen in Figure 8. The renormalization factor Z_F on the Fermi momentum is estimated to be 0.35, 0.15 and 0.25 for the three directions $\langle 100 \rangle$, $\langle 110 \rangle$ and $\langle 111 \rangle$, respectively. These values are much smaller than the theoretical results obtained so far using jellium models, which range from 0.5 [30] to 0.75 [31]. Schülke *et al.* [32] found that the value in the $\langle 100 \rangle$ direction is 0.1 ± 0.1 from the fitting to a simple model. Although their obtained value is smaller than that of our result 0.35, our value is regarded as comparable to the one in the experiment, in contrast with those predicted using jellium models. Using these occupation number

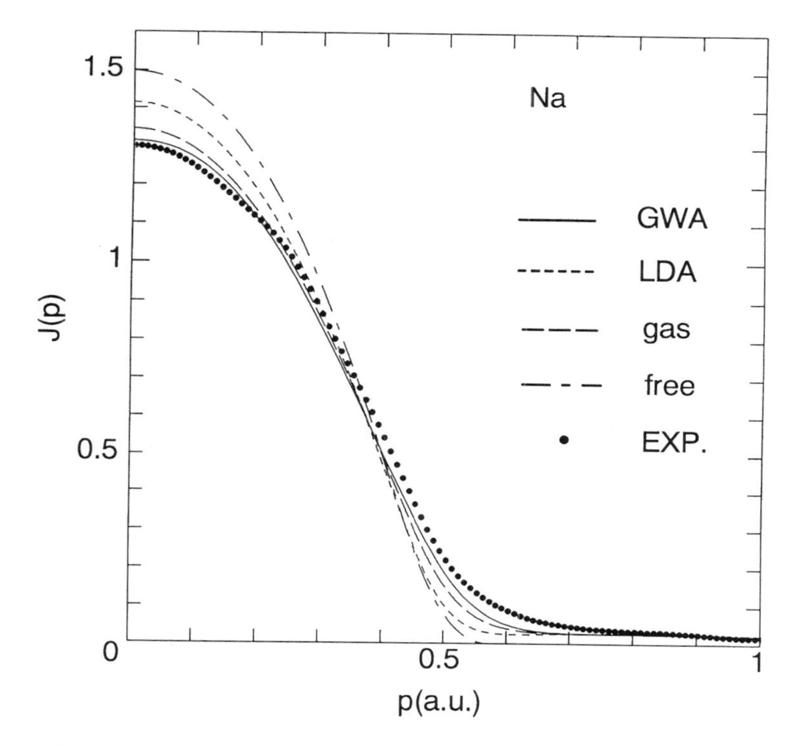

Figure 9. The valence-electron CPs of polycrystalline Na. The solid and dotted curves represent the FLAPW-GWA and FLAPW-LDA calculations, respectively. The dash-dotted and dashed curves represent the results calculated by the free-electron and FLAPW-LDA including correlation effects according to Lundqvist and Lyden [27], respectively. The dots represent the experimental result by Sakurai *et al.* [28] (after Kubo [13]).

densities N(GWA) and N(gas), CPs are calculated and shown together with the LDA results in Figure 10. In the same figure, experimental results by both Sakurai *et al.* [33] and Schülke *et al.* [32] are shown for comparison. The overall momentum resolution of the experiments of Sakurai *et al.* is 0.12 a.u. and that of Schülke *et al.* is 0.14 a.u. Calculated results are all convoluted with the momentum resolution equal to 0.12 a.u. As seen in Figure 10 the introduction of electron-correlation effects resulting from using N(gas) reduces the discrepancy between the LDA and experimental results to a certain extent. However, the reduction is smaller compared to the case of Na. On the other hand, the CPs calculated using N(GWA) lead to the drastic reduction of the discrepancy between the LDA and the experimental results as seen from Figure 10.

5. Summary and conclusions

We have studied the effects of the SIC for the filled and tightly bound bands for 'Si, diamond' and 'Cu', respectively, by utilizing the FLAPW method. In the case of Si,

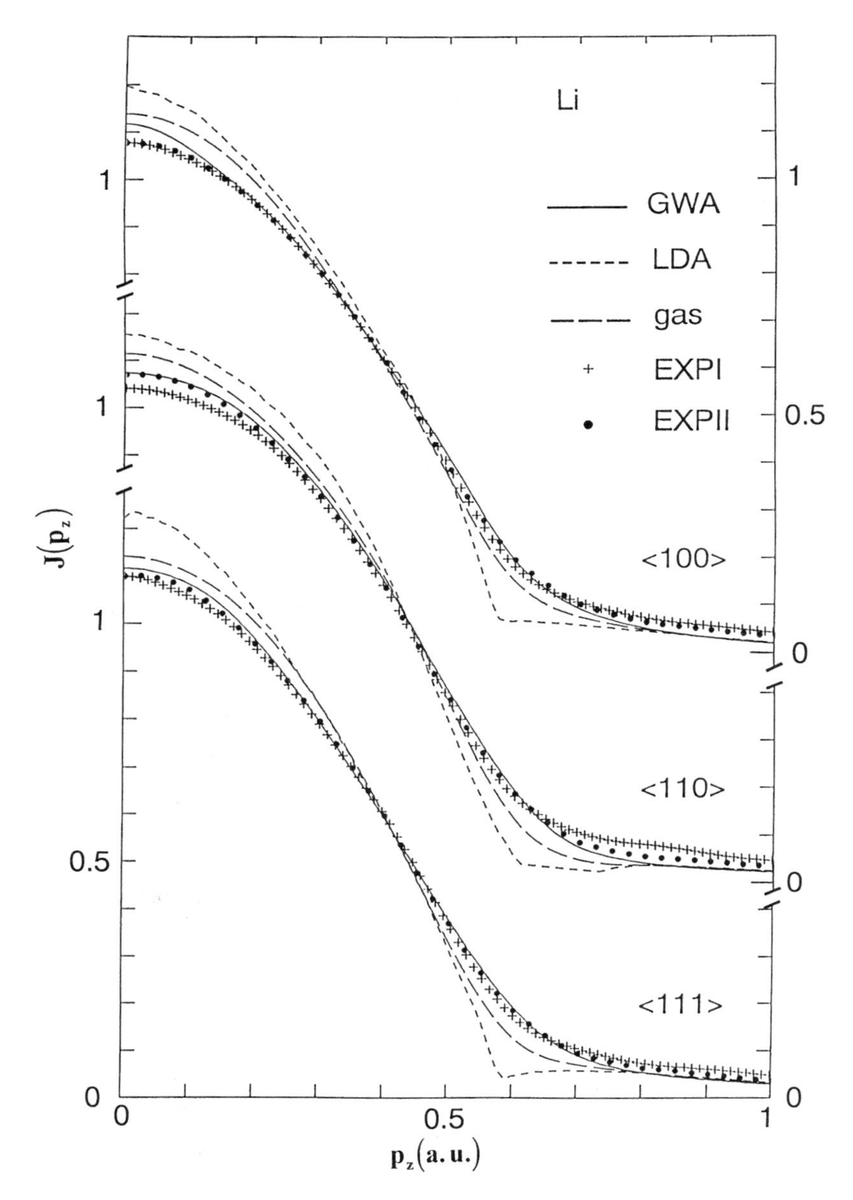

Figure 10. The valence-electron CPs of Li along the three principal symmetry directions. The solid and dotted curves represent the FLAPW-GWA and FLAPW-LDA calculations, respectively. The dashed curves represent the FLAPW-LDA calculations including correlation effects according to Lundqvist and Lyden [27]. The EXPI and EXPII represent the experimental results measured by Sakurai *et al.* [33] and Schülke *et al.* [32], respectively (after Kubo [13]).

introduction of the SIC into the FLAPW scheme changes the band structure and the band gap appreciably. CPs computed with the SIC are in better agreement with the measured profiles when their first derivatives are compared. The comparison confirms that the discrepancy between theory and experiment is the same sort as that found in other metals and alloys, suggesting that electron-correlation plays an important role. In the case of diamond, the introduction of the SIC affects the band structure, the energy gap, the wave functions and CPs. Comparison with the earlier experiment confirms an urgent need for a high resolution measurement to judge the effect of the SIC. In the case of Cu, introduction of the SIC is somewhat controversial. It has weakened the agreement between the LDA Fermi surface area and the dHvA result for the so-called neck. However, the SIC does not change the other areas so much which are mainly in the d-bands. On the other hand, the SIC has brought the LDA CP to a better agreement with the experiment. The main reason for this reduction of the discrepancy is that the SIC potential brings down and narrows the d-bands. As a result, the wave functions of the d-bands become more localized in real space. Therefore, in momentum space, they extend more in higher momenta. Although the SIC potential employed in this study is not a uniquely determined one nor rigorously formulated, the present results suggest that some kind of correction to the LDA potential is needed to explain the experimental results consistently. Furthermore, the origin of the remaining discrepancy in the shape of CP between the theory and experiment may now be ascribed to the quasi-particle nature of the electron system, in particular to the non-unity and non-zero occupation in k-space.

We have performed CP calculations of Li and Na in a quasi-particle scheme by utilizing the GWA using the wave functions and energy values of the LDA-based FLAPW computations as basis set. In the case of Na, the experimental CP is fairly well reproduced by the electron-gas model with the electron-correlation, since the lattice potential has a very weak effect on the electron states. However, the CPs calculated by using the GWA are much more reproduced than the experimental results. On the other hand, for Li, the lattice potential has a strong effect on the electron states, and the Fermi surface geometry strongly deviates from a sphere. Reflecting these characteristics of the electron states, the occupation number densities N(GWA) computed from the GWA are very different from those obtained from electron-gas models. That is, computed Z_F from the GWA is significantly smaller than that predicted from jellium models. The CPs obtained using the N(GWA)s reproduce the experimental results extremely well. These results suggest that the GWA is the most meaningful and practical way to go beyond the LDA.

Acknowledgements

The author expresses his thanks to Professor N. Shiotani and Dr. Y. Sakurai for sending their experimental results and valuable discussions. He also thanks Professor W. Schülke for providing the measured data and many fruitful discussions.

References

1. Kohn, W. and Sham, L.J. (1965) Self-consistent equations including exchange and correlation effects, *Phys. Rev.*, **140**, A1133–A1138.
2. Hohenberg, P. and Kohn, W. (1964) Inhomogeneous electron gas, *Phys. Rev.*, **136**, B864–B871.
3. Bauer, G.E. and Schneider, J.R. (1985) Electron correlation effect in the momentum density of copper metal, *Phys. Rev.*, **B31**, 681–692.
4. Shiotani, N., Tanaka, Y., Sakurai, Y., Sakai, N., Ito, M., Itoh, F., Iwazumi, T. and Kawata, H. (1993) Compton scattering study of electron momentum density in vanadium, *J. Phys. Soc. Jpn.*, **62**, 239–245.
5. Manninen, S., Honkimaki, V., Hämäläinen, K., Laukkanen, J., Blaas, C., Redinger, J., McCarthy, J. and Suortti, P. (1996) Compton-scattering study of the electronic properties of the transition-metal alloys FeAl, CoAl, and NiAl, *Phys. Rev.*, **B53**, 7714–7720.
6. Perdew, J.P. and Zunger, A. (1981) Self-interaction correction to density-functional approximations for many-electron systems, *Phys. Rev.*, **B23**, 5048–5079.
7. Hedin, L. and Lundqvist, S. (1969) Effects of electron–electron and electron–phonon interaction on the one-electron states of solids, In *Solid state Physics*, Eherenreich, H., Seitz, F. and Turnbull, D. (Eds.), Academic, New York,Vol. 23, pp. 1–181.
8. Weyrich, K.H. (1988) Full-potential linear muffin-tin-orbital method, *Phys. Rev.*, **B37**, 10269–10282.
9. Harrison, J.G. (1983) Density functional calculations for atoms in the first transition series, *J. Chem. Phys.*, **79**, 2265–2269.
10. Kubo, Y., Sakurai, Y., Tanaka, Y., Nakamura, T., Kawata, H. and Shiotani, N. (1997) Effects of self-interaction correction on Compton profiles of diamond and silicon, *J. Phys. Soc. Jpn.*, **66**, 2777–2780.
11. Kubo, Y., Sakurai, Y. and Shiotani, N. (1999) Effects of self-interaction correction on momentum density in copper, *J. Phys.: Condens. Matter*, **11**, 1683–1695.
12. Kubo, Y. (1996) Compton profiles of Li in GW approximation, *J. Phys. Soc. Jpn.*, **65**, 16–18.
13. Kubo, Y. (1997) Effects of electron correlations on Compton profiles of Li and Na in the GW approximation, *J. Phys. Soc. Jpn.*, **66**, 2236–2239.
14. Sakurai, Y. (1995) High-resolution Compton-profile measurements, *Second International Workshop on Compton Scattering and Fermiology*, Tokyo, Japan.
15. Heaton, R.A., Harrison, J.G. and Lin, C.C. (1983) Self-interaction correction for density-functional theory of electronic energy bands of solids, *Phys. Rev.*, **B28**, 5992–6007.
16. Vogel, D., Kruger, P. and Pollmann, J. (1996) Self-interaction and relaxation-corrected pseudopotential for II–VI semiconductors, *Phys. Rev.*, **B54**, 5495–5511.
17. Hamada, N. and Ohnishi, S. (1986) Self-interaction correction to the local-density approximation in the calculation of the energy band gaps of semiconductors based on the full-potential linearized augmented-plane-wave method, *Phys. Rev.*, **B34**, 9042–9044.
18. Lambrecht, W.R.L. and Andersen, O.K. (1986) Minimal basis sets in the linear muffin-tin orbital method: applications to the diamond-structure crystals C, Si, and Ge, *Phys. Rev.*, **B34**, 2439–2449.
19. Sakurai, Y. (1995) High-resolution Compton-profile measurements for silicon (private communication).
20. Reed, W.A. and Eisenberger, P. (1972) Gamma-ray Compton profiles of diamond, silicon, and germanium, *Phys. Rev.*, **B6**, 4596–4604.
21. Norman, M.R. (1984) Application of a screened self-interaction correction to transition metals: copper and zinc, *Phys. Rev.*, **B29**, 2956–2962.
22. Shoenberg, D. (1960) The de Haas–van Alphen effect in copper, silver and gold, *Phil. Mag.*, **5**, 105–110.
23. Coleridge, P.T. and Templeton, I.M. (1972) High precision de Haas–van Alphen measurements in the noble metals, *J. Phys.*, **F2**, 643–656.
24. Sakurai, Y., Kaprzyk, S., Bansil, A., Tanaka, Y., Stutz, G., Kawata, H. and Shiotani, N. (1998) Compton profiles of Cu – experiment and theory, to be published in *J. Chem. Phys.*
25. Pines, D. and Nozieres, P. (1966) *The Theory of Quantum Liquids*, Benjamin, Reading, Mass.
26. Godby, R.W., Schluter, M. and Sham, L.J. (1988) Self-energy operators and exchange-correlation potentials in semiconductor, *Phys. Rev.*, **B37**, 10159–10175.
27. Lundqvist, B.I. and Lyden, C. (1971) Calculated momentum distributions and Compton profiles of interacting conduction electrons in lithium and sodium, *Phys. Rev.*, **B4**, 3360–3370.
28. Sakurai, Y., Nanao, S., Nagashima,Y., Hyodo, T., Iwazumi, T., Kawata, H., Itoh, M., Shiotani, N. and Stewart, A.T. (1992) Electron momentum distributions in sodium and lithium, *Mater. Sci. Forum*, **105–110**, 803–806.

29. Rajput, S.S., Prasad, R., Singru, R.M., Triftshauser, W., Eckert, A., Kogel, G., Kaprzyk, S. and Bansil, A. (1993) A study of the Fermi surface of lithium and disordered lithium–magnesium alloy: theory and experiment, *J. Phys.: Condens. Matter*, **5**, 6419–6432.
30. Daniel, E. and Vosko, S.H. (1960) Momentum distribution of an interacting electron gas, *Phys. Rev.*, **120**, 2041–2044.
31. Lantto, L.J. (1980) Fermi hypernetted-chain calculations of the electron-gas correlations, *Phys. Rev.*, **B22**, 1380–1393.
32. Schülke, W., Stuz, G., Wohlert, F. and Kaprolat, A. (1996) Electron momentum-space densities of Li metal: a high-resolution Compton-scattering study, *Phys. Rev.*, **B54**, 14381–14395.
33. Sakurai, Y., Tanaka, Y., Bansil, A., Kaprzyk, S., Stewart, A.T. Nagashima, Y., Hyodo, T., Nanao, S., Kawata, H. and Shiotani, N. (1995) High-resolution Compton scattering study of Li: asphericity of the Fermi surface and electronic correlation effects, *Phys. Rev. Lett.*, **74**, 2252–2255.
34. Harrison, W.A. (1980) *Electronic Structure and the Properties of Solids*, Freeman, San Francisco, Table 10-1.

6

Interaction energy and density in the water dimer. A quantum theory of atoms in molecules: insight on the effect of basis set superposition error removal

CARLO GATTI and ANTONINO FAMULARI

Centro CNR per lo Studio delle Relazioni tra Struttura, e Reattività Chimica, via Golgi 19, 20133 Milano, Italy

1. Introduction

The study of electron density distributions resulting from molecular interactions in gas-phase complexes or in molecular crystals, is known [1, 2] to facilitate our understanding of the physical mechanisms underlying such interactions. Indeed, the action of these mechanisms is reflected in the *interaction density*, defined as the difference between the electron density distribution (EDD) of the molecular complex or crystal and that obtained by superimposing the EDDs of free molecules.

Central to the study of theoretical interaction density are the approximations one adopts in the evaluation of the corresponding *interaction energy* (E_{int}). This represents [3] only a minuscule fraction of the total energy of the global system – typically from four to seven orders of magnitude smaller. As a consequence [3] a 'correct' calculation of E_{int} requires either an inordinately and unattainable high level of precision or, which is common practice, a systematic cancelation among errors in the estimates of the various different physical contributions to E_{int}. In fact, it is reckoned [4] that 'a reliable *ab-initio* prediction of interaction potentials and energies is still a highly non-trivial task even for small atoms and molecules'. Moreover, it is also acknowledged that all the most commonly used *ab-initio* methods for computing E_{int} (supermolecular, perturbational, or hybrid) have their well-defined drawbacks and advantages [5].

This paper is a preliminary attempt towards an understanding of how the interactions densities are affected by approximations and errors introduced in the evaluation of E_{int}. The water dimer complex is investigated here, as it represents a prototype of hydrogen bonding and a sort of paradigm for molecular interactions. Owing to this and to the limited size of the system, a wealth of literature [3] has appeared on water dimer and a corresponding large spectrum of computational protocols and E_{int} estimates has been thereof proposed. Indeed, even if similar or at the limit equal E_{int} values are obtained with several methods, the resulting interaction densities may still differ among each other, since E_{int} is a delicate balance of various positive and negative energy contributions. In this respect, the study of interaction density may enhance our understanding of the performance of a given model in describing a particular molecular interaction. Not only a single, though extremely important value

Paul G. Mezey and Beverly E. Robertson (eds.), Electron, Spin and Momentum Densities and Chemical Reactivity, 93–114

like E_{int} may be checked, but the whole behavior of a scalar function in both the intra- and intermolecular regions.

One of the most important, though quite often unattended, requirement on an *ab-initio* approach to molecular interactions is that of its size consistency [6]. We refer here both to what we term *basis set size consistency* and to the more usual concept of size consistency in the evaluation of the electron-correlation contributions. The lack of basis set size consistency arises [7] from the use of an incomplete basis set and constitutes a well-known inconvenience in the evaluation of molecular interactions by a variational supermolecule approach. In fact within the complex or crystal the basis set of the subunit is improved by that of its partner(s) and vice versa, thus leading to an artificial energy lowering within the complex or crystal repeating unit. Such a bias yields to the so-called [5] basis set superposition error (BSSE), which for weak intermolecular interactions may be even comparable in magnitude to the interaction energy itself. BSSE is often *a posteriori* corrected by the counterpoise (CP) recipe [8] (or one of its many modifications) [9] that is, in its simplest formulation, the orbitals of the partner(s) are added when computing the energy of each subunit. CP corrections may however either overestimate or underestimate BSSE, while never removing it [5]. The treatment of BSSE is a problem also in the case of very simple molecular complexes. Indeed Saebø *et al.* [10] pointed out that although most of the energetic contributions to the water dimer interaction have already been computed quite accurately, one of the major goal to be reached is a clear BSSE correction.

In this paper a method [11], which allows for an *a priori* BSSE removal at the SCF level, is for the first time applied to interaction densities studies. This computational protocol which has been called SCF-MI (Self-Consistent Field for Molecular Interactions) to highlight its relationship to the standard Roothaan equations and its special usefulness in the evaluation of molecular interactions, has recently been successfully used [11–13] for evaluating E_{int} in a number of intermolecular complexes. Comparison of standard SCF interaction densities with those obtained from the SCF-MI approach should shed light on the effects of BSSE removal. Such effects may then be compared with those deriving from the introduction of Coulomb correlation corrections. To this aim, we adopt a variational perturbative valence bond (VB) approach that uses orbitals derived from the SCF-MI step and thus maintains a BSSE-free picture. Finally, no bias should be introduced in our study by the particular approach chosen to analyze the observed charge density rearrangements. Therefore, not a model but a theory which is firmly rooted in Quantum Mechanics, applied directly to the electron density ρ and giving quantitative answers, is to be adopted. Bader's Quantum Theory of Atoms in Molecules (QTAM) [14, 15] meets nicely all these requirements. Such a theory has also been recently applied to molecular crystals as a valid tool to rationalize and quantitatively detect crystal field effects on the molecular densities [16–18].

The paper is organized as follows. Section 2 summarizes the grounding of the SCF-MI and VB approaches, while Section 3 gives a brief overview of the technical details used in our computations and discusses the resulting interaction energy data. The application of our BSSE-free analysis to the study of charge density rearrangements in water dimer is presented at length in Section 4. Section 5 concludes.

2. Methods

Self-consistent field for molecular interactions

A short summary of the SCF-MI method is presented here for the simplest case of two interacting closed-shell monomers A and B. A full account of the theory is given elsewhere [11] and its generalization to interaction of an open shell with an arbitrary number of closed shell fragments has recently appeared [19].

The AB supermolecule is described by a single determinant wave function formulated in terms of doubly occupied molecular orbitals with no orthonormality constraints. For a system with $2N = 2N_\mathrm{A}+2N_\mathrm{B}$ electrons the SCF-MI wave function expressed in terms of the antisymmetrizer operator A is

$$\begin{aligned}\Psi_{\text{SCF-MI}} = [(2\mathrm{N})!]^{-1/2} A[\phi_1^\mathrm{A}(1)\bar{\phi}_1^\mathrm{A}(2)\cdots\bar{\phi}_{N_\mathrm{A}}^\mathrm{A}(2N_\mathrm{A}) \\ \times \phi_1^\mathrm{B}(2N_\mathrm{A}+1)\bar{\phi}_1^\mathrm{B}(2N_\mathrm{A}+2)\cdots\phi_{N_\mathrm{B}}^\mathrm{B}(2N_\mathrm{A}+2N_\mathrm{B})].\end{aligned} \tag{1}$$

The kernel of SCF-MI derivation is the partitioning of the basis set for the total system into two subsets:

$$\{\chi_k\}_{k=1}^{M} = \{\chi_p^\mathrm{A}\}_{p=1}^{M_\mathrm{A}} + \{\chi_q^\mathrm{B}\}_{q=1}^{M_\mathrm{B}}, \tag{2}$$

one, $\{\chi_p^\mathrm{A}\}_{p=1}^{M_\mathrm{A}}$, centered on monomer A, and the other, $\{\chi_q^\mathrm{B}\}_{q=1}^{M_\mathrm{B}}$, centered on monomer B, with $M = M_\mathrm{A} + M_\mathrm{B}$. The molecular orbitals (MO) of A are expanded in subset $\{\chi_p^\mathrm{A}\}_{p=1}^{M_\mathrm{A}}$ and those of B in subset $\{\chi_q^\mathrm{B}\}_{q=1}^{M_\mathrm{B}}$:

$$\phi_i^\mathrm{A} = \sum_{p=1}^{M_\mathrm{A}} T_{pi}^\mathrm{A}\chi_p^\mathrm{A}, \qquad \phi_i^\mathrm{B} = \sum_{q=1}^{M_\mathrm{B}} T_{qi}^\mathrm{B}\chi_q^\mathrm{B}, \tag{3}$$

that is $\Phi^\mathrm{A} = \chi^\mathrm{A}\mathbf{T}_\mathrm{A}$ and $\Phi^\mathrm{B} = \chi^\mathrm{B}\mathbf{T}_\mathrm{B}$ in matrix form. Orbitals of different fragments are left free to overlap with each other. As a consequence of the assumed partitioning, both the $(M \times N)$ matrix of MO and its variation assume a block diagonal form

$$\mathbf{T} = \begin{bmatrix} \mathbf{T}_\mathrm{A} & \mathbf{0} \\ \mathbf{0} & \mathbf{T}_\mathrm{B} \end{bmatrix}, \qquad \delta\mathbf{T} = \begin{bmatrix} \delta\mathbf{T}_\mathrm{A} & \mathbf{0} \\ \mathbf{0} & \delta\mathbf{T}_\mathrm{B} \end{bmatrix}.$$

The energy and its variation $\delta E = 0$ have apparently the standard SCF form

$$E = \mathrm{Tr}[\mathbf{D}\cdot\mathbf{h}] + \mathrm{Tr}[\mathbf{D}\cdot\mathbf{F}(\mathbf{D})], \qquad \delta E = 2\,\mathrm{Tr}[\mathbf{F}(\mathbf{D})\delta\mathbf{D}], \tag{4}$$

where $\mathbf{F}$ and $\mathbf{h}$ are the usual Fock and one-electron integral matrices expressed in the atomic orbitals basis set. However, the general stationary condition $\delta E = 0$ is also mathematically equivalent to the following coupled secular problems:

$$\begin{cases} \mathbf{F}'_\mathrm{A}\mathbf{T}_\mathrm{A} = \mathbf{S}'_\mathrm{A}\mathbf{T}_\mathrm{A}\mathbf{L}_\mathrm{A}, \\ \mathbf{T}_\mathrm{A}^\dagger\mathbf{S}'_\mathrm{A}\mathbf{T}_\mathrm{A} = \mathbf{I}_\mathrm{A}, \end{cases} \qquad \begin{cases} \mathbf{F}'_\mathrm{B}\mathbf{T}_\mathrm{B} = \mathbf{S}'_\mathrm{B}\mathbf{T}_\mathrm{B}\mathbf{L}_\mathrm{B}, \\ \mathbf{T}_\mathrm{B}^\dagger\mathbf{S}'_\mathrm{B}\mathbf{T}_\mathrm{B} = \mathbf{I}_\mathrm{B}, \end{cases} \tag{5}$$

in terms of *effective* Fock and overlap matrices $\mathbf{F}'_\mathrm{A}$, $\mathbf{F}'_\mathrm{B}$, and $\mathbf{S}'_\mathrm{A}$, $\mathbf{S}'_\mathrm{B}$:

$$\begin{aligned}
\mathbf{S}'_\mathrm{A} &= \mathbf{S}'_\mathrm{A}(\mathbf{T}_\mathrm{B}) = \mathbf{S}_\mathrm{AA} - \mathbf{S}_\mathrm{AB}\mathbf{D}_\mathrm{B}\mathbf{S}_\mathrm{BA},\\
\mathbf{S}'_\mathrm{B} &= \mathbf{S}'_\mathrm{B}(\mathbf{T}_\mathrm{A}) = \mathbf{S}_\mathrm{BB} - \mathbf{S}_\mathrm{BA}\mathbf{D}_\mathrm{A}\mathbf{S}_\mathrm{AB},\\
\mathbf{F}'_\mathrm{A} &= \mathbf{F}'_\mathrm{A}(\mathbf{T}_\mathrm{A},\mathbf{T}_\mathrm{B}) = (\mathbf{1}_\mathrm{A} \mid -\mathbf{S}_\mathrm{AB}\mathbf{D}_\mathrm{B})\mathbf{F}\left(\frac{\mathbf{1}_\mathrm{A}}{-\mathbf{D}_\mathrm{B}\mathbf{S}_\mathrm{BA}}\right),\\
\mathbf{D}_\mathrm{A}(\mathbf{T}_\mathrm{A}) &= \mathbf{T}_\mathrm{A}(\mathbf{T}^\dagger_\mathrm{A}\mathbf{S}_\mathrm{AA}\mathbf{T}_\mathrm{A})^{-1}\mathbf{T}^\dagger_\mathrm{A},\\
\mathbf{F}'_\mathrm{B} &= \mathbf{F}'_\mathrm{B}(\mathbf{T}_\mathrm{B},\mathbf{T}_\mathrm{A}) = (-\mathbf{S}_\mathrm{BA}\mathbf{D}_\mathrm{A} \mid \mathbf{1}_\mathrm{B})\mathbf{F}\left(\frac{-\mathbf{D}_\mathrm{A}\mathbf{S}_\mathrm{AB}}{\mathbf{1}_\mathrm{B}}\right),\\
\mathbf{D}_\mathrm{B}(\mathbf{T}_\mathrm{B}) &= \mathbf{T}_\mathrm{B}(\mathbf{T}^\dagger_\mathrm{B}\mathbf{S}_\mathrm{BB}\mathbf{T}_\mathrm{B})^{-1}\mathbf{T}^\dagger_\mathrm{B}.
\end{aligned} \tag{6}$$

As is apparent from the above definitions, each of these effective matrices depend on basis sets and molecular orbitals of both fragments. It is also important to observe that these matrices possess a correct asymptotic behavior as at large interfragment distances they become the usual overlap and Fock matrices of the separate fragments, while the paired secular systems uncouple and converge to the separate Roothaan equations for the single monomers. Finally, as it is usual in a supermolecular approach, the interaction energy is expressed as

$$\Delta E_\mathrm{SCF\text{-}MI} = E^\mathrm{AB}_\mathrm{SCF\text{-}MI} - E^\mathrm{A}_\mathrm{SCF} - E^\mathrm{B}_\mathrm{SCF}, \tag{7}$$

the energy of monomers being that of standard SCF wave functions. At variance with the conventional SCF supermolecular (SCF-SM) approach, the SCF-MI interaction energies exhibit an extremely rapid convergence with increasing basis set quality.

The solution of the SCF-MI equations involves the following steps:

a) construct the effective overlap and Fock matrices (Equation (6));
b) solve the generalized secular systems (Equation (5));
c) check the variation in the density matrix elements $D_\mathrm{A}$andD_B;
d) at convergence, evaluate the electronic energy (Equation (4)), otherwise go back to step (a).

The computational cost [20] of the SCF-MI algorithm is almost equal to that of standard SCF, as the time required to evaluate the effective operators is negligible and the overload caused by the doubling of secular equations to be solved is largely compensated by the reduced size of these equations. The algorithm, which has been incorporated [20] into the GAMESS-US package [21], is compatible with the usual formulation of the analytic derivatives of the SCF energy. This fact has allowed [20] the implementation of gradient optimization algorithms and of force constant matrix computations in both the direct and conventional approaches. So the SCF-MI method not only provides a complete *a priori* elimination of the BSSE, while taking into account the natural non-orthogonality of the MOs of the two interacting fragments, but also allows for a standard analytical search of minima conformations on the potential energy surface (PES) of the complex. This fact is at variance [3] with the classical CP procedure where one has to adopt point by point procedures to move on the PES and where as many as five SCF energies are to be evaluated on each point of the PES, in order to properly include [22] the treatment of geometry relaxation effects.

BSSE-free VB treatment of intermolecular forces

The correlation contribution to water dimer interaction energy and density has been evaluated [23] with a very compact multistructure VB – non-orthogonal CI – calculation. The VB approach is a natural way [24, 25] to describe the intermolecular interaction, including the effects deriving from the overlap between the orbitals of the separated fragments and the interfragment electron correlation (dispersion). The adopted wave function has the general VB form

$$\Psi_{\mathrm{AB}} = C_0 \Psi^0_{\mathrm{AB}} + \sum_{a=1}^{Na} \sum_{b=1}^{Nb} C_{ab} \Psi_{ab}^{a^* b^*} + \sum_{a=1}^{Na} C_a \Psi_a^{a^*} + \sum_{b=1}^{Nb} C_b \Psi_b^{b^*}. \tag{8}$$

It represents the configuration interaction between the SCF-MI wave function

$$\Psi^0_{\mathrm{AB}} = |\Phi_1^{\mathrm{A}} \bar{\Phi}_1^{\mathrm{A}} \Phi_2^{\mathrm{A}} \bar{\Phi}_2^{\mathrm{A}} \cdots \bar{\Phi}_{Na}^{\mathrm{A}} \Phi_1^{\mathrm{B}} \bar{\Phi}_1^{\mathrm{B}} \Phi_2^{\mathrm{B}} \bar{\Phi}_2^{\mathrm{B}} \cdots \bar{\Phi}_{Nb}^{\mathrm{B}} \rangle, \tag{9}$$

the *singly* excited *localized* configuration state functions

$$\begin{aligned} \Psi_a^{a^*} &= |\Phi_1^{\mathrm{A}} \bar{\Phi}_1^{\mathrm{A}} \cdots \Phi_a^{\mathrm{A}} \Phi_{a*}^{\mathrm{A}} \cdots \bar{\Phi}_{Na}^{\mathrm{A}} \Phi_1^{\mathrm{B}} \bar{\Phi}_1^{\mathrm{B}} \cdots \Phi_b^{\mathrm{B}} \bar{\Phi}_b^{\mathrm{B}} \cdots \bar{\Phi}_{Nb}^{\mathrm{B}} \Theta^2_{0,0} \rangle, \\ \Psi_b^{b^*} &= |\Phi_1^{\mathrm{A}} \bar{\Phi}_1^{\mathrm{A}} \cdots \Phi_a^{\mathrm{A}} \bar{\Phi}_a^{\mathrm{A}} \cdots \bar{\Phi}_{Na}^{\mathrm{A}} \Phi_1^{\mathrm{B}} \bar{\Phi}_1^{\mathrm{B}} \cdots \Phi_b^{\mathrm{B}} \Phi_{b*}^{\mathrm{B}} \cdots \bar{\Phi}_{Nb}^{\mathrm{B}} \Theta^2_{0,0} \rangle \end{aligned} \tag{10}$$

and the *doubly* excited *localized* configuration state functions

$$\Psi_{ab}^{a^* b^*} = |\Phi_1^{\mathrm{A}} \bar{\Phi}_1^{\mathrm{A}} \cdots \Phi_a^{\mathrm{A}} \Phi_{a*}^{\mathrm{A}} \cdots \bar{\Phi}_{Na}^{\mathrm{A}} \Phi_1^{\mathrm{B}} \bar{\Phi}_1^{\mathrm{B}} \cdots \Phi_b^{\mathrm{B}} \Phi_{b*}^{\mathrm{B}} \cdots \bar{\Phi}_{Nb}^{\mathrm{B}} \Theta^4_{0,0} \rangle \tag{11}$$

obtained by simultaneous single excitation localized on A and B. The singlet spin functions for the two or four electrons involved in the single or double excitation are $\Theta^2_{0,0}$ and $\Theta^4_{0,0} = C_1 \Theta^4_{0,0;1} + C_2 \Theta^4_{0,0;2}$. The configurations included in the VB wave function play a significant role in the field of intermolecular forces as they can be associated with precise physical effects (energies and associated interaction densities) and coincide with specific contributions of a perturbative approach. Namely, Ψ^0_{AB} represents the sum of the Coulombian, the exchange and the induction energy (at SCF-MI level); the $\Psi_a^{a^*}$, $\Psi_b^{b^*}$ terms in Equation (8) refine the induction energy and have been added to relax the occupied SCF-MI orbitals, which are being kept fixed during the virtual orbital determination procedure (see below); the doubly excited configurations introduce correlation between the electrons of the two fragments and are associated to the interfragment dispersion energy. The energy of VB wave function is calculated by solving the corresponding secular problem, which includes the determination of the Hamiltonian and overlap matrices between non-orthogonal VB structures. A general VB code was employed [26]. For the sake of comparison with other energy contributions to the interaction energy, the pure electrostatic and exchange term was also calculated by setting C_{ab}, C_a and C_b equal to zero in Equation (8) and constructing Ψ^0_{AB} in terms of the undistorted SCF orbitals of the isolated fragments. Such a model is hereinafter referred to as the frozen monomer (FM) model. Our correlation contributions to the interaction energy do not introduce BSSE since only localized configurations have been considered in the evaluation of the VB energy. Besides, our treatment maintains a complete size consistency as, due to Brillouin's theorem [6], the included excitations give a zero contribution to the energy of the isolated fragments. The energies for the latter to be used in Equation (7) (with VB subscripts replacing the SCF-MI ones) are therefore just the SCF energies. Particular care has been taken in the construction of the optimal virtual orbitals, in order to generate a very compact VB wave function, while maintaining a BSSE-free approach. The general procedure is described in Refs. [23, 27] and only a brief summary is reported here. Both the occupied (Φ^{A}_a, Φ^{B}_b) and virtual (Φ^{A}_{a*}, Φ^{B}_{b*}) SCF-MI orbitals are expanded only in the basis sets of their own fragment:

$$\begin{aligned} \Phi^{\mathrm{A}}_a &= \sum_{p=1}^{M_{\mathrm{A}}} \chi_p^{\mathrm{A}} \mathrm{T}^{\mathrm{A}}_{pa}, & \Phi^{\mathrm{B}}_b &= \sum_{q=1}^{M_{\mathrm{B}}} \chi_q^{\mathrm{B}} \mathrm{T}^{\mathrm{B}}_{qb}, \\ \Phi^{\mathrm{A}}_{a*} &= \sum_{p=1}^{M_{\mathrm{A}}} \chi_p^{\mathrm{A}} \mathrm{T}^{\mathrm{A}}_{pa*}, & \Phi^{\mathrm{B}}_{b*} &= \sum_{q=1}^{M_{\mathrm{B}}} \chi_q^{\mathrm{B}} \mathrm{T}^{\mathrm{B}}_{qb*}. \end{aligned} \tag{12}$$

Such constraints imply the *non-orthogonality* of the orbitals. The optimal virtual orbitals Φ^{A}_{a*} and Φ^{B}_{b*} are determined accordingly to the approximation that they separately maximize the dispersion contribution of each of the $N_{\mathrm{A}} * N_{\mathrm{B}}$ two configuration wave functions

$$\Psi'_{\mathrm{AB}} = C_0 \Psi^0_{\mathrm{AB}} + C_{ab} \Psi^{a^*b^*}_{ab}, \tag{13}$$

where $\Psi^{a^*b^*}_{ab}$ represents a doubly excited configuration in which electrons are excited from the occupied SCF-MI orbitals Φ^{A}_a and Φ^{B}_b to the virtual orbitals Φ^{A}_{a*} and Φ^{B}_{b*}, respectively. The corresponding optimum virtual orbitals are determined at the variational-perturbational level of theory by minimization of the second order expression of the energy, the final expansion (Equation (12)) for each virtual pair being achieved iteratively.

Implementation of QTAM analysis for SCF-MI and VB wave functions

It is well known that, within the framework of the MO-LCAO-HF theory, the electron density ρ at a given point $\mathbf{r}$ can be expressed as

$$\rho(\mathbf{r}) = \sum_i \lambda_i \varphi_i^* \varphi_i, \tag{14}$$

where the summation extends over the occupied molecular orbitals φ_i and λ_i represent their occupation numbers. In Equation (14) the orbitals φ_i are supposed to be orthonormal among each other, while SCF-MI orbitals (Equation (3)) are not. The PROAIM [28] code, which implements QTAM for theoretical molecular densities, evaluates ρ and its derivatives on the basis of Equation (14). Hence to interface SCF-MI wave function with PROAIM code, the final SCF-MI orbitals have been unitarily transformed by diagonalizing the global Fock matrix $\mathbf{F}$ in the basis of the Φ^A and Φ^B orbitals (Equation (3)). The same procedure was previously used to implement [20] the energy derivatives of the SCF-MI wave function in the GAMESS-US package [21]. In the case of the VB wave function, the natural orbitals and occupation numbers obtained by diagonalizing the matrix representation of the VB one-density function on the atomic basis set were used in Equation (14).

3. Computational details and interaction energy data

Water dimer

A full account on the energy computations performed, at the SCF-MI and at the SCF-MI+VB levels, is reported in Refs. [12] and [23], respectively. Computations refer to the trans-linear water dimer, with C_s symmetry, at both experimental and theoretically optimized geometries. A thorough study of the water dimer potential energy surface, providing a new SCF-MI+VB interaction potential for the molecular dynamics simulation of liquid water, can be found in Refs. [29, 30]. Computational details and results of relevance for the present study are summarized here for the trans-linear conformation only. Several basis sets have been investigated, using the geometric sequences given by the even tempered gaussian s, p basis functions generated according to Schmidt and Ruedenberg [31]. The sequence length was systematically increased up to convergence on the dimer binding energy and force constants. The isotropic part of the basis set (s, p on O and s on H atoms) was supplemented with polarization functions (d, f on O and p and d on H atoms) according to the Sadlej method [32]. Finally, to investigate the effect of extremely diffuse s, p, d, f functions on the dimer binding energy, the geometric series was prolonged by introducing additional very small exponents, down to 10^{-2}. Table 1 lists the SCF-SM (BSSE-contaminated) and SCF-MI (BSSE-free) interaction energies and optimized geometries for the water dimer, as a function of selected basis sets. The largest basis set investigated (20s10p6d6f/10s6p6d) gave a monomer energy of −76.0676 au, settling the new calculated Hartree–Fock limit for water. The corresponding dimerization energy (−3.33 kcal/mol) is to be considered as a value close to the HF limit for this quantity and has to be compared with the related SCF-SM value of −3.71 kcal/mol. Indeed, Table 1 shows that BSSE is larger than 1 kcal/mol for poor basis sets (6-31G and 6-31G**) and, more importantly, that a significant difference (0.38 kcal/mol) persists even with the biggest basis sets. For the sake of fairness, such a difference is due to both residual BSSE and incompleteness of our best basis set.

However, the error due to incompleteness should not exceed 0.20 kcal/mol in view of the most recent theoretical results on water dimer [34]. Table 1 also shows that SCF-MI energies converge much faster than SCF-SM ones with increasing basis set

Table 1. SCF-MI (*SCF-SM*) interaction energies and geometries for the C_s linear water dimer.[a]

Basis set	M	E_{int} kcal/mol	$R_{\text{O–O}}$ Å	$R_{\text{O1–H2}}$ Å
6-31G	26	−6.34 (*−7.84*)	2.916 (*2.843*)	0.953 (*0.957*)
6-31G**	50	−4.12 (*−5.54*)	3.054 (*2.980*)	0.945 (*0.948*)
TZVP[b]	112	−3.36 (*−3.95*)	3.165 (*3.028*)	0.943 (*0.944*)
TZVP++[b]	124	−3.27 (*−3.79*)	3.165 (*3.030*)	0.943 (*0.945*)
(7s4p2d/4s2p)[c]	102	−3.32 (*−3.87*)	3.164 (*3.033*)	0.944 (*0.945*)
(20s10p4d/10s4p)[d]	236	−3.27 (*−3.76*)	3.157 (*3.034*)	0.943 (*0.944*)
(20s10p5d5f/10s5p5d)[d]	480	−3.33 (*−3.71*)	3.157 (*3.034*)	0.943 (*0.944*)
(20s10p6d6f/10s6p6d)[d]	548	−3.33 (*−3.71*)	3.157 (*3.034*)	0.943 (*0.944*)

[a] Data from Ref. [12]; basis sets from Refs. [31, 32]; [b] Ref. [21]; [c] Ref. [33]; [d] Ref. [12].

size. Quite interestingly, the medium-size Millot–Stone basis set [33] (Table 1, shaded row) yields geometries and dimerization energies which are (SCF-MI results) very close to our basis set limiting values. For this reason this basis has been adopted in the VB approach and used as our 'standard' in the interaction density and QTAM analysis (see below).

The SCF-MI limit interfragment distance (3.157 Å), is significantly larger than the corresponding BSSE-contaminated estimate (3.034 Å) and that found [35] experimentally (2.98 ± 0.03 Å). Hence, if only induction (SCF-HF without BSSE) is taken into account in the intermolecular interaction, the two water molecules are kept too far apart, while BSSE seems to mimic the effect of the dispersion contribution. However, when the SCF-MI determinant is used as the reference configuration in the VB expansion, an interfragment distance of 3.00 Å is obtained [23]. In the VB calculation an active space of four MOs with one MO (oxygen $1s^2$ electrons) kept frozen was considered for each water molecule. By adopting the VB expansion given by Equation (8) and by evaluating the optimum virtual orbitals according to Equation (13), the virtual space on each fragment is equal to 16. This implies a set of 32 (16 times 2) vertical singly excited configurations and 256 (16 times 16) vertical doubly excited spatial configurations. By taking the dimension of the spin space into account, the size of the resulting VB matrix is 545. The calculated water dimer binding energy (−4.67 kcal/mol) is in very good agreement with the available experimental estimates (−5.4 ± 0.7, [36]; −5.2 ± 0.7, [37]) and so is the optimized geometry [23]. Table 2 details the contribution of the various physical effects to the water dimerization energy, using the Millot–Stone basis set. The full counterpoise procedure greatly undercorrects BSSE as its estimated E_{int} (−3.80 kcal/mol) is about 0.5 kcal/mol larger than the SCF-MI value and only slightly smaller than the SCF-SM estimate (−3.87 kcal/mol). Such an underevaluation is nearly comparable in magnitude to the effect of induction. Indeed the difference between the pure exchange plus frozen monomer electrostatic dimerization energy (FM model) and the SCF-MI dimerization energy amounts to 0.84 kcal/mol. Table 2 also shows that the estimated dispersion energy [E_{int}(VB)−E_{int}(SCF-MI)] corresponds to over 30% of E_{int}(VB) and that the two SCF procedures (SCF-SM and SCF-MI) yield interaction energy values which

Table 2. Interaction energies vs physical model for the C_s linear water dimer (Millot–Stone basis set).

Model	Energy contributions	E_{int} (kcal/mol) opt. (*exp.*) geom.
FM	Electrostatic (frozen monomers) + exchange	(*−2.14*)
SCF-SM	As above + induction + effects due to BSSE	−3.87 (*−3.84*)
SCF-FCP	As for SCF-SM but with full CP correction of BSSE	−3.80 (*−3.73*)
SCF-MI	As for SCF-SM but BSSE free	−3.32 (*−2.97*)
SCF-MI-VB	As for SCF-MI + induction refinement + dispersion	−4.67 (*−4.57*)
Exp.		*−5.2 ± 0.7*

differ by a quantity equal to about 40% of the dispersion energy. These observations strongly support the importance of a detailed study of the changes found in the interaction densities as the physical model adopted for their evaluation improves.

4. QTAM analysis of dimerization energies and densities in HOH–OH$_2$ system

Atomic energy changes

The use of QTAM provides an atomic view of the energy changes accompanying the charge rearrangements following dimerization. Atomic energies were obtained as described in Ref. [14], by integrating the negative of the kinetic energy density over the atomic basin and by scaling the resulting energy value by the factor $\gamma - 1$ ($\gamma = -V/T$, being the virial ratio), to obtain a set of atomic energies which correctly sum to the total energy E. Integration on the atomic basins of water dimer and monomer gives the changes $\Delta E(\Omega)$ in the atomic energies upon dimerization. Table 3 details such changes for the atoms of the hydrogen donor and hydrogen acceptor molecules, as a function of the adopted model, while Figure 1 shows the numbering scheme used for water dimer. A negative $\Delta E(\Omega)$ value in Table 3 indicates an energy stabilization of Ω in the dimer.

With the only exception of the FM model, which is too crude, the other considered methods give a similar qualitative picture for the atomic energy changes following dimerization, the two oxygen atoms being stabilized and H2, the hydrogen atom involved in the hydrogen bond largely destabilized (as a result of a loss of charge, see below). Conversely, when the induction mechanism is inhibited (FM model), the reverse is true, H2 being stabilized and the oxygen of the donor molecule being destabilized. A closer inspection of Table 3 yields some interesting observations. First, the dimerization energy gain is mainly due to the stabilization of the acceptor molecule, in spite of its electron population loss (see below). Dispersion stabilizes more the H-acceptor than the H-donor molecule. Indeed, the donor molecule is even slightly destabilized at SCF-SM and SCF-MI levels. Secondly, at variance with the SCF-SM case, the ratio of SCF-MI over VB atomic energy changes is nearly constant, independently on the considered atom Ω. Hence dispersion just enhances the energy stabilization or destabilization effects caused by 'true' induction mechanisms. This

Table 3. Effect of model on the changes in atomic energies $\Delta E(\Omega)$, using the Millot–Stone basis set.

Atom Ω	$\Delta E(\Omega)$ (kcal/mol), opt. geometries			
	FM[a]	SCF-SM[b]	SCF-MI[b]	VB
O1	+3.7	−11.5 (*0.89*)	−10.5 (*0.81*)	−12.8
H2	−1.7	+15.1 (*0.94*)	+13.6 (*0.84*)	+16.1
H3	+0.0	−3.6 (*1.02*)	−2.9 (*0.84*)	−3.5
Donor molecule	+2.0	+0.1	+0.2	−0.2
O4	−4.2	−13.2 (*1.00*)	−11.1 (*0.84*)	−13.2
H5	+0.1	+4.6 (*1.05*)	+3.8 (*0.85*)	4.4
Acceptor molecule	−4.1	−4.0	−3.6	−4.5

[a] FM model (see text) at exp. geom.; [b] In parenthesis the ratio with the corresponding VB model values.

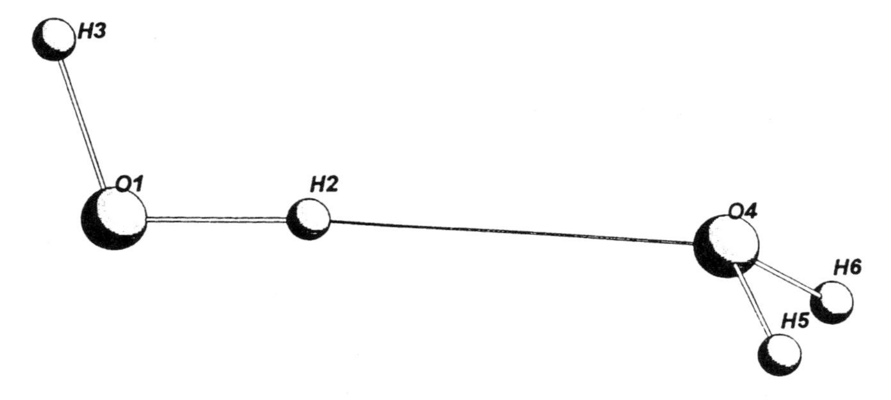

Figure 1. Numbering scheme for the trans-linear C_s water dimer.

is not the case of the standard SCF-SM method that appears to underestimate the energy changes in the part (O1 and H2 atoms) of the donor molecule more involved in the hydrogen bond, while it overestimates the energy changes of the remaining atoms. Comparison of $\Delta E(\Omega)$ values for the donor and acceptor molecules, at the two considered SCF levels, suggests that the spurious interaction energy gain due to BSSE is mainly due to an increased stabilization of the acceptor, rather than to a decreased destabilization of the donor moiety.

Charge transfer and atomic electron population changes

Upon dimerization, electron charge is transferred from the base (the H-acceptor molecule) to the acid (the H-donor molecule), in agreement with Lewis' generalized definition of an acid and a base as an electron acceptor and donor, respectively. The amount of such a charge transfer (CT) is reported in Table 4, for the two SCF models considered in this paper and as a function of the basis set size. The CTs are small and, for the SCF-SM method, are found to decrease as the basis set size increases.

Table 4. Effect of BSSE on the CT in the water dimer.

Basis set	M	CT∗100		
		SCF-SM	SCF-MI (*VB*)	(SCF-SM)/(SCF-MI)
6-31G	26	1.92	0.63	3.0
6-31G**	50	1.19	0.05	22.8
Millot–Stone	102	0.94	0.45 (*0.56*)	2.1
TZVP++	124	0.89	0.36	2.5

Table 5. Effect of model on the changes in QTAM atomic populations $\Delta N(\Omega)$, using the Millot–Stone basis set.[a]

Atom Ω	$\Delta N(\Omega) * 100$ opt. (*exp.*) geometries			
	FM	SCF-SM	SCF-MI	VB
O1	(*+1.0*)	+4.0 (*+4.7*)	+3.4 (*+5.0*)	+4.2 (*+4.7*)
H2	(*−0.8*)	−3.9 (*−4.5*)	−3.7 (*−5.4*)	−4.5 (*−4.9*)
H3	(*+0.1*)	+0.9 (*+0.9*)	+0.7 (*+0.8*)	+0.8 (*+0.8*)
Donor molecule	(*+0.3*)	+1.0 (*+1.1*)	+0.4 (*+0.5*)%	+0.5 (*+0.6*)
O4	(*−0.3*)	+1.4 (*+1.5*)	+1.5 (*+1.9*)	+1.7 (*+1.8*)
H5	(*+0.0*)	−1.2 (*−1.3*)	−0.9 (*−1.2*)	−1.1 (*−1.2*)
Acceptor molecule	(*−0.3*)	−1.0 (*−1.1*)	−0.4 (*−0.5*)	−0.5 (*−0.6*)

[a] A positive $\Delta N(\Omega)$ value indicates an electron population gain for Ω in the dimer.

The ratio of charge transfers, as obtained with the standard and the BSSE- free SCF, is always larger than two and shows a maximum for the 6-31G** basis. It appears that such a basis is large enough to allow one water molecule for a significant use of the basis functions of the partner (and vice versa) and, conversely, not big enough to make BSSE negligible. The SCF-MI charge transfer value (0.45 electrons) for the Millot–Stone basis set is half than the corresponding SCF-SM estimate and compares favorably with the VB outcome (0.56 electrons).

Atomic electron population changes $\Delta N(\Omega)$ upon dimerization are reported in Table 5 for the models considered in this study, at both experimental and optimized geometries. A detailed QTAM analysis of such electron rearrangements, in a number of hydrogen bonded systems, is reported in Ref. [38]. Here, we just investigate how the $\Delta N(\Omega)$ values vary as the theoretical level adopted for the description of the intermolecular interaction improves. A positive $\Delta N(\Omega)$ value in Table 5 indicates an electron population gain for Ω in the dimer. With the only exception of FM model, the $\Delta N(\Omega)$ values turn out to be generally larger than the charge transfer. Indeed, the remaining approaches predict a considerable redistribution of charge within the two molecules which involves an electron population loss from the tail of the base (H5+H6) and an electron population gain by the head (O1+H3) of the acid. Such a behavior is common to both experimental and optimized geometries of the water dimer. Save in FM model, the two oxygen atoms gain electron charge, the population increase for the oxygen donor being about 2.5 times greater than for the oxygen acceptor.

Table 6. Effect of BSSE on the changes in Mulliken atomic populations $\Delta N(\Omega)$.[a]

Atom Ω	$\Delta N(\Omega) * 100$ SCF-SM (SCF-MI)		
	6–31G**	Millot–Stone	TZVP++
O1	+4.6 (*+3.8*)	+2.7 (*+3.9*)	+8.3 (*+4.9*)
H2	−2.7 (*−4.7*)	−1.2 (*−4.4*)	−7.5 (*−5.2*)
H3	+1.0 (*+0.9*)	+0.5 (*+0.5*)	+0.4 (*+0.3*)
Donor molecule	+2.9 (*0*)	+1.9 (*0*)	+1.2 (*0*)
O4	+0.0 (*+2.5*)	+0.0 (*+2.3*)	+0.1 (*+3.2*)
H5	−1.4 (*−1.3*)	−1.0 (*−1.2*)	−0.6 (*−1.6*)
Acceptor molecule	−2.9 (*0*)	−1.9 (*0*)	−1.2 (*0*)

[a] A positive $\Delta N(\Omega)$ value indicates an electron population gain for Ω in the dimer.

In the case of VB and SCF-MI densities, the decrease in the electron population of the hydrogen involved in the HB is about ten times greater than and opposed to the observed CT. A similar behavior is found for the SCF-SM wave function, though the H2 atom population decrease is only five times greater than CT, as due to the overestimate of the latter. Overall, Table 5 shows that changes in electron populations upon dimerization are *qualitatively* described by the induction term only. Moreover, the effect of BSSE on $\Delta N(\Omega)$ appears rather limited. However, this observation is no longer true when $\Delta N(\Omega)$ values are computed through the standard Mulliken population analysis approach. Table 6 reports such $\Delta N(\Omega)$ values for SCF-SM and SCF-MI densities as a function of basis set size. The Mulliken CT is about two times larger than the QTAM estimate (Table 4) in the case of SCF-SM densities, while it is null for SCF-MI wave function, as due to definition of Mulliken's partitioning. The $\Delta N(\Omega)$ values at the SCF-MI level are by far more stable against basis set type than are the corresponding SCF-SM values. For instance the SCF-MI population change for H2 ranges from $-4.4 * 10^{-2}$ to $-5.2 * 10^{-2}$ electrons, while for the SCF-SM model it may differ by even one order of magnitude from basis to basis. It is also interesting to note that only for the SCF-MI wave function are the $\Delta N(\Omega)$ values obtained by Mulliken's procedure quite close to those evaluated through QTAM (Table 5). Indeed, while Mulliken's population values by themselves do not bear much physical meaning (especially for large basis sets), their variations, upon change of chemical environment, are known to be quite often reliable. Comparison of results reported in Tables 5 and 6 suggests that this holds true also for the case of the water dimer, provided the BSSE contamination is removed. The basis set instability, exhibited by the values of the SCF-SM Mulliken's population changes upon dimerization, is only caused by BSSE.

Interaction densities

The interaction density in water dimer has been the object of previous studies [39, 40]. In particular, Krijn and Feil [40] pointed out the effects of exchange repulsion and of the dominant mutual polarization of the two moieties arising from the electric fields

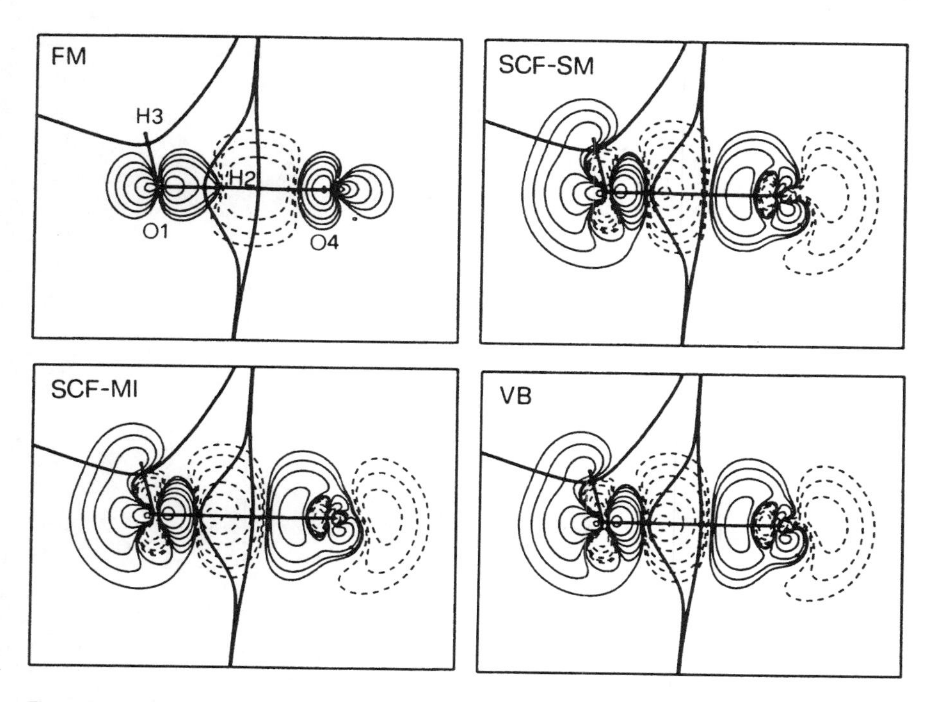

Figure 2. Effect of model on the interaction densities ($\Delta\rho = \rho_{\text{dimer}} - \rho_{\text{monomers}}$) in the water dimer ($\sigma_h$ plane, experimental geometry, Millot–Stone basis set). The values of the contours for ρ are $\pm a * 10^{-n}$, with $a = 2$, 4, 8 and n beginning at -4 and increasing in step of unity. *Dashed contours* denote negative, *solid contours* positive values of $\Delta\rho$. The pair of $\nabla\rho$ trajectories (heavy lines) which originate (bond paths) or which terminate at the bond critical points (denoted by dots) are superposed on the contour map. These latter $\nabla\rho$ trajectories mark the intersection of the interatomic surfaces with the σ_h plane, thus allowing an atomic view of interaction densities.

generated by their unperturbed charge distributions. Figure 2 displays interaction density contours for the four models used in this work, in the σ_h plane of the water dimer and at its experimental geometry. All the approaches, save the FM model, yield a similar picture for the interaction. The FM model seems unable to properly describe the polarization of the two oxygen atoms and the depletion of charge in the H2 basin. Hence, the following discussion does not refer to such a model. Figure 2 shows that the H-bond interaction is reflected throughout the complex, rather than being restricted to the region of the hydrogen bond itself. This fact was also evident from the reported analysis of the $\Delta N(\Omega)$ values. As found elsewhere [1, 41], the HB reinforces the polarity of the molecules that participate in the bond (the individual dipole moment are enhanced), the acceptor becoming a better donor and the donor becoming a better acceptor. The induced dipoles at the VB level and evaluated according to QTAM partitioning are 0.14 and 0.09 D for the donor and the acceptor molecules, respectively. The SCF-SM and SCF-MI estimates for the induced dipoles differ between each other by less than 10% and are similar to the VB values. A closer and more

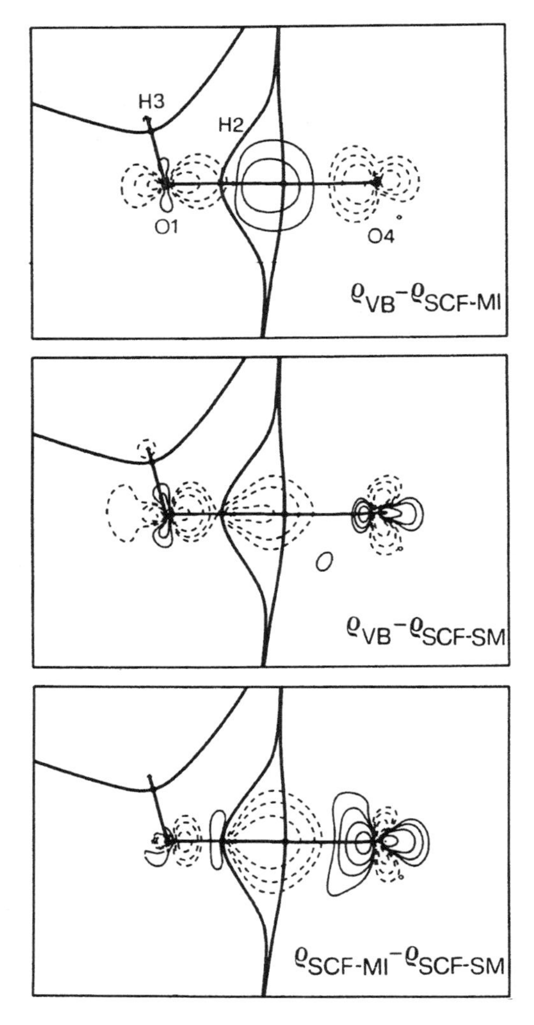

Figure 3. Differences among water dimer densities ρ, as obtained with the theoretical models explored in this work. Same symmetry plane, geometry, basis set and contour levels as that of Figure 2.

interesting inspection on the different performance of our models is given in Figure 3 where the differences between VB and either SCF-MI or SCF-SM dimer densities are displayed in the same symmetry plane and using the same $\Delta\rho$ contour levels of Figure 2. The bottom panel of Figure 3 also reports the difference between SCF-MI and SCF-SM dimer densities. Figure 3 (top) shows that by including dispersion contributions, electron charge is moved into the interfragment region, (in particular in the H2 basin) and removed from all other regions along the O–O axis. When spurious BSSE effects are introduced (Figure 3, middle) the picture becomes far less simple to interpret. However, it is clear that in this case dispersion has to remove electron charge from the interfragment region. Such an electron removal concerns all the O–O internuclear axis region, save a small part close to the lone pair of the oxygen acceptor (see below). It appears that dispersion contributions have opposite effects, whether a density describing 'true' induction or a BSSE contaminated induction, is taken as reference. Finally the difference between SCF-MI and SCF-SM densities (Figure 3, bottom) confirms that BSSE puts too much electron charge in the HB region, by removing it, in particular, from the acceptor oxygen.

The interaction densities portrayed in Figure 2 suggest that the BSSE which arise from a basis set of quite high quality (7s4p2d/4s2p on O and H, respectively) is small enough to yield electron density rearrangements *qualitatively* similar to those of the BSSE-free model. However, the density differences displayed in Figure 3 show that the BSSE effects on the density are comparable in magnitude to those induced by the dispersion contributions. As we also know (Table 2) that the small density differences shown in Figure 3 imply notable changes in the interaction energies, one wonders whether these density changes are larger or at least comparable with the estimated standard deviations (esd) for experimental ρ. High quality X-ray diffraction experiments, especially when carried out at very low temperatures, may give [42] esd for ρ in the HB regions as low as 0.001 au, a value comparable in magnitude to the innermost density contours displayed in Figure 3.

Interaction Laplacian densities

Hydrogen-bonded complexes have been discussed [14, 16, 38, 43] in terms of a generalized Lewis acid and base interaction, using the Laplacian of the electron density ($\nabla^2\rho$) as a tool for predicting their structures and studying their characteristics. As explained at length in Ref. [14], the sign of the Laplacian determines the regions which are charge depleted (positive Laplacian) or where charge concentrates. Charge concentration at a point $\mathbf{r}$ means that $\rho(\mathbf{r})$ is bigger than in an infinitesimal volume around it. The form of the Laplacian of ρ for an isolated atom reflects its shell structure since it exhibits a corresponding number of pairs of spherical shells of alternating charge concentration and charge depletion, the inner shell of each pair being always the region of charge concentration. The spherical valence shell of charge concentration (VSCC) in an isolated atom does not persist upon bonding, since local maxima and minima in $-\nabla^2\rho$ are formed within the shell, depending on the number and type of the linked atoms. It has been shown [43] that the approach of the acidic hydrogen to the base, in a HB interaction, is such as to align a $-\nabla^2\rho$ minimum in the valence shell of the acidic hydrogen atom with a base $-\nabla^2\rho$ maximum. Local minima in $-\nabla^2\rho$ are hereafter indicated as cage critical points, while local maxima are referred to as either bonded or not bonded concentrations, according to whether they occur in bonding or in lone pair regions, respectively. Figure 4(a) displays the Laplacian density in the σ_h plane of the water dimer, while Figure 4(b) shows the alignment of the H2 cage critical point (CP) of the donor water molecule with the non-bonded maximum associated to one of the lone pairs of the acceptor oxygen atom. The

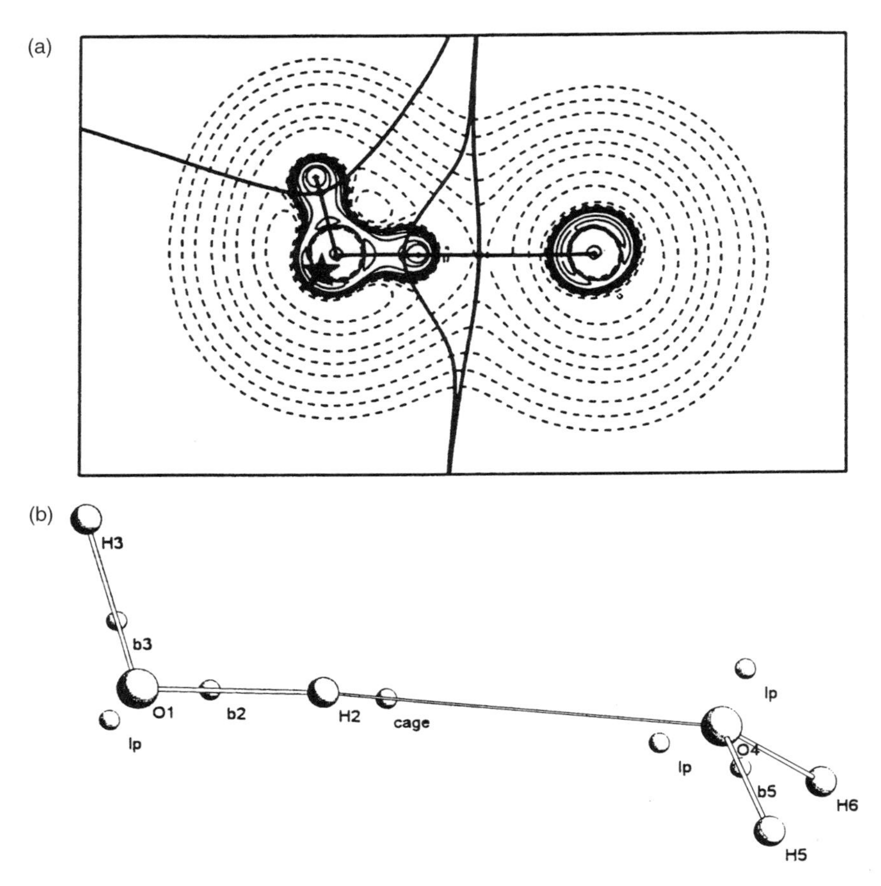

Figure 4. (a) Laplacian density (VB model, experimental geometry) in the σ_h plane of the water dimer. *Solid contours* denote negative $\nabla^2\rho$ values; (b) schematic representation of the Laplacian critical points of water which undergo the most significant changes following dimerization. Bonded maxima are denoted as b, non-bonded maxima (lone pairs) as lp. Only one O1 lp is shown in (b), while the local maximum visible in (a) and denoted with a star, corresponds to the saddle point between the two O1 lps which are symmetry related by the σ_h plane. The alignment of the H2 cage CP of the donor water molecule with the non-bonded maximum associated to one of the oxygen acceptor lps is also shown in the figure.

approach of the acid and the base involves, compared to frozen monomers, a further charge depletion of the H2 cage and a decrease of the non-bonded concentration of the acceptor oxygen pointing towards H2. Also the lone pairs of the donor oxygen (O2) become less concentrated, while the bonded concentrations of O1–H2 and O1–H3 bonds, indicated as b2 and b3 in Figure 4, respectively increase and decrease their $|-\nabla^2\rho|$ value upon dimerization. Indeed, hydrogen bonding yields a lengthening of O–H2 bond, with a parallel increase of its polarity and decrease of sharing of oxygen-bonded concentration. Just the opposite occurs to the O–H3 bond. So far we summarized the main changes induced by hydrogen bond formation on the Laplacian

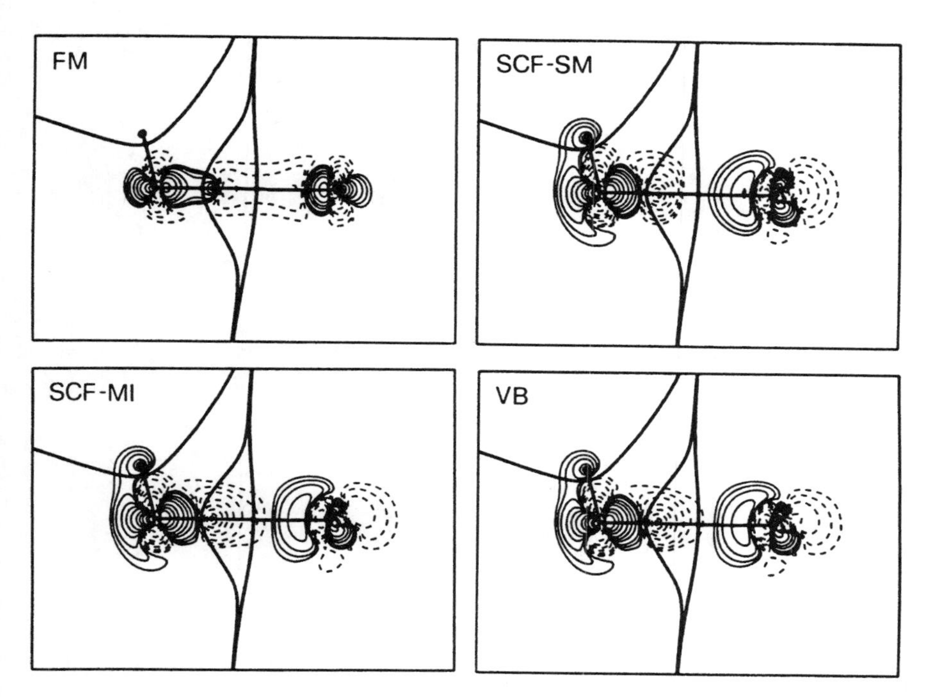

Figure 5. Effect of model on the interaction Laplacian densities $\Delta(\nabla^2\rho) = (\nabla^2\rho_{\text{dimer}} - \nabla^2\rho_{\text{monomers}})$ in water dimer (σ_h plane, experimental geometry, Millot–Stone basis set). Same contours as in Figure 2, but with n beginning at -3. *Dashed contours* denote negative, *solid contours* positive values of $\Delta(\nabla^2\rho)$. A negative contour level means that locally the dimer has less charge concentration than the pro-dimer.

distribution of isolated monomers. The question now arises whether these changes are adequately reproduced by all the approaches investigated in this study and whether the quantitative differences found among the various models are significantly smaller or, conversely, comparable in magnitude to the changes themselves. Figure 5 displays interaction Laplacian densities for the four models used in this work, while Figure 6 shows the differences $\Delta(\nabla^2\rho) = (\nabla^2\rho)_{\text{A}} - (\nabla^2\rho)_{\text{B}}$ between Laplacian densities for the dimer evaluated with models A and B (A and B being any one of the investigated models). Finally, Table 7 lists, as a function of the computational approach, the $\nabla^2\rho$ values for the Laplacian critical points shown in Figure 5, for both water monomer and dimer, at their experimental geometries. Geometry, basis set and map plane in Figure 6 is the same as that of Figure 3, while, due to the greater details given by the Laplacian function, the lowest contour level ($\pm 2 * 10^{-3}$ au) is, here, one order of magnitude larger. A negative contour level (dotted line) means (Figure 6) that locally the dimer is less charge concentrated than the pro-dimer or that the dimer evaluated according to model A is locally less charge concentrated than when evaluated with model B (not shown). Inspection of Figure 6 shows that, with the exception of the FM model, all the approaches adopted give qualitatively similar Laplacian interaction densities, a result

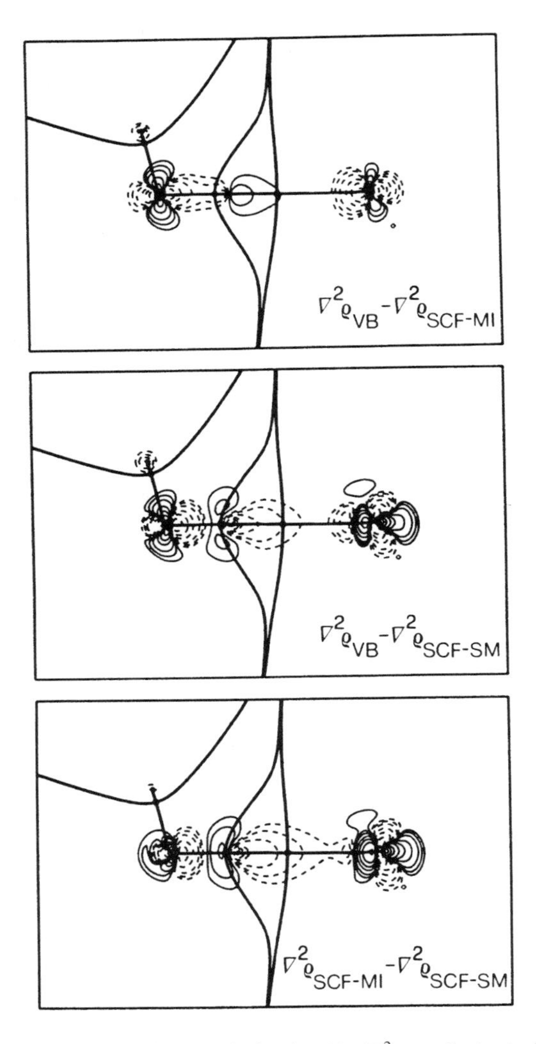

Figure 6. Differences among water dimer Laplacian densities $\nabla^2\rho$, as obtained with models explored in this work. Same symmetry plane, geometry, basis set and contour levels as that of Figure 2.

already found for the interaction densities in Figure 4. Changes in the dimer are in the expected direction, the cage on H2 being more charge depleted, the oxygen acceptor lone pair pointing towards H2 becoming less concentrated and so on. The mechanism of base–acid interaction is also evident. Charge concentration is removed from the lone pair region of O4 and moved towards the acid which enhances its acidity by further increasing the charge depletion around the H2 cage. The FM seems fully unable to describe such mechanisms, while the relevance of the quantitative differences among the remaining models can easily be appreciated. It turns out that dispersion effects

Table 7. Laplacian critical points for the water monomer and dimer.[a]

Ω	CP	$\nabla^2\rho$ au			
		FM	SCF-SM	SCF-MI	VB
Monomer					
O	lp		−5.926		
	bm		−2.715		
H	cage		+0.180		
Dimer					
O1	lp	−5.911	−5.764	−5.772	−5.771
	b2	−2.774	−2.981	−2.968	−2.954
	b3	−2.703	−2.624	−2.626	−2.630
H2	cage	+0.195	+0.199	+0.212	+0.207
O4	lp ⇒ H2	−6.037	−5.764	−5.797	−5.711
	lp	−5.916	−5.939	−5.928	−5.926
	b5	−2.701	−2.769	−2.754	−2.759

[a]Experimental geometries, Millot–Stone basis set. For labeling of critical points see Figure 4(b).

lower the charge depletion at the H2 cage point, which is exaggerated by the HF model. When BSSE is not removed the opposite is true, as BSSE underestimates the increase of charge depletion at the H2 cage point, caused by hydrogen bonding. Moreover, when compared with SCF-MI model, the dispersion effects enhance the decrease of charge concentration at the O4 lone pair, while they have to slightly increase such a concentration if BSSE is present. Such observations are made even more quantitative in Table 7. The difference between SCF-MI and SCF-SM Laplacian values at the cage point amounts to about 50% of the effect due to dimerization, while the same difference lies in the 5–20% range for the oxygen atoms lone pair concentrations. It has previously [44] been shown how the changes in the Laplacian bonded and non-bonded maxima induced by molecular association in gas phases and crystals are reflected in changes of the electric field gradient (EFG) at nuclei. From data reported in Table 7, it appears that EFG results might be affected by BSSE removal.

5. Conclusions

This study provides a detailed description of changes induced by dimerization on the electron distribution of water. The contribution of several mechanisms (electrostatic-exchange, induction, dispersion) underlying the intermolecular interaction is highlighted, and the effect of removing the BSSE contamination on the description of such mechanisms is investigated at length. Our study shows that even with a basis set of very high quality (Millot–Stone basis set), the BSSE effect is, at least for some quantities, of the same order of magnitude of changes due to dimerization. Though SCF-SM and SCF-MI provide the same qualitative picture for the charge rearrangements in water dimer, the quantitative differences between their associated electron densities are as large as 2–3 times the estimated standard deviations of good quality experimental densities in molecular crystals.

BSSE overestimates the charge transfer between monomers and accumulates too much electron charge in the hydrogen bond region by removing it in particular from the acceptor oxygen atom. Hence, dispersion contributions are found to concentrate or remove electron charge in the intermolecular region according to whether a density describing ‘true’ induction or a BSSE contaminated induction is taken as reference.

BSSE also opposes the tendency of the Hartree–Fock model to keep the interacting closed shell fragments too far apart. So, when optimized geometries are considered for the complex, BSSE is found to mimic some of those effects on the electron density distribution which would be induced by the interfragment dispersion contributions.

The SCF-MI method provides interaction energy values which converge quite rapidly with increasing basis set size. This fact makes this approach particularly recommended for large interacting moieties where basis sets of double- or triple-zeta quality are typically used and where the use of very extended basis set, like Millet–Stone’s, is precluded. The resulting BSSE effect on the interaction densities should in this case be much larger than that found for the water dimer.

In the last years, alternative approaches have been proposed for constructing electron densities for large macromolecules, like proteins, starting from smaller fragments. Methods based on discrete [14, 45, 46] or on fuzzy boundary [47–49] partitionings have been devised. The former approach leads to fragments that have been identified as proper open systems [15] and with properties defined by quantum mechanics. However, the employment of such fragments as building blocks for larger systems presents difficulties and their use is probably more suited for assessing transferability properties. The second method, though more empirical, has great advantages as far as the additivity and adjustability of fragment densities is concerned. It is our aim to explore the capability of SCF-MI method as a tool to evaluate fuzzy density fragments which reproduce interfragment interactions and do not require an *a posteriori* partitioning of the first order density matrix for their construction.

Acknowledgements

One of us (C.G.) acknowledges the support by Human Capital and Mobility programme of the European Community under Contract No. CHRX-CT93-0155. The help of Mr. M. Bandera in preparing the drawings is gratefully acknowledged.

References

1. Feil, D. (1991) X-ray diffraction and charge distribution: application to the electron density distribution in the hydrogen bond, In *The Application of Charge Density Research to Chemistry and Drug Design*, Jeffrey, G.A. and Piniella, J.F. (Eds.), NATO ASI Series B250, Plenum Press, New York.
2. de Vries, R.I.J. (1996) Exploring the limits of electron density studies by X-ray diffraction, Ph.D. Thesis, Twente University.
3. Scheiner, S. (1994) *Ab-initio* studies of hydrogen bonds: the water dimer paradigm, *Annu. Rev. Phys. Chem.*, **45**, 23–56.
4. Rybak, S., Jeziorski, B. and Szalewicz, K. (1991) Many-body symmetry-adapted perturbation theory of intermolecular interactions. H_2O and HF dimers, *J. Chem. Phys.*, **95**, 6576–6601.
5. Scheiner, S. (1991) Calculating the properties of hydrogen bonds by *Ab initio* methods, *Reviews in Computational Chemistry II*, VCH Publisher, New York.

6. Szabø, A. and Ostlund, N.S. (1989) *Modern Quantum Chemistry. Introduction to Advanced Electronic Structure Theory*, McGraw-Hill, New York.
7. Kestner, N.R. (1968) He–He Interaction in the SCF-MO approximation, *J. Chem. Phys.*, **48**, 252–257.
8. Boys, S.F. and Bernardi, F. (1970) The calculation of small molecular interaction by the differences of separate total energies. Some procedures with reduced errors, *Mol. Phys.*, **19**, 553–566.
9. van Lenthe, J.H., van Duijneveldt-van de Rijdt, J.C.G.M. and van Duijneveldt, D.B. (1987) Weakly bonded systems, *Adv. Chem. Phys.*, **69**, 521–566.
10. Saebø, S., Tong, W. and Pulay, P. Efficient elimination of basis set superposition errors by the local correlation method: Accurate *ab initio* studies of the water dimer, *J. Chem. Phys.*, **98**, 2170–2175.
11. Gianinetti, E., Raimondi, M. and Tornaghi, E. (1996) Modification of the Roothaan equations to exclude BSSE from molecular interaction calculations, *Int. J. Quantum Chem.*, **60**, 157–166.
12. Famulari, A., Raimondi, M., Sironi, M. and Gianinetti, E. (1998) Hartree–Fock limit properties of water dimer in absence of BSSE, *Chem. Phys.*, **232**, 275–287.
13. Famulari, A., Gianinetti, E., Raimondi, M., Sironi, M. and Vandoni, I. (1998) Modification of Guest and Saunders open shell SCF equations to exclude BSSE from molecular interaction calculations, *Theor. Chim. Acta*, **99**, 358–365.
14. Bader, R.F.W. (1990) Atoms in molecules. A quantum theory, *International Series of Monographs on Chemistry*, Vol. 22, Oxford University Press, Oxford.
15. Bader, R.F.W. (1994) Principle of stationary action and the definition of a proper open system, *Phys. Rev. B*, **49**, 13348–13356.
16. Gatti, C., Saunders, V.R. and Roetti, C. (1994) Crystal field effects on the topological properties of the electron density in molecular crystals: the case of urea, *J. Chem. Phys.*, **101**, 10686–10696.
17. Tsirelson, V.G., Zou, P.F. and Bader, R.F.W. (1995) Topological definition of crystal structure: determination of the bonded interactions in solid molecular chlorine, *Acta Cryst.*, **A51**, 143–153.
18. Platts, J.A. and Howard, S.T. (1996) Periodic Hartree–Fock calculations on crystalline HCN, *J. Chem. Phys.*, **105**, 4668–4674.
19. Gianinetti, E., Vandoni, I., Famulari, A. and Raimondi, M (1998) Extension of the SCF-MI method to the case of K fragments one of which is an open-shell system, *Adv. Quantum Chem.*, **31**, 251–266.
20. Famulari, A., Gianinetti, E., Raimondi, M. and Sironi, M. (1998) Implementation of gradient optimisation algorithms and force constant computations in BSSE free direct and conventional SCF approaches, *Int. J. Quantum Chem.*, **69**, 151–158.
21. Schmidt, M.W., Baldridge, K.K., Boatz, J.A., Elbert, S.T., Gordon, M.S., Jensen, J., Koseki, S., Matsunaga, N., Nguyen, K.A., Su, S.J., Windus, T.L., Dupuis, M. and Montgomery Jr., J.A. (1993) General atomic and molecular electronic structure system, *J. Comp. Chem.*, **14**, 1347–1363.
22. Xantheas, S.S. (1996) On the importance of the fragment relaxation energy terms in the estimation of the basis set superposition error correction to the intermolecular interaction energy, *J. Chem. Phys.*, **104**, 8821–8824.
23. Famulari, A., Raimondi, M., Sironi, M. and Gianinetti, E. (1998) *Ab initio* MO–VB study of water dimer in absence of BSSE, *Chem. Phys.*, **232**, 289–298.
24. Cooper, D.L., Gerratt, J. and Raimondi, M. (1987) Modern valence bond theory, *Advances Chem. Phys.*, **69**, 319–397.
25. Gerratt, J., Cooper, D.L., Karadakov, P.B. and Raimondi, M. (1997) Modern valence bond theory, *Chem. Soc. Rev.*, **26**, 87–100.
26. Raimondi, M., Campion, W. and Karplus, M. (1977) Convergence of the valence bond calculations for methane, *Mol. Phys.*, **34**, 1483–1492.
27. Raimondi, M., Sironi, M., Gerratt, J. and Cooper, D. L. (1996) Optimized spin-coupled virtual orbitals, *Int. J. Quantum Chem.*, **60**, 225–233.
28. PROAIM (1992) Mc Master University, Ontario, Canada.
29. Raimondi, M., Famulari, A., Gianinetti, E., Sironi, M., Specchio, R. and Vandoni, I. (1998) New *ab initio* VB interaction potential for molecular dynamics simulation of liquid water, *Adv. Quantum Chem.*, **32**, 263–284.
30. Famulari, A., Specchio, R., Sironi, M., and Raimondi, M. (1998) New basis set superposition error free *ab initio* MO–VB interaction potential: molecular dynamics simulation of water at critical and supercritical conditions, *J. Chem. Phys.*, **108**, 3296–3303.
31. Schmidt, M.W. and Ruedenberg, K. (1979) Effective convergence to complete orbital bases and to the atomic Hartree–Fock limit through systematic sequences of Gaussian primitives, *J. Chem. Phys.*, **71**, 3951–3962.

32. Sadlej, A.J. (1977) Molecular electric polarizabilities. Electric-field-variant (EFV) Gaussian basis set for polarizability calculations, *Chem. Phys. Lett.*, **47**, 50–54.
33. Millot, C. and Stone, A.J. (1992) Towards an accurate intermolecular potential for water, *Mol. Phys.*, **77**, 439–462.
34. Halkier, A., Koch, H., Jørgensen, P., Christiansen, O., Nielsen, I.M.B. and Helgaker, T. (1997) A systematic *ab initio* study of the water dimer in hierarchies of basis sets and correlation models, *Theor. Chem. Acc.*, **97**, 150–157.
35. Dyke, T.R., Mack, K.M. and Muenter, J.S. (1977) The structure of water dimer from molecular beam electric resonance spectroscopy, *J. Chem. Phys.*, **66**, 498–510.
36. Curtiss, L.A., Frurip, D.J. and Blander, M. (1979) Studies of molecular association in H_2O and D_2O vapors by measurement of thermal conductivity, *J. Chem. Phys.*, **71**, 2703–2711.
37. Mas, E.M. and Szalewicz, K. (1996) Effects of monomer geometry and basis set saturation on computed depth of water dimer potential, *J. Chem. Phys.*, **104**, 7606– 7614.
38. Carroll, M.T. and Bader, R.F.W. (1988) An analysis of the hydrogen bond in BASE-HF complexes using the theory of atoms in molecules, *Mol. Phys.*, **65**, 695– 722.
39. Diercksen, G.H.F. (1971) SCF-MO-LCGO studies on hydrogen bonding. The water dimer, *Theoret. Chim. Acta*, **21**, 335–367.
40. Krijn, M.P.C.M. and Feil, D. (1988), A local density-functional study of the electron density distribution in the H_2O dimer, *J. Chem. Phys.*, **89**, 5787–5793.
41. Gatti, C., Silvi, B. and Colonna, F. (1995) Dipole moment of the water molecule in the condensed phase: a periodic Hartree–Fock estimate, *Chem. Phys. Lett.*, **247**, 135– 141.
42. Destro R., Bianchi, R., Gatti, C. and Merati, F. (1991) Total electronic charge density of L-alanine from X-ray diffraction at 23 K, *Chem. Phys. Lett.*, **186**, 47–52.
43. Carroll, M.T., Chang, C. and Bader, R.F.W. (1988) Prediction of the structures of hydrogen-bonded complexes using the Laplacian of the charge density, *Mol. Phys.*, **63**, 387–405.
44. Aray, Y., Gatti, C. and Murgich, J. (1994) The electron field gradient at the N nuclei and the topology of the charge distribution in the protonation of urea, *J. Chem. Phys.*, **101**, 9800–9806.
45. Chang, C. and Bader, R.F.W. (1992) Theoretical construction of a polypeptide, *J. Phys. Chem.*, **96**, 1654–1662.
46. Popelier, P.L.A. and Bader, R.F.W. (1994) Effect of twisting a polypeptide on its geometry and electron distribution, *J. Phys. Chem.*, **98**, 4473–4481.
47. Walker, P.D. and Mezey, P.G. (1993) *J. Am. Chem. Soc.*, **115**, 12423–12430.
48. Mezey, P.G. (1995) Macromolecular density matrices and electron densities with adjustable nuclear geometries, *J. Math. Chem.*, **18**, 141–168.
49. Walker, P.D. and Mezey, P.G. (1994) *Ab initio* quality electron densities for proteins: a MEDLA approach, *J. Am. Chem. Soc.*, **116**, 12022–12032.

7

Topological analysis of X-ray protein relative density maps utilizing the eigenvector following method

KENNETH E. EDGECOMBE, ALAN ABLESON, KIM BAXTER, ANTONY CHIVERTON, JANICE GLASGOW and SUZANNE FORTIER

Molecular Scene Analysis Group, Departments of Computing and Information Science and Chemistry, Queen's University, Kingston, Ontario, Canada K7L 3N6

1. Introduction

A fundamental goal of research in the biological sciences is to understand protein structure. In theory, protein sequence information can be inferred from the fast growing volume of DNA sequence data [1] but predicting the three-dimensional structure of a protein from its sequence remains an open and important problem [2–4]. Part of the difficulty in solving this problem is due to the fact that many of the existing techniques rely on our knowledge of previously determined structures which, compared to sequence data, is relatively limited.

X-ray crystallography plays a central role in current efforts in protein structure determination. However, building an accurate and detailed protein model from crystallographic data remains a complex and time-consuming process [5]. This is due to the fact that while intensity data may be available at relatively high resolution, this is rarely the case for phase information [6]. Therefore, initial models are usually built at low to medium resolutions, where human intervention is needed for recognizing typical protein structure motifs, and then bootstrap to higher and higher resolution. Errors in the initial and subsequent models may be corrected using a refinement process in which the model is modified to minimize the difference between the experimentally observed data and the data calculated using a hypothetical crystal containing the model. The development of more sophisticated computational tools would improve the process of protein model building [7]. A goal of our research is to improve and accelerate this process through the design and development of automated tools.

In this paper we report on an approach to protein model construction that can be incorporated into a fully automated system for structure determination from crystallographic data. Our approach has the advantage of using characteristics of the experimental data to find a path through the tertiary structure of the protein without introducing bias into the data. It incorporates a spline interpolation algorithm to generate a smooth continuous density function for the protein, an eigenvector following algorithm to locate critical points in the electron density and a gradient path following algorithm to connect critical points and, thus, characterize features of the protein. The work described in this paper advances the ability to discern meaningful features of protein structure through the use of a topological analysis of the relative density at various levels of resolution. This is similar to the approach of Bader *et al.* [8] and

Paul G. Mezey and Beverly E. Robertson (eds.), Electron, Spin and Momentum Densities and Chemical Reactivity, 115–125

builds on previous work of Johnson [9] and of the Molecular Scene Analysis Group at Queen's University [10].

Bader *et al.* have developed a theory of molecular structure [8], based on the topological properties of the electron density $\rho(r)$. In this theory, a molecule may be partitioned into atoms or fragments by using zero-flux surfaces that satisfy the condition

$$\nabla\rho(r)n = 0$$

for every point on the surface of the subsystem where **n** is a unit vector to the surface. Note that points in the electron density at which the gradient of the density equals zero, that is, points at which

$$\nabla\rho(r) = 0$$

are critical points and are characterized by the signs of the eigenvalues of the Hessian of the density at that point. The sum of the signs of the eigenvalues of this 3 by 3 matrix of second derivatives of $\rho(r)$ is called the signature and is used to classify the type of critical point. There are four possible signatures for critical points of rank 3 (i.e. with three non-zero eigenvalues), designated by (rank, signature):

(3, −3) A local maximum in the density with 3 negative curvatures, called a *peak*.
(3, −1) A saddle point in the density with 2 negative and 1 positive curvatures, called a *pass*.
(3, 1) A saddle point in the density with 2 positive and 1 negative curvatures, called a *pale*.
(3, 3) A local minimum in the density with 3 positive curvatures, called a *pit*.

Points on the zero-flux surfaces that are saddle points in the density are passes or pales. Should the critical point be located on a path between bonded atoms along which the density is a maximum with respect to lateral displacement, it is known as a pass. Nuclei behave topologically as peaks and all of the gradient paths of the density in the neighborhood of a particular peak terminate at that peak. Thus, the peaks act as attractors in the gradient vector field of the density. Passes are located between neighboring attractors which are linked by a unique pair of trajectories associated with the passes. Cao *et al.* [11] pointed out that it is through the attractor behavior of nuclei that distinct atomic forms are created in the density. In the theory of molecular structure, therefore, peaks and passes play a crucial role.

The application of the theory of molecular structure to the solid state has been limited to theoretical calculations [12] or high resolution experimental data [13]. The direct application to low to medium resolution data is impractical as distinct atomic forms are for the most part impossible to characterize through the gradient vector field of the density. Although there is topology in these maps, peaks represent groups of atoms such as fragments of residues in proteins. There have been efforts, however, to utilize the topological properties of low to medium resolution crystallographic data on proteins to help deduce the structure. The program ORCRIT [14], for example, has previously been used to calculate and characterize critical points in the relative density. By choosing appropriate relative density value cutoffs, Leherte *et al.* [10] were able to trace out a spanning tree of peak–pass–peak–pass corresponding to the protein structure. The construction of a spanning tree utilizes two parameters, r_{max}, the maximum distance between critical points below which they are considered connected and, W, the weighting of the connection. The critical points themselves are located first through a simplistic comparison search of the grid, then a polynomial interpolation is performed to obtain a more accurate approximation of their position.

2. Method

In the present work, the derived relative density grid is modeled using cubic splines. This gives a smooth continuous function over which values of the relative density, gradient and Laplacian can be calculated. In order to calculate all critical points, a grid of initial starting points is chosen and a search for $(3, -3)$ critical points (peaks) initiated using the eigenvector following algorithm of Popelier [15]. Next, a check for other peaks is performed by initiating the search algorithm starting midway between those peaks already found. Following this check, a search for the passes located between proximal peaks is made. This search also uses the eigenvector following method given a starting position at the midpoint between peaks that are within 8 Å of each other. Note that for this study, only peaks and passes with relative densities greater than zero are considered. Once an acceptable pass has been located, the fifth-order Cash–Karp–Runge–Kutta gradient path tracing algorithm [16] is used to trace along the ridge of maximum density to the two peaks associated with this pass. In this way, the lines of interaction (lines of maximum density) between the peaks are traced. By tracing the lines of interaction from peak to peak, the chain of such peaks and passes representing the protein backbone, or portion thereof, may be traced. Figure 1 illustrates two portions of a protein backbone with an intraprotein interaction, such as might be traced for a disulphide bridge, and a number of peaks and passes representative of side chains. This methodology is based on Popelier's algorithm as implemented in the MORPHY program [17], but unlike the original algorithm it utilizes cubic splines and works without reference to nuclei. Note that MORPHY performs an automated topological analysis of a molecular electron density, which requires a wave function calculated from some quantum mechanical program, a Gaussian basis set and a set of nuclear coordinates as input. The nuclear coordinates are assumed to be peaks. None of these features are available for the crystallographic work, although the spline interpolation functions might be considered a basis set. The spline coefficients are fit to a density grid obtained from the XTAL program [18].

Three proteins, whose structures have been resolved and used in previous studies from this group, were chosen to serve as test cases:

Case 1: An ideal density map, that is, a calculated map from the known structure of protein BP2 (bovine pancreatic phospholipase A2) which contains 123

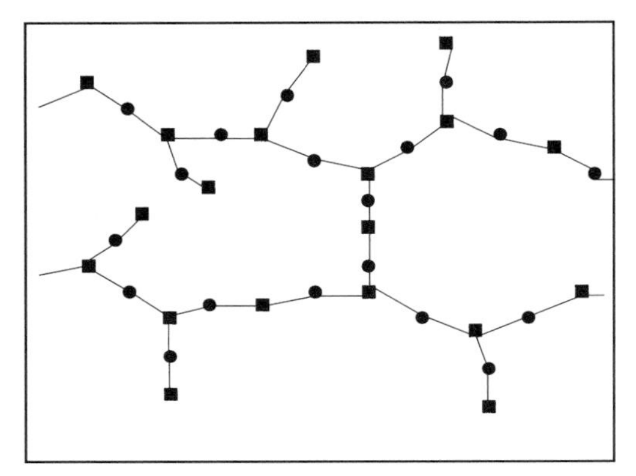

Figure 1. A two-dimensional representation that illustrates the tracing of the interaction lines to give the peak–pass–peak–pass chain representative of the protein backbone, side chains and disulphide bridge. Circles represent passes and squares peaks.

residues and in its crystalline form is a member of the $P2_12_12_1$ space group [19].

Case 2: An experimental density map for recombinant type III antifreeze protein from eel pout (AFP), which contains 66 residues and in its crystalline form is a member of the $P2_12_12_1$ space group [20].

Case 3: An experimental density map for penicillopepsin (3APP), which contains 323 residues and in its crystalline form is a member of the C2 space group [21].

All three proteins were analyzed at 3.0 Å resolution. Experiments were performed on a portion of the relative density map containing an entire connected protein. In order to discern effects of topological features just outside the boundaries of this volume, our analysis was extended 5.0 Å outside the boundaries on all sides of this volume.

3. Results and discussion

For each protein, the results were evaluated at three stages in the analysis:

(i) The assignment of peaks to residues, whether backbone or side chain atoms, utilizing a proximity criterion of 2.0 Å. The proximity criterion of 2.0 Å was chosen as it has been shown in a study involving ideal density maps of 19 proteins that over 98% of the peaks above an appropriate relative density cutoff are within that distance [22]. Note that a relative density cutoff was utilized to ensure that the backbone was adequately represented by peaks.

(ii) The assignment of peaks to residues based on the proximity (< 2.0 Å) of peaks to protein backbone atoms and the connectivity of these peaks as found through the gradient path tracing algorithm. The relative density cutoff outlined in (i) was again utilized.

(iii) A trace was made starting at the first peak (e.g. residue 1, ALA, for BP2) that simply uses the peak–pass–peak information generated by our program for the selected volume.

Case 1

(i) The positions of peaks with a relative density value greater than 0.8 were compared with the positions of the non-hydrogen atoms in the protein residues. Peaks within 2.0 Å of a residue atom were assigned to the respective residues. Only one residue, number 32 (GLY), was not assigned a peak. As well, a peak was assigned to the Ca^{2+} ion associated with the protein. Of note is the large number of side chains represented by peaks at this resolution (3.0 Å).

(ii) The location of the peaks with respect to the protein backbone atoms was considered. An assignment was made if a peak with a relative density greater than 0.8 was located within 2.0 Å of a backbone atom. Of the 123 residues present, 8 did not have a peak assigned based on this criterion. Next the results of the gradient path tracing are considered for the peaks that have been assigned. Of the two 'missing' edges, one is due to an intervening peak, that is, a third peak, not assigned to a residue, is connected to the two peaks assigned to the residues forming the edge. The other 'missing' edge is due to a side chain peak being inserted into the chain via the distance criterion. The peaks on either side of this assignment are connected by an edge to each other. Note that the lowest relative density in the trace is a pass with a density at 0.7. The fact that there are residues with no peaks fitting the distance criterion is not of concern as in each case the residues on either side are represented by peaks which are directly connected to each other by a pass.

(iii) The protein should not only have a continuous trace from start to finish, but should also display intraprotein connectivity through the disulphide bonds and the Ca^{2+} ion. Using the peak identified as residue 1 (ALA), the best trace is created using the highest passes and discounting side chains where the trace stops. Points at which there appears to be a fork are explored. From previous work, we were aware that the highest peaks usually represent disulphide bridges and heteroatoms such as Ca^{2+}, at this resolution [14]. We follow the peak–pass–peak path and trace the protein including the disulphide bridges. We complete the trace noting that there are seven disulphide bridges which connect various portions of the chain, a Ca^{2+} ion at the active site which has four passes (three large passes and one almost at the cutoff) and two bridges that are formed by side chains. The size of the peaks in the bridges allow us to identify the disulphide bridges and the Ca^{2+} ion, with the disulphides having only two passes above the cutoff and the ion three well above and one at the cutoff. With peak and pass cutoffs set at 0.8 and 0.6, respectively, the gradient path tracing used to assign peaks to passes gives an excellent trace of the protein backbone and its side chains. The two branch points where the side chains bridge between two portions of the main trace have peaks much lower than those consistent with disulphide or heteroatom peaks. One of these bridges has a peak of 1.8, consistent with an electron rich system but not as rich as a sulphur containing region ($\rho > 2$). In fact, that peak corresponds to the carboxylate group of the aspartate side chain. Note that one side

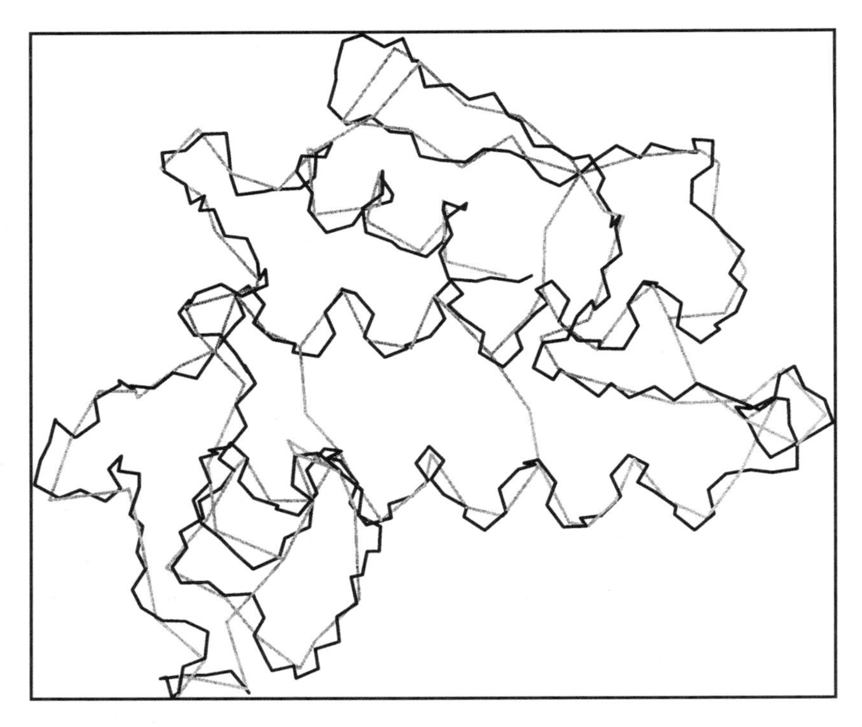

Figure 2. Correspondence of the calculated backbone trace with that of the reported backbone for BP2 [19]. As well, the calculated disulphide bridges are included to illustrate the important role they play in protein structure, binding certain regions together.

chain peak attached to peak 8 of the chain has a density of 2.2. This peak corresponds to the side chain of residue 8, methionine, and represents the sulphur atom in that chain.

Figure 2 illustrates the correspondence of the calculated trace with that of the backbone atoms. As well, the calculated disulphide bridges are included to illustrate the important role they play in protein structure, binding certain regions together. For example, the helical portions of the protein are well represented with the disulphide bridges helping to hold two helices together illustrating the important role they play in the tertiary structure of this protein.

Case 2

(i) The peaks were compared to the positions of the non-hydrogen atoms in the residues, including the side chain atoms utilizing a relative density cutoff of 0.3. Of the 66 residues, 4 do not have an atom within 2.0 Å of a peak and were not assigned a peak. They are residues 2, 30, 32 and 50.

(ii) A comparison of the location of peaks with the backbone atoms of AFP was performed. Using a density cutoff of 0.3, the peaks were assigned to backbone atoms if they were within 2.0 Å of the atoms. Of the 66 residues, 14 did not have peaks assigned. Examination of the connectivity of these peaks, utilizing the pass information from the gradient pass tracing algorithm, revealed 4 breaks in the chain, 2 of which were due to the insertion of side chain peaks included due to the distance criterion. The peaks on either side of these inserted peaks were connected directly to each other leaving only two missing edges. The lowest density value for a pass in the backbone trace is 0.19.

(iii) The tracing of the protein utilizing the peak–pass–peak information should reveal a continuous trace and also any intramolecular interactions. Starting with the peak identified with residue 1 (ALA), the backbone was traced utilizing the peak–pass–peak information. Although there are no disulphide bonds present in the protein, three bridges were detected through intraprotein interactions. Given that there were two breaks in the backbone, from residues 7 to 8 and 58 to 59, these bridges helped complete the trace. The peak density values range from 1.0 to 0.31 while the pass values range from 0.7 to 0.17. There were numerous side chains whose traces terminate after one or two peaks. Note that the highest peaks were once again those associated with sulphur atoms in the side chains of methionine.

Figure 3 illustrates the correspondence of the calculated backbone trace with that of the backbone atoms. Figure 4 illustrates the sequence of the protein and the occurrence of the two breaks in the chain.

Case 3

(i) The position of peaks with a relative density value greater than 0.5 were compared with the positions of the non-hydrogen atoms in the protein residues. Peaks within 2.0 Å of a residue atom were assigned to the respective residues. Of the 323 residues, 7 were not assigned peaks. They are residues 1, 99, 105, 109, 250, 275 and 279.

(ii) A comparison of the location of peaks with respect to the protein backbone atoms of 3APP was performed utilizing a density cutoff of 0.5. Of the 323 residues, 24 did not have a peak that was within the 2.0 Å criterion. Examining the connectivity of these peaks utilizing the pass and gradient path tracing results reveals 13 breaks in the trace, 4 of which are due to side chain peaks inserted due to the distance criterion but with the peaks on either side directly connected to each other. Another 7 breaks are due to peaks in the chain that do not meet the distance criterion but are connected to the 2 peaks where the break occurs. Of the 2 remaining breaks in the trace, there are paths to other peaks further in the chain, for example, from the peak identified as representing residue 277 there is a peak–pass–peak path to the peak representing residue 282, thus bypassing the break in the chain. A similar pattern exists for the path from residue 129 to 137, except the pattern is peak–pass–peak–pass–peak. The lowest pass in the relative density occurs between backbone peaks representing residues 12 and 13 with a density of 0.26.

(iii) This is probably the most rigorous test of the methodology as the protein should not only have a continuous trace from start to finish, but should also display intraprotein connectivity through the disulphide bond. As well, 3APP is known to resemble an approximate hexagonal close-pack in its crystal packing. The result is

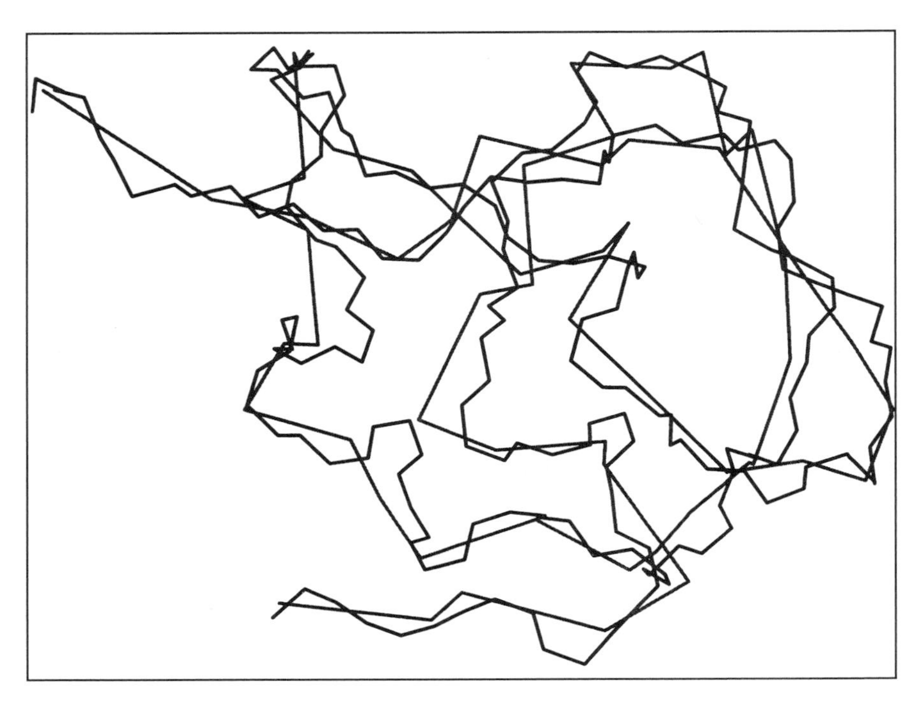

Figure 3. Correspondence of the calculated backbone trace with that of the reported backbone for AFP [20].

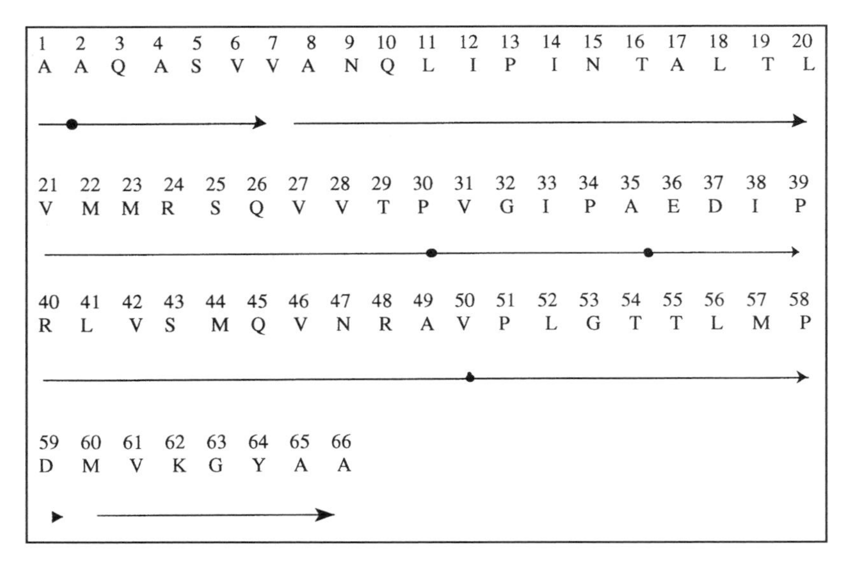

Figure 4. AFP sequence and the two breaks in the chain. The dots on the line represent residues with no associated peaks as found in stage (i) of the analysis.

that there are interprotein interactions which complicate the topology. There are also multiple intramolecular interactions due to the presence of the 6 β-pleated sheets and 6 α-helices. Using the peak identified as residue 2 (ALA), the best trace is created using the highest passes and discounting side chains where the trace stops. Points at which there appears to be a fork are explored. The resulting tree structure is extremely complicated with 41 intramolecular interactions (passes) detected, some of which have very large density values, and there are also 27 intermolecular interactions detected. In the chain itself, the lowest pass has a value of 0.25 which can be misleading considering that many of the intramolecular interactions have passes in the 0.4–1.0 range. However, at 2.8 Å resolution, it was reported that the experimental density is very weak and discontinuous in the residue regions 109–110 and 277–281 [20]. This is where we experience low passes and in the latter case a break in the continuous chain. Due to the presence of the disulphide bond and the many other intramolecular interactions we are able to complete our trace. Note that the passes to other proteins were found to have a density value range from a low of 0.46 to a high of 0.55.

Figure 5 shows the correspondence of the calculated 3APP backbone with the experimental structure.

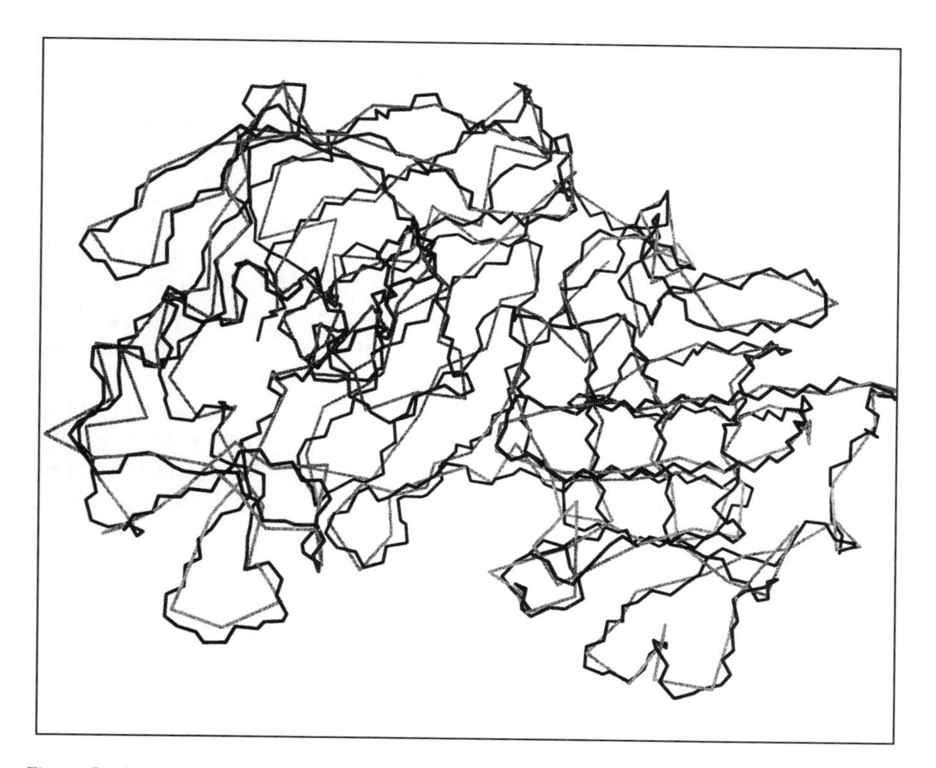

Figure 5. A portion of the calculated backbone trace and the corresponding portion of the reported backbone for 3APP [21].

Table 1. Summary of results.

	BP2	AFP	3APP
No. of peaks associated with protein[a]	944	296	1245
No. of peaks associated with protein after elimination of symmetry equivalents[a]	205	95	521
No. of residues present	123	66	323
No. of peaks in backbone trace[b]	121	56	321
No. of breaks in chain	0	2	2
Relative density range[c]	−0.95 to 2.2	−0.90 to 1.0	−1.8 to 2.3

[a] Utilizing the proximity and relative density cutoffs as discussed in case 1 and 2.
[b] As found through the eigenvector following/gradient path tracing algorithm.
[c] Note that the F_{000} term has not been included in the relative density calculation.

Comparing our results with the three known structures reveals that this methodology has correctly identified the chain, many of the side chains and the disulphide linkages. As well, there is a wealth of additional information with respect to H-bonding and other interactions. Given the range in values for the relative densities for our three test cases (see Table 1), our cutoff and lowest pass relative density values are approximately 50% or more of the values of the highest peaks. The quality of the density will obviously affect the outcome of the trace as areas where the density is discontinuous may terminate the trace if no other path is found. However, even the tracing of only portions of the backbone and side chains, combined with knowledge of the dimensions and symmetry of the unit cell should prove to be of value in further resolving the structure of the protein.

It should be noted that the present methodology provides a further advantage over the original ORCRIT program by reducing the number of peaks and/or passes that are allowed to be connected. The problem of determining the protein structure may be considered to be equivalent to determining the sequence of critical points associated with the backbone, making it useful to prune out incorrect connections. ORCRIT would allow a potential connection between any critical point (peak or pass) which lays within a certain distance of one another. The present methodology allows passes to be connected to only two peaks and peaks only to passes, with the connections determined by tracing the gradient path from pass to peak eliminating the uncertainty in the assignments. For example, for the volume studied for BP2, and considering all peaks and passes above relative densities of 0.8 and 0.6, respectively, ORCRIT's distance criteria produces 4824 connections (peak–peak, peak–pass and pass–pass), or an average of 5.05 associations per critical point. The MORDEN approach leads to only 1828 connections, or an average of 1.9 associations per critical point, thus making it easier to determine the trace corresponding to the protein structure.

Acknowledgements

The authors would like to thank Evan Steeg and Ed Willis for their contributions to the research described in this paper. We would also like to thank Zongchao Jia (Queen's

University) and Marie Fraser (University of Alberta) for providing the experimental data used in the work on AFP and 3APP, respectively. Funding for the research was provided by the Natural Science and Engineering Research Council of Canada.

References

1. Anfinsen, C.B. (1973) Principles that govern the folding of protein chains, *Science*, **181**, 223–230.
2. Reeke, Jr., G.N. (1988) Protein folding: computational approaches to an exponential time problem, *A. Rev. Comput. Sci.*, **3**, 59–84.
3. Unger, R. and Moult, J. (1993) Finding the lowest free energy conformation of a protein is a NP-hard problem: proof and implications, *Bull. Math. Biol.*, **55**, 1183–1198.
4. Fraenkel, A.S. (1993) Complexity of protein folding, *Bull. Math. Biol.*, **55**, 1199–1210.
5. Jones, T.A., Zou, J.Y., Cowan, S.W. and Kjeldgaard, M. (1991) Improved methods for building protein models in electron-density maps and the location of errors in these models. *Acta Cryst.*, **A47**, 110–119.
6. Hauptman, H. (in press) The phase problem of X-ray crystallography. In *Direct Methods for Solving Macromolecular Structures*, Fortier, S. (Ed.), Kluwer, Dordrecht.
7. Branden, C.I. and Jones, T.A. (1990) Between objectivity and subjectivity, *Nature* **343**, 687–689.
8. Bader, R.F.W. (1990) *Atoms in Molecules. A Quantum Theory*, Clarendon, Oxford.
9. Johnson, C.K. (1976) Abstract B1, *Proc. Am. Crystallogr. Assoc. Meet.*, Evanston, IL, USA. (1977) Abstract JQ6, *Proc. Am. Crystallogr. Assoc. Meet.*, Asilomar, CA, USA.
10. Leherte, L., Fortier, S., Glasgow, J. and Allen, F.H. (1994) Molecular scene analysis: application of a topological approach to the automated interpretation of protein electron-density maps, *Acta Cryst.*, **D50**, 155–166 and references therein.
11. Cao, W.L., Gatti, C., MacDougall, P.J. and Bader, R.F.W. (1987) On the presence of non-nuclear attractors in the charge distributions of Li and Na clusters, *Chem. Phys. Lett.*, **141**, 380–385.
12. Mei, C., Edgecombe, K.E., Smith, Jr., V.H. and Heilingbrunner, A. (1993) Topological analysis of the charge density of solids: *bcc* sodium and lithium, *Int. J. Quantum Chem.*, **48**, 287–293; Gatti, C., Saunders, V.R. and Roetti, C. (1994) Crystal field effects on the topological properties of the electron density in molecular crystals: the case of urea, *J. Chem. Phys.*, **101**, 10 686–10 696; Tsirelson, V.G., Zou, P.F., Tang, T.H. and Bader, R.F.W. (1995) Topological definition of crystal structure: determination of the bonded interactions in solid molecular chlorine, *Acta Cryst.*, **A51**, 143–153 and references therein.
13. Destro, R., Bianchi, R., Gatti, C. and Merati, F. (1991) Total electronic charge density of L-alanine from X-ray diffraction at 23 K, *Chem. Phys. Lett.*, **186**, 47–52; Iversen, B.B., Larsen, F.K., Souhassou, M. and Takata, M. (1995) Experimental for the existence of non-nuclear maxima in the electron-density distribution of metallic beryllium. A comparative study of the maximum entropy method and the multipole refinement method, *Acta Cryst.*, **B51**, 580–591 and references therein.
14. Johnson, C.K. (1977) ORCRIT. The Oak Ridge Critical Point Network Program. Chemistry Division, Oak Ridge National Laboratory, USA.
15. Popelier, P.L.A. (1994) A robust algorithm to locate automatically all types of critical points in the charge density and its Laplacian, *Chem. Phys. Lett.*, **228**, 160–164.
16. Press, W.H., Flannery, B.P., Teukolsky, S.A. and Vetterling, W.T. (1996) *Numerical Recipes*, 2nd edition, Cambridge Press, Cambridge.
17. Popelier, P.L.A. (1996) MORPHY, a program for an automated 'Atoms in Molecules' Analysis, *Comput. Phys. Comm.*, **93**, 212.
18. Hall, S.R. and Stewart, J.M. (1990) *XTAL* 3.0 *User's Manual*, Lamb Printers, Perth.
19. Dijkstra, B.W., Kalk, K.H., Hol, W.G.J. and Drenth, J. (1981) Structure of bovine pancreatic phospholipase A2 at 1.7 Å resolution, *J. Mol. Biol.*, **147**, 97–123.
20. Jia, Z., DeLuca, C.I., Chao, H. and Davies, P.L. (1996) Structural basis for the binding of a globular antifreeze protein to ice, *Nature*, **384**, 285–288.
21. James, M.N.G. and Sielecki, A.R. (1983) Structure and refinement of penicillopepsin at 1.8 Å resolution, *J. Mol. Biol.*, **163**, 299–361.
22. Baxter, K. (1997) Private communication, Department of Computing and Information Science, Queen's University, Kingston.

8

The number of independent parameters defining a projector: proof in matrix representation and resolution of previously conflicting arguments

ARNAUD J.A. SOIRAT* and LOU MASSA

Chemistry Department, Graduate Center and Hunter College, The City University of New York, 695 Park Ave., New York, NY 10021, USA

1. Introduction

Projectors often arise in attempts to describe experiments within the structure of Quantum Mechanics. For example, in the case of the coherent scattering of X-rays by crystals the ideal measured intensities are given by the square of the structure factors

$$F(\mathbf{k}) = \int P(\mathbf{r})e^{\mathrm{i}\mathbf{k}\cdot\mathbf{r}}\,\mathrm{d}\mathbf{r} \tag{1}$$

where $\mathbf{k}$ is the scattering vector, $\mathbf{r}$ is a position vector, and P is the spinless-electron density. In order to obtain a quantum interpretation of the measured structure factors, it is natural to expand the molecular orbitals, $\underline{\underline{\phi}}$, of the system studied in an orthonormal basis, $\underline{\underline{\psi}}$, thus

$$\underline{\underline{\phi}}(\mathbf{r}) = \underline{\underline{C}}\,\underline{\underline{\psi}}(\mathbf{r}), \tag{2}$$

where the matrix $\underline{\underline{C}}$ contains expansion coefficients. The density associated with an independent particle model, i.e. with a Slater determinant wave function, then becomes

$$P(\mathbf{r}) \equiv 2\sum_{k=1}^{N} \underline{\underline{\phi}}_k(\mathbf{r})\underline{\underline{\phi}}_k^*(\mathbf{r}) \equiv 2\,\mathrm{Tr}\underline{\underline{C}}^+\underline{\underline{C}}\,\underline{\underline{\psi}}(\mathbf{r})\underline{\underline{\psi}}^+(\mathbf{r}). \tag{3}$$

By defining the population matrix

$$\underline{\underline{P}} = \underline{\underline{C}}^+\underline{\underline{C}}, \tag{4}$$

the electron density may be written as

$$P(\mathbf{r}) \equiv 2\,\mathrm{Tr}\underline{\underline{P}}\,\underline{\underline{\psi}}(\mathbf{r})\underline{\underline{\psi}}^+(\mathbf{r}). \tag{5}$$

The elements of $\underline{\underline{P}}$ may now be considered to be experimental parameters obtained simply by an experimental fit to the measured X-ray structure factors (Equation (1)).

---

*Currently with Aluminium Pectiney, France

Paul G. Mezey and Beverly E. Robertson (eds.), Electron, Spin and Momentum Densities and Chemical Reactivity, 127–145

However, to ensure that the electron density thus obtained is N-representable by a single determinant of N doubly occupied molecular orbitals it is necessary and sufficient that $\underline{\underline{P}}$ be a normalized, hermitian, projector [1], i.e.

$$\underline{\underline{P}}^2 = \underline{\underline{P}}; \quad \underline{\underline{P}}^+ = \underline{\underline{P}}; \quad \mathrm{Tr}\,\underline{\underline{P}} = N. \tag{6}$$

Thus the following question arises quite naturally: how many independent experimental conditions are required to entirely determine a normalized, hermitian, projector $\underline{\underline{P}}$?

This important question has been addressed several times in the literature, with various authors reaching apparently conflicting results, or, in some cases, apparently agreeing on the results but for conflicting reasons.

The first derivation treating the general case of the projection from an M-dimensional space (spanned by the functions $\underline{\psi}$) onto an N-dimensional subspace (spanned by the molecular orbitals $\underline{\phi}$) was due to Clinton Galli and Massa (CGM) [1] who reached the conclusion that for a complex $\underline{\underline{P}}$ matrix, the number of complex constraints required to fix the projector is $N(M - N)$. Later on, Pecora [2] considered the problem and followed essentially the lines of argument in [1], but criticized their counting of hermiticity conditions and hence reached the different conclusion that for a complex $\underline{\underline{P}}$, the number of real constraints required to fix the projector is $2NM - N(N + 1)$, while the number of complex constraints is half this latter number. Still later, Levy and Goldstein on the one hand agreed with the criticism by Pecora, but on the other hand, by means of a different line of argument, reached a result apparently similar to that of CGM and in disagreement with Pecora; they found, indeed, that for a real $\underline{\underline{P}}$ the number of real conditions to fix the projector is $N(M - N)$.

These various conflicting results, summarized in Table 1, leave the question addressed unresolved. Also, Refs. [1, 3] display the results in different ways, making their comparison less than obvious. In attempting to make comparisons, one might assume two things:

1. The number of complex constraints to fix a complex $\underline{\underline{P}}$ would be the same as real constraints to fix a real $\underline{\underline{P}}$;
2. Twice as many real conditions would be needed to fix a complex $\underline{\underline{P}}$ as complex conditions.

However, both assumptions are in general not valid[1], thus complicating the comparison of results in the various papers and, in some cases, causing errors in the extrapolation of one result from another within a paper, as we shall see.

For the special case of a projection from an M-dimensional space onto an $N =$ one-dimensional subspace, Fano [4], Roman [5], and Blum [6] have obtained the number of real constraints required to fix a complex projector as $K_{\mathrm{C,R}} = 2M - 2$.

[1]For example, according to Hamermesh (see Ref. [11]), the number of real conditions to uniquely determine an $(N \times N)$ (complex) unitary matrix is N^2 , while the number of real conditions to uniquely fix a (real) orthogonal matrix of same dimensions is not $N^2/2$ but $N(N + 1)/2$.

Table 1. Number K of independent parameters in a projector: comparison of different published formulae.

Reference	Complex $\underline{\underline{P}}$ and complex constraints	Complex $\underline{\underline{P}}$ and real constraints	Real $\underline{\underline{P}}$ and real constraints
[1]	$N(M-N)$	NA	NA
[2]	$NM - N(N+1)/2$	$2NM - N(N+1)$	$NM - N(N+1)/2$
[3]	NA	NA	$N(M-N)$

NA = Not applicable.

There is no doubt that this result is correct, as a close examination of their derivation would suggest. Interestingly enough, the results of all reduce properly to this result in the one-dimensional case, although there are disagreements in the N-dimensional case. (Note: We use the symbol K to represent a count of the parameters which fix a matrix. A first subscript, C or R, is attached to indicate whether the matrix is complex or real, and a second subscript, C or R, is attached to indicate whether the parameters counted are complex or real. For example, $K_{\mathrm{C,R}}$ signifies the number of real parameters required to fix a complex matrix.)

To clarify this problem, our approach will be the following: first, we shall devote ourselves to finding a formula for K, independently of any of the three existing derivations made for the most general N-dimensional case; then we shall compare our answer to the previously published results.

Since the one-dimensional result for K is definitely correct, our approach will be to generalize Fano–Roman's [4, 5] derivation to the N-dimensional case. This shall be our goal in the coming section.

2. Generalization of Fano–Roman's derivation to the N-dimensional case

2.1. *Number of real conditions to fix a complex $\underline{\underline{P}}$*

Consider an $(M \times M)$ complex $\underline{\underline{P}}$ matrix which satisfies the conditions in Equation (6). According to a well-known theorem on projectors [7], for the above equations to hold, it is necessary and sufficient that

$$\begin{aligned} \underline{\underline{P}}^{+} &= \underline{\underline{P}}, \\ \text{Rank } \underline{\underline{P}} &= N, \\ \lambda_i &= 1 \quad \forall i = 1, \ldots, N, \end{aligned} \tag{7}$$

where λ_i's refer to the eigenvalues.

Note that this system properly reduces to that of Fano–Roman [4, 5], in the one-dimensional case, i.e.

$$\begin{aligned} \underline{\underline{P}}^{+} &= \underline{\underline{P}}, \\ \text{Rank } \underline{\underline{P}} &= 1, \\ \text{Tr } \underline{\underline{P}} &= 1. \end{aligned} \tag{8}$$

Hence, the problem of counting how many experimental conditions are required to fix the matrix $\underline{\underline{P}}$ satisfying Equation (6) is equivalent to that for the case satisfying Equation (7). We treat here the problem of imposing Equations (7), because, as we shall see, we can display a clear counting procedure for them. This approach represents a generalization of that used by Fano and Roman for the specific case $N = 1$.

We proceed as follows: for a complex $\underline{\underline{P}}$, we count the number of real parameters which completely define the entire matrix; from this number, we subtract the number of real conditions imposed by N-representability (i.e. hermiticity, rank N and unit eigenvalues). The remaining number of parameters represents the number of real (experimental) conditions required to complete the definition of the projector considered. Such a number is the solution to the problem posed in this paper. Later on, we shall consider the two other cases previously mentioned, that is, complex independent parameters of a complex $\underline{\underline{P}}$, and real independent parameters of a real $\underline{\underline{P}}$.

2.1.1. Number of real parameters contained in complex $\underline{\underline{P}}$

The $\underline{\underline{P}}$ matrix, made of complex elements, p_{ij}, may be written as

$$\underline{\underline{P}} \equiv (p_{ij}) \equiv (\mathrm{Re}_{ij} + \mathrm{i}\mathrm{Im}_{ij}), \tag{9}$$

so that each element is explicitly defined by two real numbers, Re_{ij} and Im_{ij}. For a $\underline{\underline{P}}$ matrix of dimensions $(M \times M)$, the total number of real parameters defining the matrix is therefore

$$N_{\mathrm{total,(C,R)}} = 2M^2. \tag{10}$$

This is the number whose reduction by parameters fixed by the N-representability constraints yields a count of the remaining parameters which must be fixed by additional experimental constraints, such as those of Equation (1).

2.1.2. Hermiticity constraint

In order to properly count the number of real conditions arising from the hermiticity constraint $\underline{\underline{P}}^{+} = \underline{\underline{P}}$, it is first necessary to determine the number of diagonal and off-diagonal elements in this matrix: the $\underline{\underline{P}}$ matrix being of dimension $M \times M$, there are M diagonal complex elements, and consequently a total of $(M^2 - M)$ off-diagonal complex elements, or $M(M-1)/2$ complex elements in each off-diagonal triangle.

The hermiticity constraint may, then, be transcribed into the following equivalent conditions on the $\underline{\underline{P}}$ matrix elements:

Diagonal elements

$$\forall i = 1, \ldots, M, \quad p_{ii}^{*} = p_{ii} \Leftrightarrow \mathrm{Im}_{ii} = 0$$
$$\Rightarrow M \text{real conditions}; \tag{11}$$

Off-diagonal elements

$$\forall i \neq j, \quad p_{ij}^{*} = p_{ji} \Leftrightarrow \begin{cases} \mathrm{Re}_{ij} = \mathrm{Re}_{ji}, \\ \mathrm{Im}_{ij} = -\mathrm{Im}_{ji}, \end{cases}$$

$$\Rightarrow 2\frac{M(M-1)}{2} \text{ real conditions.} \tag{12}$$

Therefore, by summing the above results, we are led to a total number of

$$N_{\text{hermiticity},(\mathrm{C,R})} = M^2 \text{ real conditions.} \tag{13}$$

The reader may note that this result is in accordance with Fano–Roman, who treat only the case N=1, for the hermiticity constraint is the same no matter what the dimension of the subspace.

2.1.3. Number of real parameters fixed by rank

The rank of any $(P \times Q)$ matrix $\underline{\underline{A}} = (a_{ij})$, for which $a_{11} \neq 0$, may be computed using the following algorithm [8]:

$$\text{Rank } \underline{\underline{A}} = 1 + \text{Rank } \underline{\underline{D}}, \tag{14}$$

where $\underline{\underline{D}}$ is a $((P-1) \times (Q-1))$ matrix of the form

$$\underline{\underline{D}} = \begin{pmatrix} d_{22} & \dots & d_{2N} \\ d_{32} & \dots & d_{3N} \\ \cdot & \dots & \cdot \\ \cdot & \dots & \cdot \\ \cdot & \dots & \cdot \\ d_{M2} & \dots & d_{MN} \end{pmatrix} \tag{15}$$

whose elements are the (2x2) subdeterminants

$$d_{ij} = \begin{vmatrix} a_{11} & a_{1j} \\ a_{i1} & a_{ij} \end{vmatrix}. \tag{16}$$

Before using this algorithm, we note the following theorem.

Theorem Let $\underline{\underline{A}}$ be an hermitian matrix. Then, the matrix $\underline{\underline{D}}$ arising from the algorithm for calculating the rank of a matrix , i.e.,

$$\text{Rank } \underline{\underline{A}} = 1 + \text{Rank } \underline{\underline{D}},$$

where the $\underline{\underline{D}}$ elements retain their previous meaning, is hermitian.

This theorem is essential to the proper counting of the number of conditions arising from the constraint Rank $\underline{\underline{P}} = N$, as we shall now see.

In order to use the above algorithm for computing the rank of $\underline{\underline{P}}$, p_{11} must be different from 0. However, this is no restriction, since it is always possible to reach this condition for a non-zero matrix by using an appropriate elementary row operation (if necessary) transformation which always leaves the rank of the matrix unchanged. We can therefore assume, without any loss of generality, that the condition $p_{11} \neq 0$ is always satisfied. The $\underline{\underline{P}}$ matrix being of dimension $(M \times M)$, the computation of its rank following this algorithm will yield the relation

$$\text{Rank } \underline{\underline{P}} = 1 + \text{Rank } \underline{\underline{D}}, \tag{17}$$

where $\underline{\underline{D}}$ retains its previous meaning, except for its dimensions which are here $((M-1) \times (M-1))$. The same procedure can be used iteratively, and after the Nth iteration, one obtains

$$\text{Rank } \underline{\underline{P}} = N + \text{Rank } \underline{\underline{X}}, \tag{18}$$

where $\underline{\underline{X}}$ is an $((M-N) \times (M-N))$ matrix.

The constraint Rank $\underline{\underline{P}} = N$ is then equivalent to

$$\text{Rank } \underline{\underline{P}} = N \Leftrightarrow N + \text{Rank}\underline{\underline{X}} = N \Leftrightarrow \text{Rank}\underline{\underline{X}} = 0. \tag{19}$$

However, the rank of any matrix other than a zero matrix cannot be 0, while the rank of a zero matrix is defined to be 0 [9]. The following equivalence is, thus,

$$\text{Rank}\underline{\underline{P}} = N \Leftrightarrow \underline{\underline{X}} = \underline{\underline{0}}. \tag{20}$$

We shall now use the theorem previously mentioned. Since $\underline{\underline{P}}$ is hermitian during this rank computation, so too are $\underline{\underline{D}}$ and $\underline{\underline{X}}$. The above constraint on $\underline{\underline{X}}$, Equation (20), along with its hermiticity property, leads to the following number of conditions on its elements, and therefore ultimately on the $\underline{\underline{P}}$ elements:

Diagonal elements

$$\forall i = 1, \ldots, (M-N), \quad x_{ii} = 0 \;\Rightarrow\; \text{Re}[x_{ii}] = 0 \tag{21}$$

$(\text{Im}[x_{ii}] = 0$ being obvious since x_{ii} is real)

$\Rightarrow \quad (M-N)$ real conditions.

Off-diagonal elements

$$\forall i \neq j, \quad x_{ij} = 0. \tag{22}$$

But $x_{ji} = x_{ij}^*$, and, thus, only the constraints on, say, the upper off-diagonal triangle of $\underline{\underline{X}}$ is to be counted:

$$\Rightarrow \quad 2\frac{(M-N)(M-N-1)}{2} \text{ real conditions.}$$

As a conclusion, summing the above results shows that the constraint Rank $\underline{\underline{P}} = N$ yields a total number of

$$N_{\text{rank},(\text{C,R})} = (M-N)^2 \text{ real conditions.} \tag{23}$$

One may notice that this last formula properly reduces to the $(M-1)^2$ real conditions found by Roman [5] in the $N = 1$ case.

2.1.4. Number of real parameters fixed by unit eigenvalues

Since $\underline{\underline{P}}$ has been already constrained to be hermitian, it is legitimate to assume, without any loss of generality that $\underline{\underline{P}}$ is always diagonalizable into, say, $\underline{\underline{P}}'$, by a unitary transformation of the basis elements [10]. The diagonal elements of $\underline{\underline{P}}'$, then called its eigenvalues, are real. The rank constraint on $\underline{\underline{P}}$ (which is basis independent) further reduces the number of non-zero eigenvalues to N. Let λ_i $(i = 1, \ldots, N)$, be these non-zero eigenvalues.

Hence, it is always possible to find a unitary transformation into a basis in which the matrix is diagonal, and can be written as

$$\underline{\underline{P}}' = \begin{pmatrix} \lambda_1 & & \cdots & & 0 \\ & \ddots & & & \\ \vdots & & \lambda_N & & \vdots \\ & & & \ddots & \\ 0 & & \cdots & & 0 \end{pmatrix} \tag{24}$$

where $\lambda_i \in \Re^*, \forall i = 1, \ldots, N$.

Imposing unit magnitudes upon the eigenvalues, we have

$$\lambda_1 = \lambda_2 = \cdots = \lambda_N = 1 \quad \Rightarrow \quad N \text{ real conditions,} \tag{25}$$

which would allow one to recover a diagonalized $\underline{\underline{P}}$ matrix of the form

$$\underline{\underline{P}}' = \begin{pmatrix} \underline{\underline{1}}_N & \underline{\underline{0}} \\ \underline{\underline{0}} & \underline{\underline{0}} \end{pmatrix}. \tag{26}$$

The basis in which $\underline{\underline{P}}$ is of the above form is made of the collection of occupied and unoccupied eigenstates $\underline{\phi}_k$, i.e. $\{\underline{\phi}_{k(\text{occupied})}, \underline{\phi}_{k(\text{unoccupied})}\}$. Ultimately, the very process of projection allows one to select the N occupied ones, and it is not necessary to consider the unoccupied ones (at least for the ground-state description of the system).

However, to determine the number of real pieces of information required to fix the projection from an M-dimensional space onto an N-dimensional subspace spanned, not by the particular $\{\underline{\phi}_{k(\text{occupied})}\}$ basis in which $\underline{\underline{P}}$ is diagonal, but by any basis of the subspace, it is necessary to subtract the number of real parameters required to fix a particular basis in the N-dimensional subspace from the total $K_{\text{C,R}}$; such a number corresponds to the N^2 real conditions that are necessary to fix a unitary transformation [11] in the subspace. But, as the phases of the eigenstates, $\underline{\phi}$, are arbitrary as far as the physical state is concerned [4, 12], this latter number is reduced by N, the number of eigenstates belonging to the projection space. Hence, the number of independent real parameters in the unitary transformation which fixes the basis spanning the subspace defined by $\underline{\underline{P}}$ when reduced by the number of arbitrary phases is equal to $(N^2 - N)$.

Summing the above results, the unit magnitudes of the non-zero eigenvalues of $\underline{\underline{P}}$ yields a total number of

$$N_{\text{unit eigenvalues},(\text{C},\text{R})} = N^2 \text{ real conditions.} \tag{27}$$

2.1.5. Summation of various contributions to K

We now obtain the solution to our problem of enumeration by subtracting from the total number of parameters in $\underline{\underline{P}}$ those fixed independently by hermiticity, rank, and unit eigenvalues.

Thus,

$$K_{\text{C,R}} = N_{\text{total},(\text{C},\text{R})} - (N_{\text{hermiticity},(\text{C},\text{R})} + N_{\text{rank},(\text{C},\text{R})} + N_{\text{unit eigenvalues},(\text{C},\text{R})}),$$

i.e.,

$$K_{\text{C,R}} = 2M^2 - (M^2 + (M - N)^2 + N^2),$$

or,

$$K_{\text{C,R}} = 2N(M - N) \text{ real conditions.} \tag{28}$$

This number is the answer to the question originally posed. This is the number of real conditions required to fix experimentally a complex, normalized, hermitian, projection matrix. For example, this number of experimental structure factors, Equation (1), would suffice to fix $\underline{\underline{P}}$, Equation (6).

2.2. Number of real (complex) conditions to fix a real (complex) $\underline{\underline{P}}$

By simply using arguments analogous to those used in deriving Equation (28), one may find that if the projector is real, then

$$K_{\text{R,R}} = N(M - N) \text{ real conditions.} \tag{29}$$

while, for a complex $\underline{\underline{P}}$,

$$K_{\text{C,C}} = N(M - N) \text{ complex conditions.} \tag{30}$$

Having derived K by a method independent of those used in [1–3], we now compare our result to those obtained in these previous papers. From the perspective of our present counting procedure, we hope to shed light on the previous derivations. We shall take them up in the order in which they appeared in the literature.

3. Further investigation of the previous derivations of K

3.1. CGM's derivation

In counting the number of orthonormalization conditions on $\underline{\underline{C}}$, CGM apparently did not assume the hermiticity of the scalar product in the subspace, but rather chose to impose it. Their calculation of K ran along the following lines: a complex projector, which is hermitian and normalized, may be factored into [13]

$$\underline{\underline{P}} = \underline{\underline{C}}^{+}\underline{\underline{C}}, \tag{31}$$

where the complex $\underline{\underline{C}}$ is $(N \times M)$, and

$$\underline{\underline{C}}\,\underline{\underline{C}}^{+} = \underline{\underline{1}}_{N}. \tag{32}$$

Counting the parameters in $\underline{\underline{P}}$ is now converted into counting the parameters in **C**, which defines $\underline{\underline{P}}$. Thus, the number of elements in $\underline{\underline{C}}$, NM, is reduced by the N^2 orthonormalization conditions arising from Equation (32). They found, then, that

$$K_{\mathrm{C,C}} = NM - N^2 = N(M - N) \text{ complex conditions.} \tag{33}$$

Notice that, in counting the orthogonalization conditions, the 'upper triangle' and the 'lower triangle' in $\underline{\underline{C}}\,\underline{\underline{C}}^{+}$ have separately been set to zero. It is for this reason that we interpret the derivation as imposing hermiticity on the subspace rather than assuming it. If hermiticity were assumed, then the vanishing of the 'upper triangle' of $\underline{\underline{C}}\,\underline{\underline{C}}^{+}$ would automatically require the vanishing of the 'lower triangle', and both would not be counted as independent orthogonalization conditions.

However, as we show in an appendix, when counting the number of conditions arising from Equation (32), one does not have to impose hermiticity on the inner product but can take it for granted. The reader may find it, then, quite interesting to understand why the CGM derivation obtains, nevertheless, the right answer for K. Rather than counting the number of independent parameters in $\underline{\underline{P}}$, CGM instead counted those in $\underline{\underline{C}}$. In doing so, they apparently overcounted by one of the off-diagonal triangles ($2(N^2 - N)/2$ real parameters) by choosing to impose the hermiticity of the inner product in the occupied subspace. But, since the counting is based on the $\underline{\underline{C}}$ matrix, one has to correct for the fact that actually less information is required to uniquely determine the $\underline{\underline{P}}$ matrix. That is to say, one has to subtract the number of conditions necessary to fix a unitary transformation apart from the arbitrarily chosen phases of the basis elements. Such a correction would have given

$$K_{\mathrm{C,R}}(\text{for } \underline{\underline{P}}) = K_{\mathrm{C,R}}(\text{for } \underline{\underline{C}}) - (N^2 - N). \tag{34}$$

Apparently, the oversight of this last correction exactly compensates for the previous overcounting associated with imposing hermiticity on the subspace. Therefore, an exact compensation of errors has yielded precisely the correct result.

3.2. Pecora's derivation

Following the approach of CGM based on McWeeny's [13] decomposition of $\underline{\underline{P}}$ into $\underline{\underline{C}}^+\underline{\underline{C}}$, Pecora arrived at results different from CGM.

After factoring of the $\underline{\underline{P}}$ matrix, Pecora considers that 'the constraints are summarized by the equation $\underline{\underline{C}}\underline{\underline{C}}^+ = \underline{\underline{1}}_N$, and $\underline{\underline{P}}$ is completely determined by $\underline{\underline{C}}$'. This analysis, too, is based on counting the number of complex conditions on the complex elements of $\underline{\underline{C}}$.

It seems the key step in this derivation, which differs from the analysis of CGM, is the following. In the system of equations resulting from the constraint $\underline{\underline{C}}\underline{\underline{C}}^+ = \underline{\underline{1}}_N$, Pecora considers that '$N(N-1)$ of [them] are simply complex conjugates of each other', yielding a total number of complex conditions equal to $N(N+1)/2$. This is, in fact, equivalent to considering the $\underline{\underline{C}}\underline{\underline{C}}^+$ matrix as hermitian, i.e.,

$$(\underline{\underline{C}}\underline{\underline{C}}^+)^+ = \underline{\underline{C}}\underline{\underline{C}}^+. \tag{35}$$

More fundamentally, what Pecora seems to assume – although never explicitly saying so – is the following property. Since the condition $\underline{\underline{C}}\underline{\underline{C}}^+ = \underline{\underline{1}}_N$ is actually the orthonormalization constraint on the ϕ_k's since $\underline{\underline{C}}\underline{\underline{C}}^+ = (\langle\Phi_k|\Phi_m\rangle)$, it is supposed that the scalar product between any two wavefunctions ϕ_k is hermitian. That is to say, it is assumed that the subspace on which the projection is made is a Hilbert subspace.

Based on this assumption, the result for the number of complex conditions to uniquely determine $\underline{\underline{P}}$, is then given to be

$$K_{\mathrm{C,C}} = NM - \frac{N(N+1)}{2}. \tag{36}$$

Finally, Pecora generalizes to

$$K_{\mathrm{C,R}} = 2NM - N(N+1) = 2K_{\mathrm{C,C}}, \tag{37}$$

obtaining $K_{\mathrm{C,R}}$, the number of real parameters in a complex $\underline{\underline{P}}$, by simply doubling $K_{\mathrm{C,C}}$, the number of complex parameters.

However, if one were to exactly follow what seem to be Pecora's assumptions about the scalar product being hermitian, one would get a different result from Pecora when counting the number of real conditions on the complex $\underline{\underline{P}}$ matrix, arising from the constraint $\underline{\underline{C}}\underline{\underline{C}}^+ = \underline{\underline{1}}_N$. In fact, when the $\underline{\underline{C}}\underline{\underline{C}}^+$ matrix is considered to be hermitian, the normalization condition on the N complex diagonal elements of $\underline{\underline{C}}\underline{\underline{C}}^+$ yields N real conditions and not $2N$ as Pecora seemed to tacitly suppose. This is due to the fact that the diagonal elements are already known to be real since $\underline{\underline{C}}\underline{\underline{C}}^+$ is hermitian, and hence, $\mathrm{Im}_{ii} = 0$ is not a separate constraint.

The orthogonalization condition on the off-diagonal elements correctly yields $2\{N(N-1)/2\}$ real conditions. The assumption of the hermiticity property of the scalar product in the subspace of N dimensions, would lead finally to, $K_{\mathrm{C,R}} = 2NM - N^2$ in the case of a complex $\underline{\underline{P}}$, and not $2K_{\mathrm{C,C}}$, as had been claimed.

However, if one completely determines $\underline{\underline{C}}$, and therefore the accompanying $\underline{\phi}$'s, one has to do so apart from their phases which are known to be physically meaningless, as previously said. This further decreases the number of conditions to uniquely determine $\underline{\underline{C}}$, apart from the $\underline{\underline{\phi}}$'s phases, by a number N and yields:

$$K_{\mathrm{C,R}} = 2NM - N^2 - N = 2NM - N(N+1). \tag{38}$$

Interestingly enough, this turns out to be the very result that Pecora claims for the following reason: the overcounting of the number of real conditions on the diagonal elements of the assumed hermitian $\underline{\underline{C}}\,\underline{\underline{C}}^+$ matrix (N in number) exactly compensates for the oversight of the conditions relative to the arbitrary phase of each $\underline{\underline{\phi}}$. Here, we have a second case in which an exact compensation of errors has occurred.

Of course, it is apparent that $K_{\mathrm{C,R}}$ is given by differing formulas in the work of CGM and Pecora. The formula of CGM correctly answers the question posed in this paper. However, as we shall see later, the formula of Pecora correctly answers quite a different, but related question. For now, we turn to the remaining paper which is to be considered.

3.3. *Levy and Goldstein's derivation*

Levy and Goldstein chose to tackle the problem in a different way. They based their reasoning, in one of their derivations, on the orthogonal decomposition of the space spanned by $\{|\psi_j\rangle\}$, into $\mathbf{S}$ and $\mathbf{S}^{\perp}$, respectively the occupied and unoccupied subspaces.

As is known [14], there is a one-to-one correspondence between each subspace and its accompanying projection.

Based on this notion, Levy and Goldstein then developed a formula for the real number of pieces of information necessary to fix uniquely $\Psi_{\mathrm{det}}(1, 2, \ldots, N)$ described in a real function basis. They wrote: 'the number of independent parameters in Ψ_{det} is equal to the number of equations required to fix the $\underline{\underline{\phi}}$ subspace. We now assert that this number is $N(M-N)$ because there are $N(M-N)$ orthogonality relations between the $\underline{\underline{\phi}}$ and the $\underline{\underline{\phi}}^{\perp}$ orbitals:

$$\langle \Phi_i | \Phi_j \rangle = 0, \quad i = 1, 2, \ldots, N; \quad j = 1, 2, \ldots, (M-N)'. \tag{39}$$

Although it is clear that there are $N(M-N)$ orthogonality relations between the $\underline{\underline{\phi}}$'s and the $\underline{\underline{\phi}}^{\perp}$'s, it is not clear why this is exactly equal to K, unless one has additional knowledge of $\underline{\underline{\phi}}$'s and the $\underline{\underline{\phi}}^{\perp}$'s, but such knowledge would be incorporated in additional constraints which would have to be counted and would presumably alter the expression for K that was obtained.

Indeed, one may give the following counter-argument to Levy and Goldstein's assumption: for simplicity, take the case of a projection from a three-dimensional space ($M = 3$) onto a two-dimensional subspace ($N = 2$). $\mathbf{S}^{\perp}$ is then a one-dimensional subspace, and any $\underline{\underline{\phi}}^{\perp} \in \mathbf{S}^{\perp}$ spans the subspace and therefore completely specifies it. Figure 1 somewhat clarifies our hypotheses.

According to Levy and Goldstein, the $\mathbf{S}$ subspace is then completely specified by the $N(M-N) = 2$ orthogonality relations between the Φ's and the $\Phi^{\perp}$. Let Φ_1 and

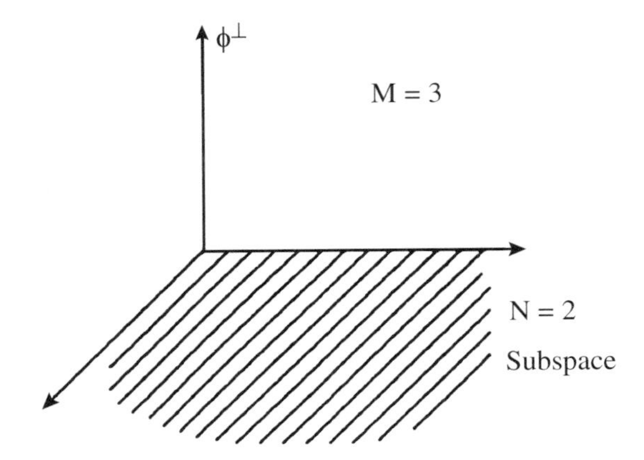

Figure 1. Orthogonal decomposition of a three-dimensional Hilbert space: geometrical representation of the two orthogonal subspaces.

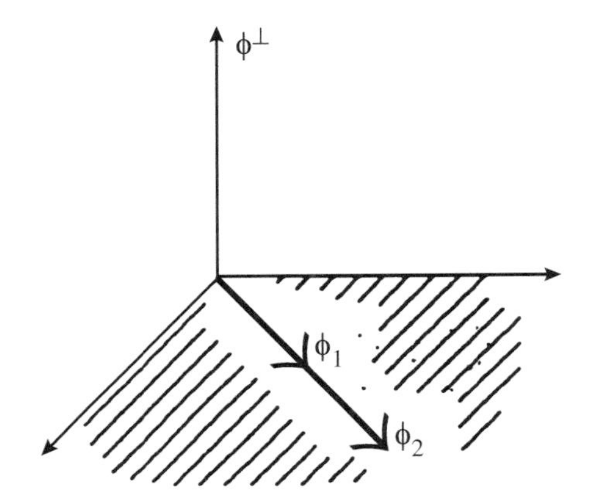

Figure 2. Orthogonal decomposition of a three-dimensional Hilbert space: case of two collinear vectors in the two-dimensional subspace.

Φ_2 be the two elements of **S**. One has then the following relations:

$$\Phi^{\perp} \perp \Phi_1; \qquad \Phi^{\perp} \perp \Phi_2.$$

However, Φ_1 and Φ_2 could be such that $\Phi_1 = k\Phi_2$, as described by Figure 2.

In such a case, the two vectors being collinear, do not form a basis of the **S** subspace and, consequently, do not entirely define the subspace they belong to.

In the case where $M = 4$ and $N = 3$, one could have the following situation:

$$\Phi^{\perp} \perp \Phi_1; \qquad \Phi^{\perp} \perp \Phi_2; \qquad \Phi^{\perp} \perp \Phi_3$$

but where $\Phi_3 = \Phi_1 + \Phi_2$. Here again, the set $\{\Phi_1, \Phi_2, \Phi_3\}$ does not span the **S** subspace.

As a consequence, such examples show that the orthogonality relations (between vectors in different subspaces) alone, do not fix the **S** subspace. To do so, one would need some previous additional information on the basis which spans **S** and $\mathbf{S}^{\perp}$. That is to say, one would need to constrain the set of recovered Φ's to form a basis of the occupied subspace. This would then make additional orthogonality constraints within the subspace to take into account in the search for a K formula.

The formula in [3] gives the correct answer for $K_{\mathrm{R,R}}$ to determine real $\underline{\underline{P}}$. Apparently, coincidences occur in all three derivations reviewed in this section.

4. Discussion

A problem arises in considering the result given by Pecora for $K_{\mathrm{C,R}}$. Indeed, this result does not seem to be correct since it does not properly reduce to 0 when $M = N$, but to $(N^2 - N)$, instead.

However, when $M = N$, the projection operation is done onto the whole space and is, thus, the identity transformation; $\underline{\underline{P}}$ is then the unit matrix and, as a consequence, no information is needed to determine it, leading to $K = 0$ in such a case.

Based on this argument, Levy and Goldstein correctly implied – in their footnote #7 – that Pecora's formula was wrong, and did not discuss it further. Earlier, we criticized Pecora's derivation, but we point out here an interpretation under which Pecora's formula is correct.

A normalized, hermitian projector $\underline{\underline{P}}$ can always be diagonalized, according to the following procedure [10]:

$$\underline{\underline{P}} = \underline{\underline{U}}^{+}\underline{\underline{P}}'\underline{\underline{U}}, \tag{40}$$

where

$$\underline{\underline{U}}^{+}\underline{\underline{U}} = \underline{\underline{U}}\,\underline{\underline{U}}^{+} = \underline{\underline{1}}_{M} \tag{41}$$

and

$$\underline{\underline{P}}' = \begin{pmatrix} \underline{\underline{1}}_N & \underline{\underline{0}} \\ \underline{\underline{0}} & \underline{\underline{0}} \end{pmatrix}. \tag{42}$$

As McWeeny [13] showed by the reverse transformation of Equation (40), $\underline{\underline{P}}$ can always be factored into

$$\underline{\underline{P}} = \underline{\underline{C}}^{+}\underline{\underline{C}}, \tag{43}$$

where the $(N \times M)$ $\underline{\underline{C}}$ matrix is made of the first N rows of the unitary matrix $\underline{\underline{U}}$, while $\underline{\underline{C}}^{+}$ of the first N columns of $\underline{\underline{U}}^{+}$.

However, the decomposition of $\underline{\underline{P}}$ into $\underline{\underline{C}}^{+}\underline{\underline{C}}$ is not unique, since, as Pecora wrote, any $\underline{\underline{C}} = \underline{\underline{V}}\,\underline{\underline{C}}$ (where $\underline{\underline{V}}$ is understood to be an $(N \times N)$ unitary matrix) will generate the same $\underline{\underline{P}}$ matrix, which 'is just a basic fact of Quantum Mechanics or, more generally, linear eigenvalue theory restated in population matrix language'. Hence, the only way one might speak of the uniqueness of $\underline{\underline{C}}$ is within a unitary transformation.

Moreover, as previously mentioned, the CGM (and [3]) formula properly reduces to $K = 0$ when $M = N$. We shall now examine the decomposition of $\underline{\underline{P}}$ into $\underline{\underline{C}}^+\underline{\underline{C}}$ in such a case.

From the definition of $\underline{\underline{C}}$, it is clear that the rectangular $\underline{\underline{C}}$ matrix of the previous case becomes now the square $\underline{\underline{U}}$ matrix, so that $\underline{\underline{P}}$ can always be written, when $M = N$, as

$$\underline{\underline{P}} = \underline{\underline{U}}^+\underline{\underline{U}}, \quad \text{where } \underline{\underline{U}}^+\underline{\underline{U}} = \underline{\underline{U}}\,\underline{\underline{U}}^+ = \underline{\underline{1}}_M. \tag{44}$$

Of course, there is an infinity of unitary transformations in the space we are dealing with, that satisfies this equation.

Now suppose that we were to determine one particular complex $\underline{\underline{U}}$ matrix out of the infinity. We stated earlier that the number of real independent conditions to uniquely determine $\underline{\underline{U}}$, apart from the phases of each of the N eigenstates $\underline{\underline{\phi}}_k$, is: $K_{\mathrm{U(C,R)}} = N^2 - N$.

If one realizes that this last number $K_{\mathrm{U(C,R)}}$ is precisely the difference between Pecora's and CGM's formulae for K_{P},

$$2MN - N^2 - N = 2N(M - N) + (N^2 - N), \tag{45}$$

one may interpret Pecora's formula as follows: $2MN - N^2 - N$ is the number of independent real conditions to uniquely determine the projection transformation from the M-dimensional space onto the N-dimensional subspace ($2N(M-N)$ conditions), as well as the N basis vectors of the subspace, apart from their phases, arbitrarily chosen (additional $(N^2 - N)$ conditions). That is to say, such a K_{P} allows one to *uniquely decompose* $\underline{\underline{P}}$ into $\underline{\underline{C}}^+\underline{\underline{C}}$.

Figure 3 summarizes this interpretation, in the easily visualizable case of a projection from an $M = 3$ dimensional space onto an $N = 2$ dimensional subspace. This interpretation clarifies the relationship between the CGM (and [3]) formula and Pecora's, and also the following:

1. Pecora noticed that 'the phase information of $\underline{\underline{C}}$ is lost in the original constraints' [i.e. $\underline{\underline{P}}^2 = \underline{\underline{P}}$; $\mathrm{Tr}\,\underline{\underline{P}} = N$], but found it 'not at all clear'. Here, we showed in which way one might take into account the loss of the phase information in $\underline{\underline{C}}$: when calculating the number of conditions to uniquely determine $\underline{\underline{C}}$, one has to impose, over and above the constraints arising from fixing the projector, the conditions to determine a particular unitary transformation in the N-dimensional subspace, apart from the phases of the basis functions which are physically meaningless in the context of Quantum Mechanics.
2. The formulae obtained for K by CGM (and [3]) and [2], reduce to the proper result for the case where $N = 1$. Indeed, this is because, in such a case, there is only one possible orientation of the basis vector in the one-dimensional subspace (the subspace being fixed), and its phase is physically meaningless.

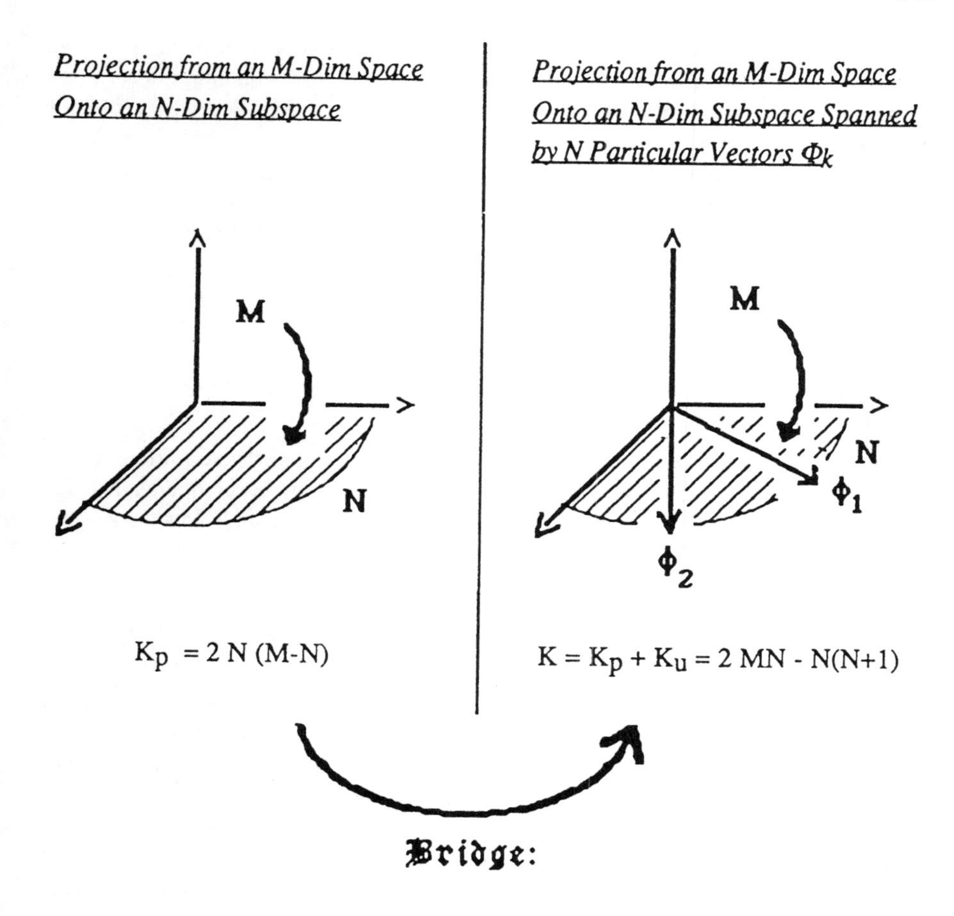

Figure 3. Number K of independent parameters in a projector: geometrical interpretation of Pecora's vs CGM's formulae.

5. Conclusion

In this paper, we have answered the fundamental question of determining how many independent experimental measurements (or theoretical conditions) are needed to fix a projector. Conflicts which appear in the previous literature treating this question, and that we have simply noted earlier [15], have here been resolved. In particular, we have explained how to properly interpret the K-formulae in [1–3].

One way of using experimental conditions that determine a projector is provided by the methods of Quantum Crystallography [16–20], where a quantum description of the X-ray diffraction experiment is realized. In practice, one makes a least squares fitting to experimental X-ray structure factors, consistent with the constraints of N-representability, using the Clinton–Massa algorithm [16]. Ultimately, this procedure allows one to recover quantum mechanically valid, exact density delivering, reduced density matrices which are projectors N-representable by a single Slater determinant [17–20]. These projectors through density functional theory, also contain in themselves the only information required for their own correction to include correlation effects.

Appendix: Proof of the inner product hermiticity of a subspace of an hermitian space

In their derivation, CGM tacitly assumed that one has to impose the hermiticity of the scalar product defined in the subspace to ensure the subspace to be a Hilbert subspace.

To decide whether or not this is a legitimate assumption, we shall now answer the following questions:

1. Is the space spanned by (ψ_j) a Hilbert space?
2. If yes, does the projection process preserve the hermiticity character of the scalar product?

For convenience of argumentation, we from now on use the function representation of our formalism, which restrains the generality of the results only in the sense that the $\mathbf{L}^2$ space is a particular example of separable Hilbert space; the generalization to any separable Hilbert space is, however, straightforward.

In the most general formulation of our formalism expressed in the function representation, we first make the choice of a basis $\{\psi_j(l)\}$ which spans an M-dimensional space. This basis can be any set of M linearly independent complex functions, normalized or not, as long as it satisfies one single condition. Since we are dealing with the description of the state of the system by Quantum Mechanics, $\psi_j(l)$'s must be well-behaved functions and therefore must be chosen among the elements of the $\mathbf{L}^2$ space [21]. Any choice of a basis not satisfying this latter condition would violate the Quantum Mechanics formalism – described in the continuous representation. The basis functions being elements of the $\mathbf{L}^2$ space, our space is as a consequence an M-dimensional complex separable Hilbert space, written $\mathbf{h}$, and therefore possesses by definition an inner product structure satisfying the corresponding axioms [22]. In particular, the inner product is hermitian, i.e. for any choice of basis which is a linearly independent subset of $\mathbf{L}^2$ it is always true that

$$\langle \psi_i \mid \psi_j \rangle = \langle \psi_j \mid \psi_i \rangle^*$$

$$\Leftrightarrow \int \psi_i^*(l)\psi_j(l)\,\mathrm{d}l = \left(\int \psi_j^*(l)\psi_i(l)\,\mathrm{d}l \right)^* . \qquad \text{(A-1)}$$

Hence the first question is answered affirmatively.

Now we shall examine the second one, which may be resolved in different ways.

1. Our first way of answering the last question will be based on the fundamental theorems on Hilbert space [14]. Indeed, the theorem on separability tells us that any subspace of **h** is also a separable Hilbert space. As a consequence, the inner product defined on, say, the occupied subspace is hermitian irrespectively of the choice of the basis $\{\psi_j(l)\}$, as long as this latter satisfies the fundamental requirements of Quantum Mechanics. One should therefore not have to impose this property as a constraint when counting the number of conditions arising from the constraint $\underline{\underline{C}}\,\underline{\underline{C}}^+ = \underline{\underline{1}}_N$ but, on the contrary, can take it for granted.

2. A second way of resolving this question is provided by examining the constraint itself. Indeed, the condition $\underline{\underline{C}}\,\underline{\underline{C}}^+ = \underline{\underline{1}}_N$ is equivalent to requiring the orthonormalization of the basis functions $\Phi_k(l)$ of the occupied subspace. That is to say, $\Phi_k(l)$'s must satisfy

$$\int \Phi_k^*(l)\Phi_m(l)\,\mathrm{d}l = \delta_{km}. \tag{A-2}$$

However, since $\Phi_k(l)$ is described in the space basis by

$$\Phi_k(l) = \sum_{j=1}^{M} c_{kj}\psi_j(l), \tag{A-3}$$

the orthonormalization condition can be written as

$$\sum_{jn=1}^{M} c_{kj*}c_{mn} \int \psi_j^*(l)\psi_n(l)\,\mathrm{d}l = \delta_{km}, \tag{A-4}$$

and it is obvious that the hermiticity property of the inner product of the space is conferred to the subspace. This is in fact nothing more than the above answer restated in terms of the exact expression of the inner product.

3. Finally, one may suggest a third way of solving this problem by further investigating McWeeny's theorem of decomposition [13]. Consider first a general matrix $\underline{\underline{P}}$ of M^2 dimensions. If this $\underline{\underline{P}}$ matrix is of rank r, $r \leq M$ it is then always decomposable into a product of two rectangular matrices of respective shapes, $(M \times r)$ and $(r \times M)$ [23]. Now consider each of the three constraints on $\underline{\underline{P}}$:

a) $\underline{\underline{P}}^+ = \underline{\underline{P}}$: then, it is always possible to find a unitary transformation so that $\underline{\underline{P}} = \underline{\underline{U}}^+\underline{\underline{P}}'\underline{\underline{U}}$; $\underline{\underline{P}} = \underline{\underline{C}}^+\underline{\underline{C}}$ if and only if $\underline{\underline{P}}^2 = \underline{\underline{P}}$, since then

$$\underline{\underline{P}}' = \begin{pmatrix} \underline{\underline{1}} & \underline{\underline{0}} \\ \underline{\underline{0}} & \underline{\underline{0}} \end{pmatrix}.$$

However, so far, $\underline{\underline{C}}$ is a rectangular matrix of dimensions $(r \times M)$ where r is the rank of $\underline{\underline{P}}$.

b) Rank $\underline{\underline{P}} = N$: then,

$$\underline{\underline{P}}' = \begin{pmatrix} \underline{\underline{1}}_N & \underline{\underline{0}} \\ \underline{\underline{0}} & \underline{\underline{0}} \end{pmatrix}$$

and $\underline{\underline{C}}$ becomes an $(N \times M)$ matrix.

c) $\underline{\underline{P}}^2 = \underline{\underline{P}}$:

$$\begin{aligned} &\Leftrightarrow \quad (\underline{\underline{C}}^+\underline{\underline{C}})(\underline{\underline{C}}^+\underline{\underline{C}}) = \underline{\underline{C}}^+\underline{\underline{C}}, \\ &\Leftrightarrow \quad \underline{\underline{C}}^+(\underline{\underline{C}}\,\underline{\underline{C}}^+)\underline{\underline{C}} = \underline{\underline{C}}^+\underline{\underline{C}} \quad \Leftrightarrow \quad \underline{\underline{C}}\,\underline{\underline{C}}^+ = \underline{\underline{1}}_N. \end{aligned} \tag{A-5}$$

Therefore, it is clear that one may rewrite the $\underline{\underline{P}}$ constraints as

- if $\underline{\underline{P}}^+ = \underline{\underline{P}}$ then $\underline{\underline{P}} = \underline{\underline{C}}^+\underline{\underline{C}}$, where $\underline{\underline{C}}$ is $(N \times M)$;
- if and only if $\underline{\underline{C}}\,\underline{\underline{C}}^+ = \underline{\underline{1}}_N$;
- i.e. if and only if $\underline{\underline{P}}^2 = \underline{\underline{P}}$ and Rank $\underline{\underline{P}} = N$.

It appears thus, that writing $\underline{\underline{P}}$ as $\underline{\underline{C}}^+\underline{\underline{C}}$ where $\underline{\underline{C}}\,\underline{\underline{C}}^+ = \underline{\underline{1}}_N$ is already taking account of the first constraint of hermiticity of $\underline{\underline{P}}$, and also includes the two conditions $\underline{\underline{P}}^2 = \underline{\underline{P}}$ and Rank $\underline{\underline{P}} = N$. That is to say, the hermiticity constraint is implicitly taken into account as soon as one decomposes $\underline{\underline{P}}$ as $\underline{\underline{C}}^+\underline{\underline{C}}$, the two other constraints being completely summarized by $\underline{\underline{C}}\,\underline{\underline{C}}^+ = \underline{\underline{1}}_N$. It seems therefore not necessary to superimpose the hermiticity constraint on $\underline{\underline{C}}\,\underline{\underline{C}}^+ = \underline{\underline{1}}_N$, for it has already been done.

As a conclusion to this part, when counting the number of conditions arising from $\underline{\underline{C}}\,\underline{\underline{C}}^+ = \underline{\underline{1}}_N$, one does not have to impose the inner product to be hermitian but can take it for granted.

References

1. Clinton, W.L., Galli, A.J. and Massa, L.J. (1969) *Phys. Rev.*, **177**, 7.
2. Pecora, L.M. (1986) *Phys. Rev. B*, **33**, 5987.
3. Levy, M. and Goldstein, J.A. (1987)*Phys. Rev. B*, **35**, 7887.
4. Fano, U. (1957) *Rev. Mod. Phys.*, **29**, 75.
5. Roman, P. (1965) *Advanced Quantum Theory*, Addison-Wesley: New York.
6. Blum, K. (1981) *Density Matrix Theory and Applications*, Plenum Press, New York.
7. Graybill, F.A. and Marsaglia, G. (1957) *Ann. Math. Statist.*, **28**, 678.
8. Gerstein, L.J. (1988) *Amer. Math. Month.*, **95**(10), 950.
9. Pettofrezo, A.J. (1966) *Matrices and Transformations*, Prentice Hall, Englewood Cliffs (Dover, New York, 1978).
10. Cohen-Tannoudji, C., Diu, B. and Laloe. F. (1977) *Quantum Mechanics,* Vols. I and II (Engl. Transl.), Wiley-Interscience, New-York.
11. Hamermesh, M. (1962) *Group Theory and its Applications to Physical Problems*, Addison-Wesley, New York (Dover, New York, 1990).
12. Dirac, P.A.M. (1958) *The Principles of Quantum Mechanics (4th edn.)*, Oxford University, New York.
13. McWeeny, R. (1960) *Rev. Mod. Phys.*, **32**, 335.
14. Istratescu, V.I. (1987) *Inner Product Structures, Theory & Applications*, D. Reidel, Boston.
15. Soirat, A.J.A. and Massa, L. (1994) *Phys. Rev. B*, **50–55**, 3392.
16. Clinton, W.L. and Massa, L.J. (1972) *Phys. Rev. Lett.*, **29**, 1363.
17. Massa, L.J. and Clinton, W.L. (1972) *Trans. Amer. Crystallogr. Assn.*, **8**, 149.
18. Clinton, W.L., Frishberg, C.A., Massa, L.J. and Oldfield, P.A. (1973) *Int. J. Quantum Chem. Symp.*, **7**, 505.

19. Frishberg, C.A., Goldberg, M.J. and Massa, L.J. 'Quantum model of the coherent diffraction experiment: recent generalizations and applications'. In Ref. [23] of this bibliography, p. 101.
20. Massa, L., Goldberg, M.J., Frishberg, C., Boehme, R.F. and La Placa, S.J. (1985) *Phys. Rev. Lett.*, **55**, 622.
21. Prugovecki, E. (1981) *Quantum Mechanics in Hilbert Space (2nd edn.)*, Academic, New York.
22. Halmos, P.R. (1957) *Introduction to Hilbert Space (and the Theory of Spectral Multiplicity) (2nd edn.)*, Chelsea, New York.
23. Graybill, F.A. (1961) *An Introduction to Linear Statistical Models*, McGraw-Hill, New York.

9

Kinetic equation, optical potential, tensor theory and structure factor refinement in high-energy electron diffraction

LIAN-MAO PENG

Beijing Laboratory of Electron Microscopy, Institute of Physics and Center for Condensed Matter Physics, Chinese Academy of Sciences, P.O. Box 2724, Beijing 100080, People's Republic of China

1. Introduction

In this chapter we will present an account of methods used in high-energy electron diffraction for describing the movement of fast electrons in a solid and for retrieving crystal structure factors from energy-filtered experimental diffraction data. By high-energy electron we mean electrons in the energy range of a few keV to a few MeV. For if the incident beam energy is smaller than about 1 keV the incident electron will hardly be distinguishable from one of the solids. The treatment of the scattering processes is then complicated by exchange effects due to the mixture of the incident electron and the electrons of the solid, and by complicated virtual inelastic scattering effects [1]. On the other hand, if the energy is very high, say greater than 10 MeV, Bremsstrahlung losses become severe and in addition the specimen can be seriously damaged by electron induced atomic displacements [2].

High-energy electrons may be scattered elastically or inelastically by a solid. In an *elastic collision* the solid remains in its original state so that the incident electron does not lose any energy, i.e. $\phi_f = \phi_i$ (here the subscripts 'f' and 'i' denote the final state and the initial state respectively). On the other hand in an *inelastic collision* the incident electron loses an amount of energy ΔE equal to $E_f - E_i$, and the solid is excited from the initial state ϕ_i to a final state ϕ_f. Without loss of generality we can partition the interaction potential into time-independent and time-dependent parts. The time-independent part of the potential gives rise to elastic scattering, while the time-dependent part gives rise to inelastic scattering. Techniques utilizing elastically scattered electrons can be used to study the electron distribution and atomic structure of solids and utilizing inelastically scattered electrons can be used to probe the dynamics of solids [3, 4].

The general problem of high-energy electron diffraction by a solid may be formulated self-consistently on the basis of a kinetic equation for the one-particle density matrix $\rho(\mathbf{r}, \mathbf{r}', t, t') = \psi(\mathbf{r}, t)\psi(\mathbf{r}', t')$, $\psi(\mathbf{r}, t)$ being the wave function of a high-energy electron propagating in the solid [5]. This approach provides a general treatment of spatial and temporal coherence of electrons and takes account of both elastic and inelastic scattering [6, 7]. It can be shown using the kinetic equation that the problem of multiple elastic and inelastic scattering by a solid is entirely determined by two universal functions, that is (1) the *Coulomb potential* averaged over the motion of the crystal particles, i.e. the crystal electrons and nuclei and

Paul G. Mezey and Beverly E. Robertson (eds.), Electron, Spin and Momentum Densities and Chemical Reactivity, 147–168

(2) the *mixed dynamic form factor* of inelastic excitations [8, 9]. In a general electron diffraction experiment, the scattering cross section contains information about both these functions simultaneously.

For high-energy electron diffraction the influence of the time-dependent part of the potential, giving rise to inelastic scattering, is usually much smaller than that of the time-independent part. To a good approximation the effect of the inelastic scattering on the elastic scattering may be represented by regarding the interaction potential between the incident electron and the solid to be complex. This complex potential is usually called the *optical potential*, by analogy with the long-standing use of a complex refractive index for discussing the optical properties of partially absorbing media [10, 11]. After an inelastic collision the solid is excited to a higher energy state and the incident electron is removed from an *elastic channel* and enters an *inelastic channel*. Since for high-energy electrons the probability that the inelastically scattered electron will reappear in the elastic channel is very small [12], as far as the elastic scattering is concerned, the inelastically scattered electron can be considered to have been *absorbed* by the crystal, and the inelastic scattering events contribute only an imaginary addition to the time-independent part of the potential [13].

Using the effective *optical potential* the general kinetic equation reduces to a one-body Schrödinger equation. To an excellent approximation, the real part of the optical potential equals the averaged Coulomb potential and the imaginary part represents the first order correction resulting from inelastic diffuse scattering. Experimentally the real part of the optical potential may be measured using, for example, the technique of convergent-beam electron diffraction (CBED), and the structure factors of crystals may be retrieved [14–23].

For high-energy electron diffraction there exist three main inelastic scattering mechanisms. These are, respectively, the collective excitation of the valence electrons (*plasmon excitation*) which has an energy of the order of 10–40 eV, single electron excitations with energies up to few thousand eV, and the excitation of lattice vibrations (*phonon excitation*) with energies typically of 10^{-2} eV. It has been shown, see for example Rez [12] and Whelan [24, 25], that for all but the direct transmitted beams the contribution from phonon excitation or thermal diffuse scattering (TDS) is an order of magnitude larger than contributions from plasmon and single electron excitations. For high-order reflections the imaginary part of the optical potential may be calculated accurately using an Einstein model of TDS [26–31]. For low-order reflections the calculated imaginary part of the structure factors are less accurate. These low-order structure factors may, however, be taken to be the fitting parameters in the structure factor refinement procedure although they may not correspond directly to real physical quantities.

To a first order approximation, the scattering potential of a crystal may be represented as a sum of contributions from isolated atoms, having charge distributions of spherical symmetry around their nuclei. In a real crystal the charge distribution deviates from the spherical symmetry around the nucleus and the difference reflects the charge redistribution or bonding in the crystal. The problem of experimental measurement of crystal bonding is therefore a problem of structure factor refinement, i.e. accurate determination of the difference between the true crystal structure factors and that of isolated atoms. The structure factors may in principle be extracted from energy-filtered experimental diffraction data by varying crystal structure factors and minimizing a merit function which measures the difference between experiments and theoretical models. There are at least two major drawbacks for the direct application of the minimization scheme. First, the procedure may not be able to return a unique set of parameters that give the minimum to the merit function in multiple parameter space and, second, the procedure is numerically very expansive. Both problems may be solved to a large extent by the use of a perturbation approach called the *tensor theory* of electron diffraction [20, 22, 32–34]. The validity of this approach will be discussed and its application to structure factor refinement will be demonstrated using experimental results from a single crystal of silicon.

The plan of this chapter is as follows. Section 2 outlines the general equations that govern the movement of high-energy electrons in a solid, and Section 3 describes the concept and computation of the optical potential and the reduction of the general kinetic equation to a one-body Schrödinger equation for the elastic wave field. Section 4 presents the tensor theory of high-energy electron diffraction for the description of energy-filtered electron diffraction data, and Section 5 gives its application to crystal structure factor refinement. The summary and conclusions are given in Section 6.

2. Kinetic equation

The dynamical elastic and inelastic scattering of high-energy electrons by solids may be described by three fundamental equations [5]. The first equation determines the wave amplitude $G_0(\mathbf{r}, \mathbf{r}', E)$, or the *Green function*, at point $\mathbf{r}$ due to a point source of electrons at $\mathbf{r}'$ in the averaged potential $\langle V(\mathbf{r})\rangle$:

$$\left[E + \frac{\hbar^2}{2m}\nabla^2 - \langle V(\mathbf{r})\rangle\right] G_0(\mathbf{r}, \mathbf{r}', E) = \delta(\mathbf{r} - \mathbf{r}'), \tag{1}$$

where the time-averaged interaction potential $\langle V(\mathbf{r})\rangle$ is made over the motion of the crystal particles and is defined as

$$\langle V(r)\rangle = \frac{1}{Z}\sum_n \exp(-\varepsilon_n/k_B T)\langle n|V(\mathbf{r}, \mathbf{r}_1, \ldots)|n\rangle, \tag{2}$$

where $Z = \sum_n \exp(-\varepsilon_n/k_B T)$ is the partition function, ε_n the nth eigenvalue of the crystal Hamiltonian H_{cr}, i.e. $H_{cr}|n\rangle = \varepsilon|n\rangle$, $|n\rangle$ being the nth eigenstate of the crystal system.

The second equation determines the wave amplitude $G(\mathbf{r}, \mathbf{r}', E)$ at $\mathbf{r}$ due to a point source of electrons at $\mathbf{r}'$, with the influence of the fluctuating part of the interaction included

$$\left[E + \frac{\hbar^2}{2m}\nabla^2 - \langle V(\mathbf{r})\rangle\right] G(\mathbf{r}, \mathbf{r}', E) - \int \mathrm{d}\mathbf{x} \int \mathrm{d}\omega\, \overline{s}(\mathbf{r}, \mathbf{x}, \omega) G_0(\mathbf{r}, \mathbf{x}, E - \hbar\omega) G(\mathbf{x}, \mathbf{r}', E) = \delta(\mathbf{r} - \mathbf{r}'), \tag{3}$$

where $\overline{s}(\mathbf{r}, \mathbf{x}, \omega)$ is the Van Hove *dynamical form factor* [8] which is defined as

$$\overline{s}(\mathbf{r}, \mathbf{r}', \omega) = \frac{1}{Z} \sum_{n,n_1} \exp(-\varepsilon_n / k_B T) \delta V_{n,n_1}(\mathbf{r}) \delta V_{n_1,n}(\mathbf{r}') \delta \left(\omega - \frac{\varepsilon_{n_1} - \varepsilon_n}{\hbar} \right), \tag{4}$$

in which $\delta V_{n,n_1}(\mathbf{r})$ represents the fluctuating part of the interaction

$$\delta V_{n,n_1}(\mathbf{r}) = \langle n | V(\mathbf{r}, \mathbf{r}_1, \ldots) | n_1 \rangle - \delta_{n,n_1} \langle V(\mathbf{r}) \rangle, \tag{5}$$

with $\langle V(\mathbf{r}) \rangle$ being given by Equation (2).

The third equation is the *kinetic equation*, which describes the evolution of the one-particle density matrix $\rho(\mathbf{r}, \mathbf{r}', E)$ of the electron in the process of multiple elastic and inelastic scattering in a solid

$$\rho(\mathbf{r}, \mathbf{r}', E) = \rho_0(\mathbf{r}, \mathbf{r}', E) + \int \mathrm{d}\mathbf{x}\, \mathrm{d}\mathbf{x}'\, G(\mathbf{r}, \mathbf{x}, E) G^*(\mathbf{r}', \mathbf{x}', E) \times \left[\int \mathrm{d}\omega\, \overline{s}(\mathbf{x}', \mathbf{x}, \omega) \rho(\mathbf{x}, \mathbf{x}', E + \hbar\omega) \right], \tag{6}$$

that is the one-particle density matrix is a sum of the 'coherent' wave ρ_0 and waves inelastically scattered at $(\mathbf{x}, \mathbf{x}')$, the propagation of which to $(\mathbf{r}, \mathbf{r}')$ is described by the product of two Green's functions $G(\mathbf{r}, \mathbf{x}, E) G^*(\mathbf{r}', \mathbf{x}', E)$. The one-particle density matrix $\rho(\mathbf{r}, \mathbf{r}', E)$ is indeed the *spectral one-particle density matrix* which is related to the usual bilinear combination of two wave functions

$$\rho(\mathbf{r}, \mathbf{r}', t, t') = \psi(\mathbf{r}, t) \psi^*(\mathbf{r}', t)$$

by the Fourier transformation

$$\rho(\mathbf{r}, \mathbf{r}', E) = \frac{1}{2\pi\hbar} \int_{-\infty}^{\infty} \mathrm{d}(t - t') \exp\left[-\frac{E}{\hbar}(t - t') \right] \rho(\mathbf{r}, \mathbf{r}', t, t').$$

In the simplest case where the interaction potential does not depend on time we have

$$\psi(\mathbf{r}, t) = \psi_\mu(\mathbf{r}) \exp\left(-\mathrm{i} \frac{E_\mu}{\hbar} t \right),$$

where $\psi_\mu(\mathbf{r})$ is a wave function of the continuous spectrum, and the spectral one-particle density matrix $\rho(\mathbf{r}, \mathbf{r}', E)$ is given by

$$\rho(\mathbf{r}, \mathbf{r}', E) = \psi_\mu(\mathbf{r}) \psi_\mu^*(\mathbf{r}') \delta(E - E_\mu).$$

In general the spectral one-particle density matrix $\rho(\mathbf{r}, \mathbf{r}', E)$ describes the mutual coherence of the wave field of high-energy electrons at the points $\mathbf{r}$ and $\mathbf{r}'$. For the simplest case of time-independent interaction potential the diagonal elements of

$\rho(\mathbf{r}, \mathbf{r}', E) = |\psi_\mu(\mathbf{r})|^2$, i.e. the element is proportional to the probability of finding the electron with energy E at the point $\mathbf{r}$. In a general case the kinetic equation describes the evolution of $\rho(\mathbf{r}, \mathbf{r}', E)$ due to the process of multiple elastic and inelastic scattering. The distribution of electrons over a solid angle and energy is related to the double differential cross section

$$\frac{d^2\sigma}{d\Omega\, dR} = \frac{k}{k_0}\left(\frac{k\cos\theta}{2\pi}\right)^2 \times \lim_{z\to\pm\infty} \int d^2R\, d^2R'\, \exp(-i\mathbf{q}_t \cdot \mathbf{R} + i\mathbf{q}_t'\mathbf{R}')\rho(\mathbf{R}, z, \mathbf{R}', z, E), \quad (7)$$

where $k = \sqrt{2mE/\hbar^2}$, θ is the angle between the wave vector of the scattered electron and the z-axis of the chosen system of coordinates, $\mathbf{R} = (x, y)$ and $\mathbf{q}_t = \mathbf{q}_{x,y}$. The positive and negative signs correspond to the forward and backward scattering, respectively.

In summary, the movement of a high-energy electron in a solid may be described by a set of three Equations (1), (4) and (6). From these equations we may conclude that for high-energy electron diffraction the problem of multiple elastic and inelastic scattering by a solid is entirely determined by two functions, i.e. (1) the Coulomb interaction potential averaged over the motion of the crystal particles $\langle V(\mathbf{r})\rangle$ and (2) the mixed dynamic form factor $\overline{s}(\mathbf{r}, \mathbf{r}', E)$ of inelastic excitations of the solid.

3. Optical potential

In this section we consider the problem of scattering of a well-collimated beam of high-energy electrons of energy E_0 by a crystal. The incident electron wave function then has the form of a plane wave

$$\psi_0(\mathbf{r}) = \exp(i\mathbf{k}_0 \cdot \mathbf{r}),$$

where $\mathbf{k}_0$ is the wave vector of the incident electron. Neglecting the effect of the time-dependent part of the interaction potential, the movement of the incident high-energy electron in a solid is governed by Equation (1). Let ψ_{k_0} be the wave function of the fast electron, we have

$$\left[-\frac{\hbar^2}{2m}\nabla^2 + \langle V(\mathbf{r})\rangle\right]\psi_{k_0}(\mathbf{r}) = E_0\psi_{k_0}, \quad (8)$$

E_0 being the energy of the fast electron. To a good approximation, the effect of inelastically scattered electrons on the elastic electron wave field may be treated via a first order perturbation method. From Equation (4) we have

$$\left[-\frac{\hbar^2}{2m}\nabla^2 + V(\mathbf{r})\right]\psi_{k_0} = E_0\psi_{k_0}, \quad (9)$$

where V is called the *optical potential* [13, 35, 36] and is given by

$$V^{\mathrm{op}} \approx V^{(0)} + V^{(1)},$$

with

$$V^{(0)} = \langle V(\mathbf{r}) \rangle$$

being the averaged potential and

$$V^{(1)} = \left\langle (V - \langle V \rangle) \frac{1}{E_k + E_\alpha + \mathrm{i}\varepsilon - h_0 - H_0} (V - \langle V \rangle) \right\rangle, \tag{10}$$

being the first order correction due to diffuse scattering. Recent quantitative electron diffraction work has shown that this approximation works with high precision [16, 21, 23, 37, 38]. In Equation (10) $h_0 = -(\hbar^2/2m_0)\nabla^2$ is the free Hamiltonian of the incident electron, H_0 is the Hamiltonian for all the electrons and nuclei of the crystal, and E_α is the αth eigenvalue of the crystal Hamiltonian, i.e. $H_0\phi_\alpha = E_\alpha\phi_\alpha$, ϕ_α being the αth eigenstate of the crystal.

For TDS and to a good approximation we may assume that the atomic electrons follow adiabatically the motion of nucleus and that all atomic electrons are in their ground states [39]. The interacting potential is then given by

$$\begin{aligned} V(\mathbf{r}, \ldots, \mathbf{r}_n, \ldots) &= \sum_n \left\{ -\frac{Z_n e^2}{|\mathbf{r} - \mathbf{r}_n|} + \int \frac{e^2}{|\mathbf{r} - \mathbf{R}' - \mathbf{r}_n|} \rho_n^0(\mathbf{R}')\,\mathrm{d}\mathbf{R}' \right\} \\ &= \sum_n \int \mathrm{d}\mathbf{R}\, \varphi_n(\mathbf{r} - \mathbf{R}) \delta(\mathbf{R} - \mathbf{r}_n), \end{aligned} \tag{11}$$

in which Z_n is the atomic number of the nth atom and $\rho_n^0(\mathbf{R}')$ is its corresponding electron density in its ground state, the summation on n is over all atoms in the crystal, and $\varphi_n(\mathbf{r})$ is given by

$$\varphi_n(\mathbf{r}) = -\frac{Z_n e^2}{r} + \int \frac{e^2}{|\mathbf{r} - \mathbf{R}'|} \rho_n^0(\mathbf{R}')\,\mathrm{d}\mathbf{R}'. \tag{12}$$

Let $\mathbf{r}_n = \mathbf{R}_n + \mathbf{u}_n$, where $\mathbf{R}_n$ denotes the equilibrium position of the nth atom and $\mathbf{u}_n$ represents the thermal displacement of the atom from its thermal equilibrium position, we have for the averaged potential

$$\langle V(\mathbf{r}) \rangle = \sum_n \int \mathrm{d}\mathbf{R}\, \varphi_n(\mathbf{r} - \mathbf{R}) \langle \delta(\mathbf{R} - \mathbf{R}_n - \mathbf{u}_n) \rangle. \tag{13}$$

The Fourier coefficients of the averaged potential is given by

$$\begin{aligned} V_g &= \frac{1}{V} \int \langle V(\mathbf{r}) \rangle \exp(\mathrm{i}\mathbf{g} \cdot \mathbf{r})\,\mathrm{d}\mathbf{r} \\ &= \frac{h^2}{8\pi^2 m_0} \cdot \frac{4\pi}{\Omega} \sum_i f_i^{\mathrm{B}}(s) T_i(\mathbf{g}) \exp(-\mathrm{i}\mathbf{g} \cdot \mathbf{R}_i), \end{aligned} \tag{14}$$

in which V and Ω are the volume of the crystal and a unit cell respectively, $f_i^{\mathrm{B}}(s)$ is the *Born atomic scattering amplitude* [40] with $\mathbf{s} = \mathbf{g}/4\pi$, $T_i(\mathbf{g})$ is the *temperature factor* [41] of the ith atom, and the summation on i is over a unit cell. In Equation (14) the Born atomic scattering amplitude is related to $\varphi_i(\mathbf{r})$ via the following relation:

$$f_i^{\mathrm{B}}(\mathbf{g}) = -\frac{2\pi m_0}{h^2}\int \varphi_i(\mathbf{r})\exp(-i\mathbf{g}\cdot\mathbf{r})\,\mathrm{d}\mathbf{r}, \tag{15}$$

and for a harmonic crystal the temperature factor is given by

$$T_i(\mathbf{g}) = \langle \exp(-\mathbf{g}\cdot\mathbf{u}_i)\rangle = \exp\left\{-\tfrac{1}{2}\langle(\mathbf{g}\cdot\mathbf{u}_i)^2\rangle\right\}. \tag{16}$$

Let $\mathbf{a}_1, \mathbf{a}_2, \mathbf{a}_3$ be the real space lattice vectors and $\mathbf{b}_1, \mathbf{b}_2, \mathbf{b}_3$ be the reciprocal space lattice vectors. We have then the following relations:

$$a_i \cdot b_j = \delta_{ij}.$$

In terms of these vectors a real space displacement vector $\mathbf{u}$ can be expressed as

$$\mathbf{u} = u_1\mathbf{a}_1 + u_2\mathbf{a}_2 + u_3\mathbf{a}_3, \tag{17}$$

and a reciprocal space vector $\mathbf{g}$ as

$$\mathbf{g} = h\mathbf{b}_1 + k\mathbf{b}_2 + l\mathbf{b}_3, \tag{18}$$

giving

$$\begin{aligned}\langle(\mathbf{g}\cdot\mathbf{u})^2\rangle = {} & h^2\langle u_1^2\rangle + k^2\langle u_2^2\rangle + l^2\langle u_3^2\rangle + 2hk\langle u_1u_2\rangle + 2hl\langle u_1u_3\rangle \\ & + 2kl\langle u_2u_3\rangle.\end{aligned} \tag{19}$$

In matrix notation the above expression can be written as

$$(\mathbf{g}\cdot\mathbf{u})^2 = (h, k, l)\begin{pmatrix}u_1\\u_2\\u_3\end{pmatrix}(u_1, u_2, u_3)\begin{pmatrix}h\\k\\l\end{pmatrix} = \mathbf{G}^{\mathrm{T}}(\mathbf{X}\mathbf{X}^{\mathrm{T}})\mathbf{G},$$

where $\mathbf{G}$ and $\mathbf{X}$ are 3×1 column vectors and their transpose are given by

$$\mathbf{G}^{\mathrm{T}} = (h, k, l), \qquad \mathbf{X}^{\mathrm{T}} = (u_1, u_2, u_3).$$

The temperature factor (16) then becomes

$$T_i(\mathbf{g}) = \exp\{-\mathbf{G}^{\mathrm{T}}\beta_i\mathbf{G}\}, \tag{20}$$

in which the matrix $\beta = 1/2(\mathbf{X}\mathbf{X}^{\mathrm{T}})$ is a symmetric matrix

$$\beta = \frac{1}{2}\langle\mathbf{X}\mathbf{X}^{\mathrm{T}}\rangle = \frac{1}{2}\begin{pmatrix}\langle u_1^2\rangle & \langle u_1u_2\rangle & \langle u_1u_3\rangle\\ \langle u_2u_1\rangle & \langle u_2^2\rangle & \langle u_2u_3\rangle\\ \langle u_3u_1\rangle & \langle u_3u_2\rangle & \langle u_3^2\rangle\end{pmatrix}, \tag{21}$$

and is usually referred to as the *mean-square displacement matrix*. In X-ray crystallography the general anisotropic vibration parameters are usually given as the elements of a **U** matrix which are related to that of the β matrix by the following relation:

$$\beta_{ij} = 2\pi^2 U_{ij} b_i b_j. \tag{22}$$

Explicitly the anisotropic temperature factor is given by

$$\begin{aligned} T(\mathbf{g}) = \exp\{-2\pi^2[&U_{11}(hb_1)^2 + U_{22}(kb_2)^2 + U_{33}(lb_3)^2 \\ &+ 2U_{12}(hb_1)(kb_2) + 2U_{13}(hb_1)(lb_3) + 2U_{23}(kb_2)(lb_3)]\}. \end{aligned} \tag{23}$$

Experimentally U_{ij} may be obtained by fitting quantitatively the calculated X-ray beam intensities with the experimentally measured X-ray intensities using a general anisotropic temperature factor [42].

We now consider the first order correction to the average potential, i.e. $V^{(1)}$. In real space representation, substituting Equations (11) and (13) into (10) gives [36]

$$\begin{aligned} V^{(1)}(\mathbf{r}, \mathbf{r}') = \sum_{ij} \Big\langle \iint & \mathrm{d}\mathbf{R}\,\mathrm{d}\mathbf{R}'\, \varphi_i(\mathbf{r} - \mathbf{R})[\delta(\mathbf{R} - \mathbf{r}_i) - \langle \delta(\mathbf{R} - \mathbf{r}_i)\rangle] \\ & \times \left\langle \mathbf{r} \left| \frac{1}{E_k + E_\alpha + \mathrm{i}\varepsilon - h_0 - H_0} \right| \mathbf{r}' \right\rangle \varphi_j(\mathbf{r}' - \mathbf{R}') \\ & \times \Big[\delta(\mathbf{R}' - \mathbf{r}_j) - \big\langle \delta(\mathbf{R}' - \mathbf{r}_j)\big\rangle\Big]\Big\rangle. \end{aligned} \tag{24}$$

For thermal diffuse scattering since the energies of phonons are much smaller than the energy of the incident electrons, we may neglect E_α and H_0 in (24). Neglecting the effect of *virtual diffuse scattering* [12] and using the *high-energy approximation* [43] we obtain

$$\begin{aligned} V^{(1)}(\mathbf{g}, \mathbf{h}) &= -\mathrm{i}\frac{h^2}{8\pi^2 m_0} \cdot \frac{4\pi}{\Omega} \sum_i f_i^{\mathrm{TDS}}(\mathbf{s}) \exp[-\mathrm{i}(\mathbf{g} - \mathbf{h}) \cdot \mathbf{R}_i] \\ &= V^{(1)}(\mathbf{g} - \mathbf{h}), \end{aligned} \tag{25}$$

where

$$\begin{aligned} f_i^{\mathrm{TDS}}(\mathbf{s}) = \frac{2h}{m_0 v} \int & \mathrm{d}\mathbf{s}'\, f_i\left(\left|\frac{\mathbf{s}}{2} + \mathbf{s}'\right|\right) f_i\left(\left|\frac{\mathbf{s}}{2} - \mathbf{s}'\right|\right) \\ & \times \left\{T_i(\mathbf{s}) - T_i\left(\frac{\mathbf{s}}{2} + \mathbf{s}'\right) T_i\left(\frac{\mathbf{s}}{2} - \mathbf{s}'\right)\right\}, \end{aligned} \tag{26}$$

where v is the velocity of the electron. For isotropic thermal vibrations of the crystal lattice we have

$$\beta = \frac{1}{2}\begin{pmatrix} \langle u^2\rangle & 0 & 0 \\ 0 & \langle u^2\rangle & 0 \\ 0 & 0 & \langle u^2\rangle \end{pmatrix},$$

and therefore

$$T(\mathbf{g}) = \exp\left\{-\tfrac{1}{2}\langle u^2\rangle g^2\right\} = \exp(-Bs^2), \tag{27}$$

in which $B = 8\pi^2\langle u^2\rangle$ is the usual Debye–Waller temperature B-factor of the atom. Substitution of Equation (27) into Equation (26) gives

$$\begin{aligned} f_i^{\text{TDS}}(\mathbf{s}) = & \frac{2h}{m_0 v}\int \mathrm{d}\mathbf{s}'\, f_i\left(\left|\frac{\mathbf{s}}{2}+\mathbf{s}'\right|\right) f_i\left(\left|\frac{\mathbf{s}}{2}-\mathbf{s}'\right|\right) \\ & \times \exp(-B_i s^2)\left\{1-\exp\left[-2B_i(s'^2-s^2/4)\right]\right\} \end{aligned} \tag{28}$$

and this is the Hall and Hirsch formula [26]. In real space we have

$$V^{(1)}(\mathbf{r}) = \int V^{(1)}(\mathbf{s})\exp(\mathrm{i}\mathbf{s}\cdot\mathbf{r})\,\mathrm{d}\mathbf{s}, \tag{29}$$

and this expression clearly shows that $V^{(1)}(\mathbf{r})$ is a *local potential*, i.e. it depends only on one site coordinate $\mathbf{r}$.

4. Tensor theory

In this section we will discuss perturbation methods suitable for high-energy electron diffraction. For simplicity, in this section we will be concerned with only periodic structures and a transmission diffraction geometry. In the context of electron diffraction theory, the perturbation method has been extensively used and developed. Applications have been made to take into account the effects of weak beams [44, 45]; inelastic scattering [46]; higher-order Laue zone diffraction [47]; crystal structure determination [48] and crystal structure factors refinement [38, 49]. A formal mathematical expression for the first order partial derivatives of the scattering matrix has been derived by Speer *et al.* [50], and a formal second order perturbation theory has been developed by Peng [22, 34].

It is assumed from the outset that the crystal potential may be written as a sum of two parts:

$$V(\mathbf{r}) = V_0(\mathbf{r}) + \Delta V(\mathbf{r}),$$

in which $V_0(\mathbf{r})$ is a known potential, hereafter we will refer to the structure giving rise to this potential as the *reference structure*. The second term in the above expression $\Delta V(\mathbf{r})$ is a small quantity which may be regarded as a perturbation on $V_0(\mathbf{r})$.

Considering only forward scattering by a crystal, the one-body Schrödinger wave equation may be transformed into a first order eigenequation [44, 51]

$$\frac{2k_0 S_g}{(1+g_z/k_{0_z})}B_g^{(j)} + \sum_{h\neq g}\frac{U_{g-h}B_h^{(j)}}{\sqrt{1+g_z/k_{0_z}}\sqrt{1+h_z/k_{0_z}}} = 2k_{0_z}\gamma^{(j)}B_g^{(j)}, \tag{30}$$

in which $S_g = [K^2 - (\mathbf{k} + \mathbf{g})^2]/2K_z$ is the usual excitation error measuring the distance between the reflection $\mathbf{g}$ and the Ewald sphere, $K^2 = k_0^2 + U_0$, U_g is the gth Fourier component of the crystal scattering potential field, $\gamma^{(j)}$ and $\{B_g^{(j)}\}$ are the eigenvalue and corresponding eigenvector of the jth Bloch wave, respectively.

In matrix notation, Equation (30) can be rewritten as

$$(\mathbf{S} + \mathbf{U})\mathbf{B} = \mathbf{B}\gamma, \tag{31}$$

in which the matrices $\mathbf{S}$ and γ are diagonal matrices with

$$\{\mathbf{S}\}_{gg} = \frac{2k_0 S_g}{\sqrt{1 + g_z/k_{0_z}}}, \qquad \{\gamma\}_{jj} = 2k_{0_z}\gamma^{(j)}, \tag{32}$$

and the elements of the matrices $\mathbf{U}$ and $\mathbf{B}$ are given by

$$\{\mathbf{U}\}_{gh} = \frac{U_{g-h}}{\sqrt{1 + g_z/k_{0_z}}\sqrt{1 + h_z/k_{0_z}}}, \qquad \{\mathbf{B}\}_{gi} = B_g^{(i)}. \tag{33}$$

Similarly we may define a right-hand eigenvector $\overline{\mathbf{B}}$ satisfying

$$\overline{\mathbf{B}}(\mathbf{S} + \mathbf{U}) = \gamma\overline{\mathbf{B}} \tag{34}$$

and it can be easily shown that $\mathbf{B}\overline{\mathbf{B}} = \overline{\mathbf{B}}\mathbf{B} = \mathbf{I}$, i.e.

$$\sum_i \overline{B}_g^{(i)} B_h^{(i)} = \delta_{gh}, \qquad \sum_g \overline{B}_g^{(i)} B_g^{(j)} = \delta_{ij}.$$

When the interaction potential $V_0(\mathbf{r})$ is subjected to a small variation $\Delta V_0(\mathbf{r})$, both the eigenvalues and eigenvectors of the initial system change their values. If the perturbation is small enough, the changes in both the eigenvalues and eigenvectors may be obtained by the use of the perturbation theory. Following the standard procedures of quantum mechanics, the changes may be expressed in a tensor form [34], by analogy with the tensor theory of low-energy of electron diffraction [52]

$$\Delta\gamma^{(j)} = {}^1\mathbf{u}^{(j)} \cdot \Delta\mathbf{U} + \Delta\mathbf{U} \cdot {}^2\mathbf{u}^{(j)} \cdot \Delta\mathbf{U}, \tag{35}$$

$$\Delta B_g^{(j)} = {}^1\varepsilon_g^{(j)} \cdot \Delta\mathbf{U} + \Delta\mathbf{U} \cdot {}^2\varepsilon_g^{(j)} \cdot \Delta\mathbf{U}, \tag{36}$$

$$\Delta\overline{B}_g^{(j)} = {}^1\overline{\varepsilon}_g^{(j)} \cdot \Delta\mathbf{U} + \Delta\mathbf{U} \cdot {}^2\overline{\varepsilon}_g^{(j)} \cdot \Delta\mathbf{U}, \tag{37}$$

in which $\Delta\mathbf{U} = \{\Delta U_g\}$ and

$$\{{}^1\mathbf{u}^{(j)}\}_l = \mathbf{u}_l^{(jj)}, \qquad \{{}^2\mathbf{u}^{(j)}\}_{l,k} = \sum_{j' \neq j} \frac{u_l^{(jj')} u_k^{(j'j)}}{\gamma_0^{(j)} - \gamma_0^{(j')}}, \tag{38}$$

with

$$u_l^{(jj')} = \frac{1}{2k_{0_z}} \sum_g \frac{\overline{B}_g^{(j)} B_{g-l}^{(j')}}{\sqrt{1 + g_z/k_{0_z}}\sqrt{1 + (g_z - l_z)/k_{0_z}}} \tag{39}$$

and

$$\left\{{}^{1}\varepsilon_{g}^{(j)}\right\}_{l} = \sum_{j' \neq j} \frac{B_{0_g}^{(j')} u_{l}^{(j'j)}}{\gamma_0^{(j)} - \gamma_0^{(j')}}, \qquad \left\{{}^{1}\overline{\varepsilon}_{g}^{(j)}\right\}_{l} = \sum_{j' \neq j} \frac{\overline{B}_{0_g}^{(j')} u_{l}^{(jj')}}{\gamma_0^{(j)} - \gamma_0^{(j')}}, \tag{40}$$

$$\left\{{}^{2}\varepsilon_{g}^{(j)}\right\}_{l,k} = \sum_{j' \neq j} \frac{B_{0g}{}^{(j')}}{\gamma_0^{(j)} - \gamma_0^{(j')}} \left\{ -\frac{u_{l}^{(j'j)} u_{k}^{(jj)}}{\gamma_0^{(j)} - \gamma_0^{(j')}} - \sum_{i \neq j} \frac{u_{l}^{(j'i)} u_{k}^{(ij)}}{\gamma_0^{(j)} - \gamma_0^{(i)}} \right\} - \frac{1}{2} \sum_{j' \neq j} \frac{u_{l}^{(jj')} u_{k}^{(j'j)}}{\left[\gamma_0^{(j)} - \gamma_0^{(j')}\right]^2} B_{0g}{}^{(j)}, \tag{41}$$

$$\left\{{}^{2}\overline{\varepsilon}_{g}^{(j)}\right\}_{l,k} = \sum_{j' \neq j} \frac{\overline{B}_{0g}{}^{(j')}}{\gamma_0^{(j)} - \gamma_0^{(j')}} \left\{ -\frac{u_{l}^{(jj)} u_{k}^{(jj')}}{\gamma_0^{(j)} - \gamma_0^{(j')}} + \sum_{i \neq j} \frac{u_{l}^{(ji)} u_{k}^{(ij')}}{\gamma_0^{(j)} - \gamma_0^{(i)}} \right\} - \frac{1}{2} \sum_{j' \neq j} \frac{u_{l}^{(jj')} u_{k}^{(j'j)}}{\left[\gamma_0^{(j)} - \gamma_0^{(j')}\right]^2} \overline{B}_{0g}{}^{(j)}. \tag{42}$$

The diffracted beam amplitude is given by

$$F_g = \sum_{j} \left(\overline{B}_{0_0}^{(j)} + \Delta \overline{B}_{0_0}^{(j)}\right) \left(B_{0_g}^{(j)} + \Delta B_{0_g}^{(j)}\right) \exp\left[\mathrm{i}\left(\gamma_0^{(j)} + \Delta\gamma^{(j)}\right) z\right]. \tag{43}$$

At the entrance surface $z = 0$, the above expression gives $F_g = \delta_{g0}$, i.e. there exists only the incident beam above the crystal in the vacuum region.

For given vectors and matrices $\mathbf{u}$ and ε, the calculation of the diffracted beam amplitude is an operation of the order of $n(p + p^2)$, where n is the number of Bloch waves having appreciable excitation amplitudes, and p is the number of varying crystal structure factors. As will be shown in the following section that for a typical zone axis incidence, the number of Bloch waves having appreciable excitation amplitude is usually less than 20. For simple cases, the number of fitting parameters is usually less than 30. This situation should be compared with the case of a full dynamical calculation scaling as $O(N^3)$, with N being the total number of beams involved. For a typical zone-axis incidence this number is usually larger than 100. For each calculation the tensor theory is therefore about 50 times more efficient than the full dynamical diffraction theory. For a typical numerical minimization routine, for example the NAG routine E04GBF using quasi-Newton algorithm [53], the number of multiplications performed per iteration of the routine is approximated $pm^2 + O(p^3)$, m being the number of data points. For a refinement procedure involving 30 parameters the present scheme is therefore many thousand times faster than the standard procedure. In the following section we will be concerned with the validity of the present tensor theory, the computation of the tensor expressions, and its application to crystal structure factor refinement from energy-filtered experimental CBED patterns.

5. Results

Shown in Figure 1 is an energy-filtered experimental CBED pattern obtained from a silicon sample and along the [110] zone axis, using a primary beam energy of 195.35 keV. The CBED pattern was obtained by focusing a convergent electron beam, defined by a circular aperture, onto the specimen (for a detailed discussion of this method of electron diffraction, see Spence and Zuo [37]). Each diffraction spot of the conventional electron diffraction pattern is then spread into a circular disk, and each point in the disk corresponds to a particular angle of incidence. The variation of intensity across each disk represents the variation of the diffracted beam intensity associated with that disk as a function of the incident angle. A graphical representation of this variation is called a *rocking curve*. CBED patterns are essentially two-dimensional rocking curves from a very small illuminated area, which in the present study is of the order of 1.4 nm. It is then reasonable to assume that the CBED pattern is obtained from an area of uniform crystal thickness and orientation. The pattern obtained in this way is therefore well defined, from an area free from crystal defects and from effects due to bending, and is well suited for comparison with theoretical calculations.

Figure 1 shows, among other disks, the transmitted (000) disk, two (002) type disks, and four $(1\bar{1}1)$ type disks. It should be pointed out that the (002) type reflections are kinematically forbidden, and the appearance of the (002) type disks in the

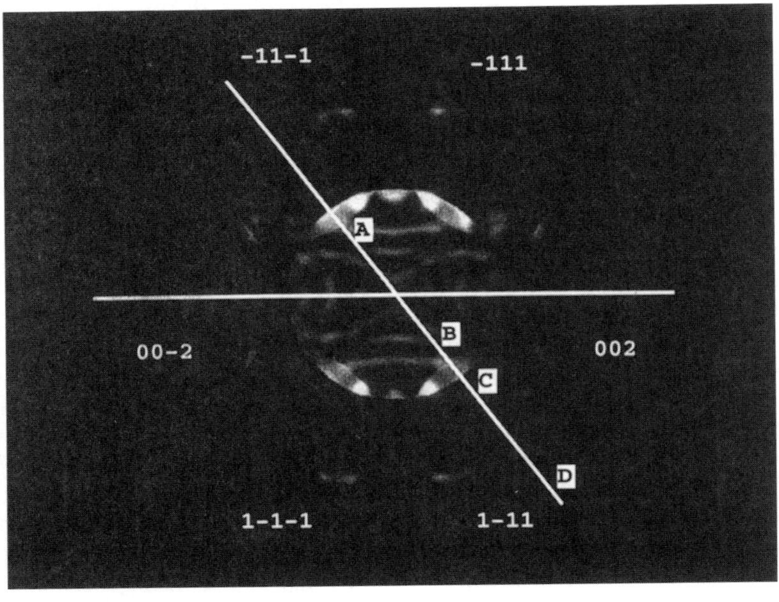

Figure 1. Energy filtered experimental Si[110] zone axis CBED pattern. The pattern was obtained for a primary beam energy of 195.35 keV, an energy window of 10 eV and an electron probe size of 1.4 nm, using a Philips CM200/FEG electron microscope.

experimental CBED pattern is mainly due to the multiple diffraction processes, for example via the scattering of $(1\bar{1}1)$ followed by $(\bar{1}11)$. The raw experimental CBED pattern was recorded using a 1024×1024 slow-scan CCD camera. We have therefore in a single CBED pattern more than one million data points available. Since the number of structural parameters for a single crystal of silicon is much smaller than the number of available data points, we choose to extract certain well-defined lines of data from the two-dimensional CBED pattern and for simplicity we will hereafter simply call these line scans CBED rocking curves.

Shown in Figure 2 are calculated CBED rocking curves, corresponding to a line scan along $[1\bar{1}1]$ direction as shown in Figure 1. The first data point of Figure 2 corresponds roughly to point A of Figure 1, the 98th data point to point B, 99th data point to point C and the 196th point to point D. This figure shows that the effect of the number of Bloch waves used in dynamical diffraction calculations on the calculated CBED rocking curves.

All tensor expressions (35)–(42) involve summation over Bloch waves, i.e. summation over j. For a dynamical diffraction calculation involving N beams, the number of Bloch waves resulting from Equation (30) equals the number of beams, i.e. N. It should be noted, however, that not all of these Bloch waves will be strongly excited within the crystal and contribute to the electron wave field. The excitation amplitudes of the Bloch waves in the crystal are given by $\{\overline{B}_0^{(j)}\}$. Extensive numerical calculations show that in a typical dynamical diffraction calculation, although typically more than

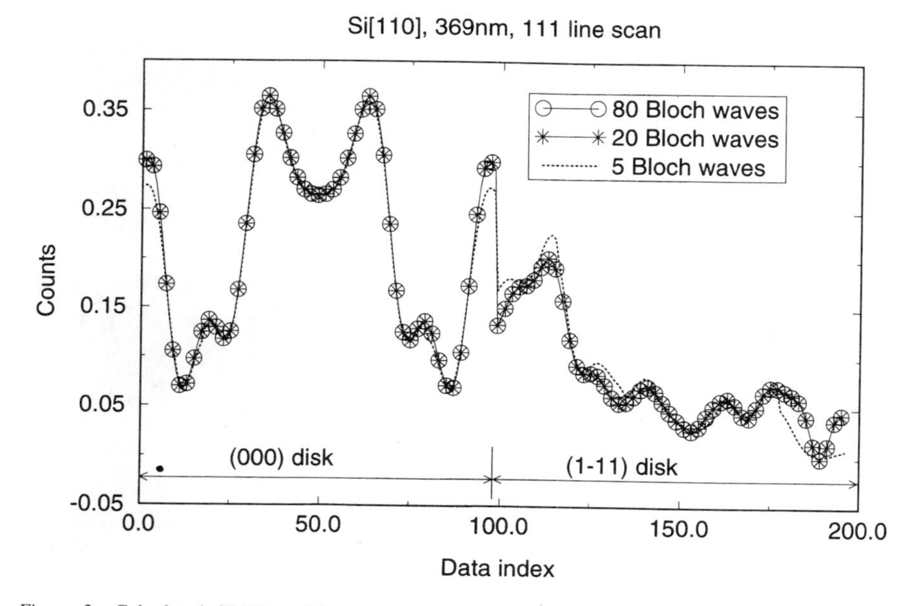

Figure 2. Calculated CBED rocking curves for Si[110], a primary beam energy of 193.35 keV and a crystal thickness of 369 nm. The three curves shown in the figure were calculated using 80 Bloch waves (circle+solid line) 20 Bloch waves (star solid line) and 5 Bloch waves (dotted line) and the curves correspond to the line of Figure 1 along A–D.

100 beams are needed for a zone-axis incidence, only less than 30 most strongly excited Bloch waves are required to achieve a convergent result for calculating the diffracted beam amplitudes F_g. Shown in Figure 2 are three rocking curves calculated using 5, 20 and 80 most strongly excited Bloch waves. This figure shows clearly that results obtained using 20 and 80 Bloch waves are indistinguishable, suggesting that in the present case only about 20 Bloch waves have been excited in the crystal with appreciable amplitude and have contributed to the electron wave field.

Among all the tensor components, the computation of the second order tensors $^2\varepsilon$ and $^2\overline{\varepsilon}$ are most time consuming. According to the order of complexity, we may divide our approximate theories as (1) first order or linear theory, using only the linear tensors, $^1\mathbf{u}$, $^1\varepsilon$ and $^1\overline{\varepsilon}$ for calculating the correction to both the eigenvalues and eigenvectors; (2) quasi-second order theory, treating the correction in eigenvalues using both the first and second order tensors and correction in eigenvector using only the first order tensor, i.e. $^1\varepsilon$ and $^1\overline{\varepsilon}$; (3) full second order theory, using full tensor expressions for treating both corrections in eigenvalues and eigenvectors. Roughly speaking, both the linear theory and the quasi-second order theory are methods which scale as M, M being the number of strongly excited Bloch waves within the crystal. The full second order theory scales as M^2. Shown in Figure 3 are five rocking curves calculated using the full dynamical theory (solid curve), using the full tensor expressions for both the eigenvalues and eigenvectors (circle and dotted line), using second order tensor expression for eigenvalue and first order expressions for eigenvectors (star and dotted

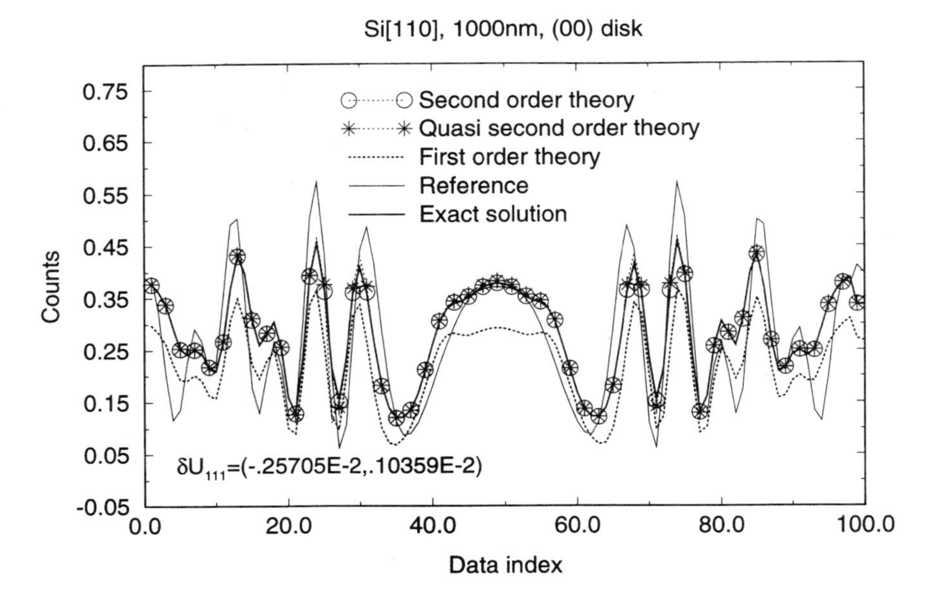

Figure 3. Calculated CBED rocking curves within the (000) disk. The calculations were made for a Si[110] zone axis, a primary beam energy of 193.35 keV and a crystal thickness of 1000 nm. The curves shown in the figure correspond to the line scan A–B of Figure 1.

line), using first order expressions for both the eigenvalues and eigenvectors (dotted line). The reference structure is taken to be that composed of neutral atoms, and the perturbation be the difference between the (111) structure factors of a real single crystal of silicon and that of the reference structure. The crystal thickness used in the calculation was 1000 nm. It is seen that for this crystal thickness while the first order theory differs substantially from the exact solution, both the full second order tensor theory and the quasi- second order theory give excellent results over the whole range of incidence. In what follows we shall therefore use only the quasi-second order theory.

In principle the validity of an approximate theory depends on the crystal thickness. The parameter most widely used for estimating the validity of the kinematical or a single diffraction theory is the distinction distance, see for example Hirsch *et al.* [46]. For an averaged interaction potential, the extinction distance is roughly proportional to the inverse of the potential. For a single crystal of silicon, the extinction distance is of the order of 50 nm, suggesting that for a crystal thickness comparable to that value multiple diffraction processes will begin to dominate the diffraction processes and a kinematical diffraction theory will no longer be valid. Since the charge redistribution or the formation of bonding in a real crystal introduces a change in the crystal structure factors which is typically less than 5% of the total crystal structure factors, one would expect that the corresponding extinction distance for the first order perturbation treatment of the bonding effect be about 20 times that due to the whole crystal potential, i.e. of the order of $20 \times 50\,\text{nm} = 1000\,\text{nm}$. For the second order tensor theory the distance would be twice that value. We would then expect that the validity of the first order theory to be about one-third of the extinction distance, i.e. for the first order theory the validity is about 330 nm, and that for the second order theory 700 nm. Shown in Figures 4–6 are three sets of rocking curves for a crystal thickness of (a) 250 nm (Figure 4); (b) 500 nm (Figure 5); and (c) 1000 nm (Figure 6). These figures clearly show that for a crystal thickness smaller than a few hundreds of angstroms, see for example Figure 4, both the first and second order theory works well. For a crystal thickness of larger than, say 500 nm, see Figure 5, only the second order theory provides an adequate description of the perturbation caused by the crystal bonding. Figure 6 shows that for a crystal thickness as large as 1000 nm the second order tensor theory remains accurate for describing the effect of bonding. Noticing that a typical crystal thickness used in CBED experiments lies from a few hundred to less than 5000 Å, we would expect the second order theory to provide a generally valid description of the bonding effect in real crystals. It should also be pointed out that the perturbation theory may also be used in combination with the method of iteration, i.e. a new reference structure which is closer to the true solution may always be defined as the solution return from the previous application of the tensor theory, and the accuracy of the solution may therefore be improved via iteration.

One of the most important questions in quantitative electron diffraction work concerns whether or not the solution obtained is unique. It may be shown that in a general situation the solution obtained is not unique [54]. In the study of crystal bonding, however, since we have a fairly good starting point, i.e. the isolated atoms approximation of the crystal, we will show that the solution obtained from quantitative

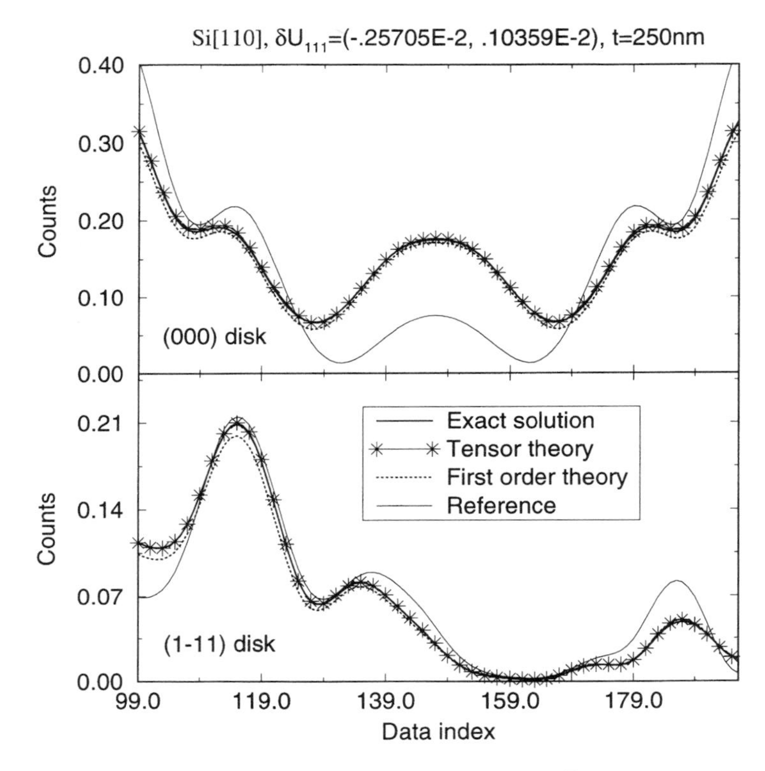

Figure 4. Calculated CBED rocking curves within the (000) and the $(1\bar{1}1)$ disks in a Si[110] zone axis CBED pattern. All curves shown in the figure were calculated for a crystal thickness of 250 nm, and a primary beam energy of 196.35 keV., and correspond to the line scan A–B of Figure 1.

electron diffraction is unique, with the only uncertainty being that due to the statistic noise always present in real experiments.

The goodness-of-fit between the experimental and theoretically calculated CBED rocking curves is described by a *merit function*, and in the present study we use the chi-square merit function defined as

$$\chi^2 = \sum_k \frac{1}{\sigma_k^2} \left[I_{\text{exp}}^k - I_{\text{cal}}^k \right]^2, \tag{44}$$

in which σ_k^2 denotes the variance of the kth experimental data point, I_{exp} and I_{cal} refer to experimental and calculated diffracted beam intensities respectively. The variance of the experimental data σ_k may be estimated using experimentally measured values of detector quantum efficiency for different beam intensities i.e. I_{exp} see for example [23]. In a simple first order perturbation theory, the x^2 function depends on the structural parameters, i.e. ΔU_g, quadratically. Within the validity of the full tensor theory, the dependence is quadrennial. In both cases a unique minimum exists in the

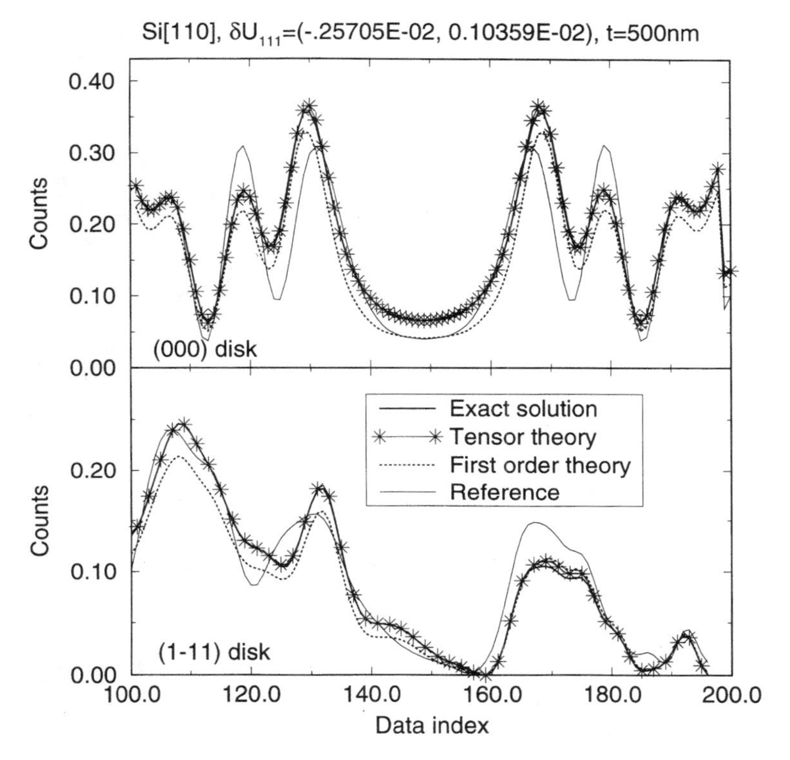

Figure 5. Calculated CBED rocking curves. This figure is essentially the same as Figure 4, except that all calculations were made for a crystal thickness of 500 nm.

x^2 surface, since the third order terms will affect only the degree of asymmetry of the x^2 surface around its minimum and the fourth order term affect only the degree of peakedness or flatness of the x^2 surface rather than introduce additional minimums. Shown in Figure 7 is a plot of the x^2 function as a function of δU_{111}, i.e. variation in the $\{111\}$ structure factors. The three curves are obtained using the full dynamical theory (circles) and making expansions using the quasi-tensor expressions around $\delta U_{111} = -0.0025$ and $-0.005\,\text{Å}^{-1}$, respectively. The neutral atom approximation gives $U_{111} = 0.050136\,\text{Å}^{2}$. A variation of $\delta U_{111} = -0.0025$ and $-0.005\,\text{Å}^{-1}$ therefore represent 5% and 10% variations in U_{111}. It should be noted that both values are larger than that caused by the bonding effect in a single crystal of silicon. Figure 7 shows that for any variation in U_{111} of less than about 5%, the tensor theory will always be able to return a very accurate value of U_{111} giving rise to the unique minimum in the x^2 surface. For a variation of up to 10% in U_{111}, although the tensor theory cannot return accurate solution, the returned solution is nevertheless much closer to the true solution, and the solution may be taken to be the new starting point

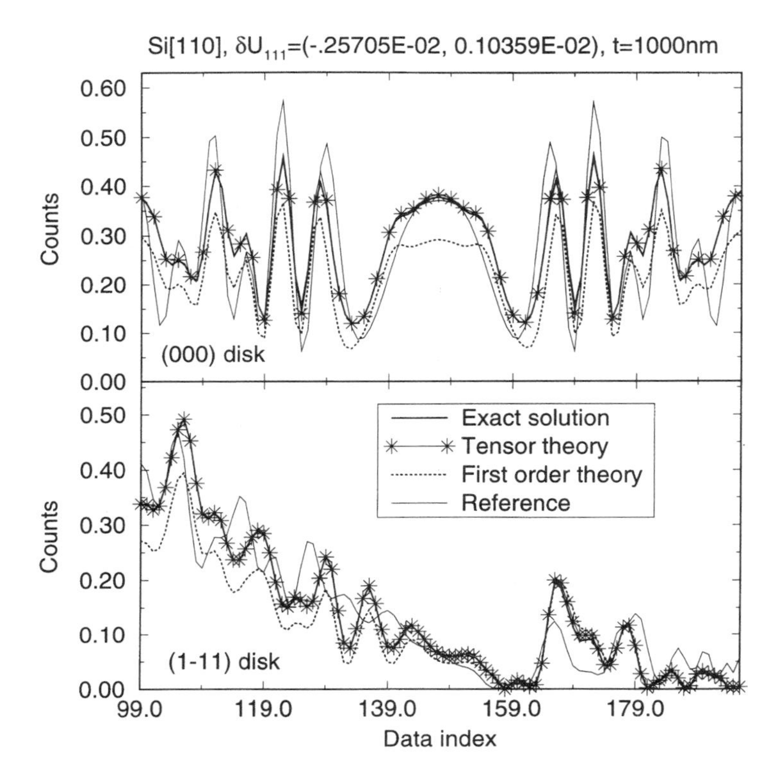

Figure 6. Calculated CBED rocking curves. This figure is essentially the same as Figure 4, except that all calculations were made for a crystal thickness of 1000 nm.

and a more accurate solution can be obtained by repeated applications of the tensor theory.

Shown in Figure 8 is essentially the same plot as Figure 7, but one of the expansions is made using the first order theory. This figure shows that although the first order theory is not adequate for an accurate description of the x^2 surface even for a 5% variation around the true solution, the first order theory nevertheless results in a solution which is closer to the true solution compared with its starting point and that the method of iteration may be applied to improve the accuracy of the solution.

Shown in Figure 9 are experimental CBED rocking curves extracted from Figure 1 along (a) $[1\bar{1}1]$ and (b) [002] directions, the corresponding fitted rocking curves and the residual between the experimental and fitted rocking curves. The fitting was made using the quasi-second order tensor theory, using a primary beam energy of 195.35 keV and a crystal thickness of 369 nm. It was found that for the present study no iteration is required, and the direct application of the quasi-second order theory returns a minimum x^2 values of 1.4, which is very close to the ideal value of 1.0, suggesting that systematic errors introduced by other factors that have not been considered here have been minimized.

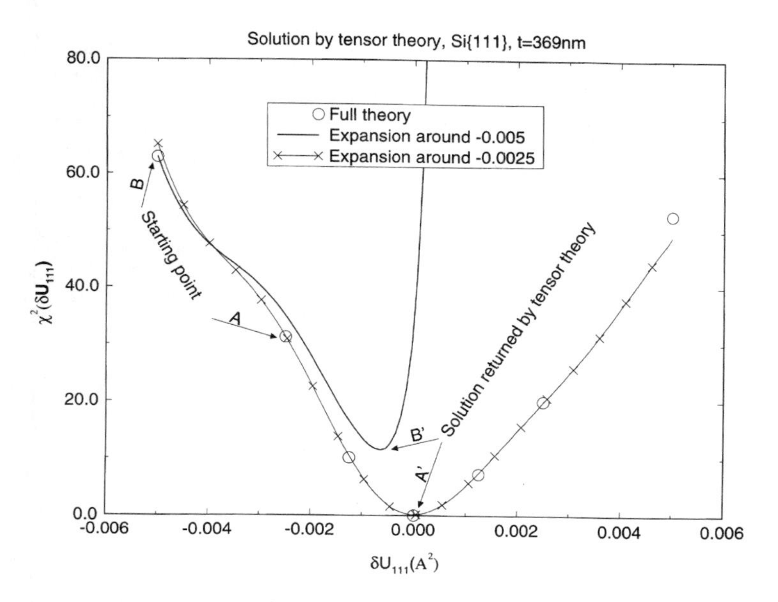

Figure 7. One-dimensional plot of x^2 as a function of δU_{111}. The three curves in the figure are exact plots (circle) calculated using the full dynamical theory and approximate expansions using the full tensor theory around $\delta U_{111} = -0.005\ \text{Å}^{-1}$ (solid line) and around $\delta U_{111} = -0.0025\ \text{Å}^{-1}$ (cross and solid line) respectively.

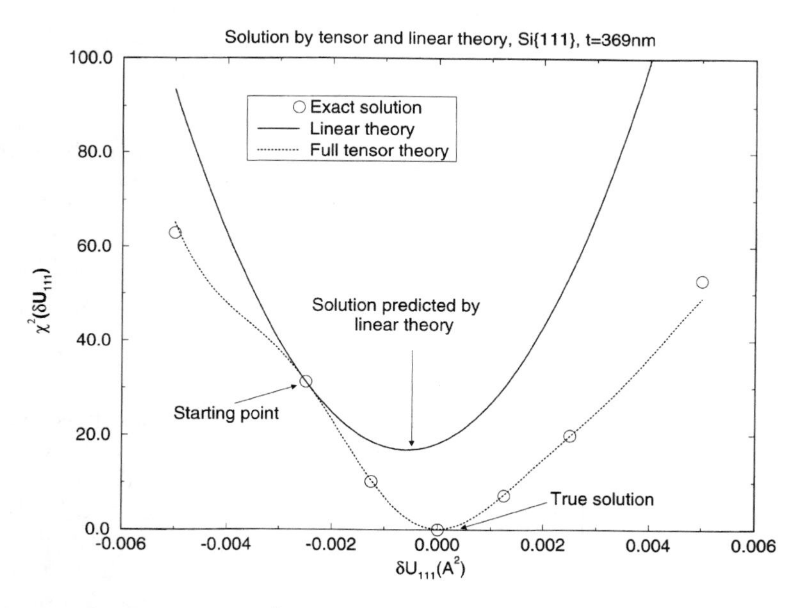

Figure 8. One dimensional plot of x^2 as a function of δU_{111}. The three curves in the figure are exact plots calculated using full dynamical theory (circle) full tensor expansion around $\delta U_{111} = -0.0025\ \text{Å}^{-1}$ (solid line) and line tensor expansion around $\delta U_{111} = -0.0025\ \text{Å}^{-1}$ (cross and solid line).

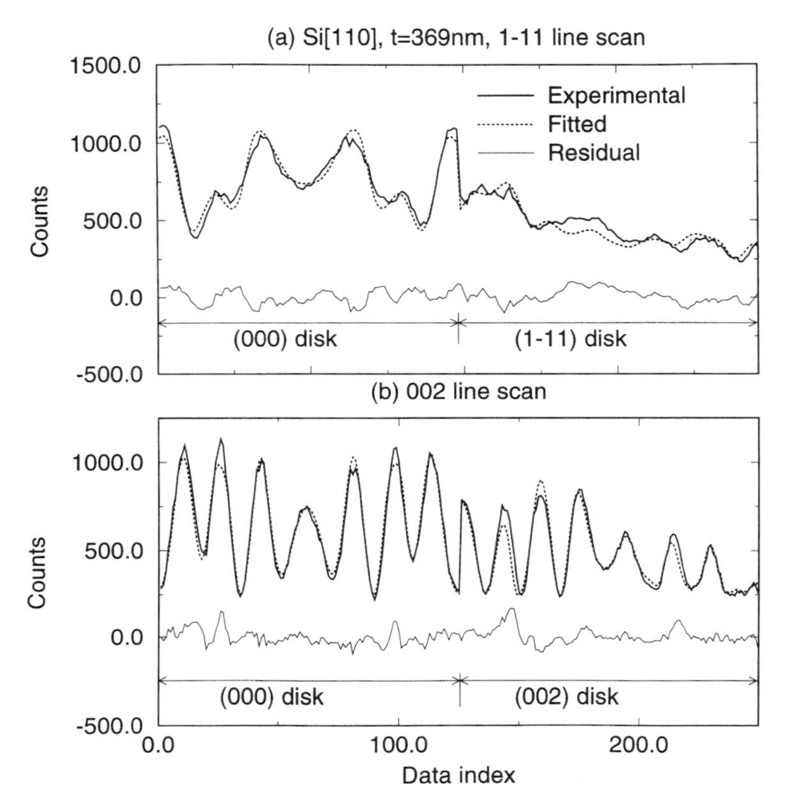

Figure 9. Energy-filtered experimental and fitted Si[110] CBED rocking curves for (a) a line scan along the $[1\bar{1}1]$ direction and (b) a line scan along the [002] direction (see Figure 1). The calculations were made for a primary beam energy of 195.35 keV and a crystal thickness of 369 nm.

6. Conclusion

In summary, in this chapter we have presented methods of different complexity and therefore validity for describing the process of elastic and inelastic scattering of high-energy electrons in a solid. For the general description of the multiple scattering events of both the elastically and inelastically scattered electrons, the kinetic equation should be used. When considering only the elastically scattered electrons, the general kinetic equation reduces to a one-body Schrödinger equation with the interaction potential being regarded as an effective optical potential. For accurate structure factor refinement or the measurement of the bonding effect in a crystal the tensor theory may be used and its accuracy may be improved by the method of iteration.

Acknowledgements

The author wishes to thank Drs. J.M. Zuo, S.L. Dudarev, X.F. Duan and G. Ren for stimulating discussions and collaboration. This work is supported by the National Natural Science Foundation of China (Grant No. 19425006) and by the Chinese Academy of Sciences.

References

1. Pendry, J.B. (1974) *Low Energy Electron Diffraction*, Academic Press, New York.
2. Reimer, L. (1989) *Transmission Electron Microscopy – Physics of Image Formation and Microanalysis*, 2nd edn. Springer-Verlag, Berlin.
3. Cowley, J.M. (Ed.) (1993) *Electron Diffraction Techniques*, Oxford University Press, Oxford.
4. Egerton, R.F. (1986) *Electron Energy Loss Spectroscopy in the Electron Microscope*, Plenum Press, New York.
5. Dudarev, S.L., Peng, L.-M. and Whelan, M.J. (1993) Correlations in space and time and dynamical diffraction of high energy electrons by crystals, *Phys. Rev. B*, **48**(18), 13408–13429.
6. Dudarev, S.L., Peng, L.-M. and Whelan, M.J. (1992). On the damping of coherence in the small-angle inelastic scattering of high-energy electrons by crystals, *Phys. Letts. A*, **170**, 111–115.
7. Peng, L.-M., Dudarev, S.L. and Whelan, M.J. (1993). Evidence for the damping of coherence in inelastic scattering of high-energy electrons by crystals, *Phys. Letts.*, **A175**, 461–464.
8. Van Hove, L. (1954) Correlations in space and time and Born approximation scattering in systems of interacting particles, *Phys. Rev.*, **95**(1), 249–262,.
9. Kohl, H. and Rose, H. (1985) Theory of image formation by inelastically scattered electrons in the electron microscope. In *Adv. Electronics and Electron Physics*, Hawkes, P.W. (Ed.), Vol. 65, Academic Press, New York, pp. 173–227.
10. Hodgson, P.E. (1963) *The Optical Model of Elastic Scattering*, Oxford University Press, New York.
11. Yoshioka. H (1957) Effect of inelastic waves on electron diffraction, *J. Phys. Japan*, **12**(6), 618–628.
12. Rez, P. (1976) *The Theory of Inelastic Scattering in the Electron Microscopy of Crystals*, PhD thesis, St. Catherine's College, Oxford, Oxford.
13. Dederichs, P.H. (1972) Dynamical diffraction theory by optical potential methods, *Solid State Phys.*, **27**, 125.
14. Smart, D.J. and Humphreys, C.J. (1978) The crystal potential in electron diffraction and in band theory, *Inst. Phys. Confs. Ser.*, **41**, 145–149.
15. Fox, A.G., and Fisher, R.M. (1986) Accurate structure factor determination and electron charge distributions of binary cubic solid solutions, *Phil. Mag. A*, **53**, 815–832.
16. Zuo, J.M., Spence, J.C.H., and O'Keeffe, M. (1988) Bonding in GaAs, *Phys. Rev. Letts.*, **61**(3), 353–356.
17. Spence, J.C.H. (1993) On the accurate measurement of structure factor amplitude and phases by electron diffraction, *Acta. Cryst. A*, **49**, 231–260.
18. Zuo, J.M., O'Keeffe, M., Rez, P. and Spence, J.C.H. (1997) Charge density of MgO: implications of precise new measurements for theory, *Phys. Rev. Letts.*, **78**(25), 4777–4780.
19. Høier, R., Bakken, L.N., Marthinsen, K. and Holmestad, R. (1993) Structure factor determination in non-centrosymmetric crystals by a two-dimensional CBED-based parameter refinement method, *Ultramicroscopy*, **49**, 159–170.
20. Peng, L.-M. and Zuo, J.M. (1995) Direct retrieval of crystal structure factors in THEED, *Ultramicroscopy*, **57**, 1–9.
21. Saunders, M., Bird, D.M., Zaluzec, N.J., Burgess, W.G., Preston, A.R. and Humphreys, C.J. (1995) Measurement of low-order structure factors for silicon from zone-axis CBED patterns, *Ultramicroscopy*, **60**, 311–323.
22. Peng, L.-M. (1997) Direct retrieval of crystal and surface structure using high energy electrons, *MICRON*, **28**(2), 159–173.
23. Ren, G., Zuo, J.M. and Peng, L.-M. (1997) Accurate measurements of crystal structure factors using a FEG electron microscope, *MICRON*, **28**.

24. Whelan, M.J. (1965) Inelastic scattering of fast electrons by crystals – I. Interband excitations, *Applied Phys.*, **36**, 2099–2103.
25. Whelan, M.J. (1965) Inelastic scattering of fast electrons by crystals – II. Phonon scattering, *Applied Phys.*, **36**, 2103–2110.
26. Hall, C.R. and Hirsch, P.B. (1965) Effect of thermal diffuse scattering on propagation of high energy electron through crystals, *Proc. R. Soc. London A*, **286**, 158–177.
27. Rossouw, C.J. and Bursill, L.A. (1985) Interpretation of dynamical diffuse scattering of fast electrons in rutile, *Acta Cryst. A*, **41**, 320–327.
28. Bird, D.M. and King, Q.A. (1990) Absorptive form factor for high-energy electron diffraction, *Acta Cryst. A*, **46**, 202–208.
29. Dudarev, S.L., Peng, L.-M. and Whelan, M.J. (1995) On the Doyle-Turner representation of the optical potential for RHEED calculations, *Surface Sci.*, **330**, 86–100.
30. Peng, L.-M., Ren, C., Dudarev, S.L. and Whelan, M.J. (1996) Robust parameterization of elastic and absorptive electron atomic scattering factors, *Acta Cryst. A*, **52**, 257–276.
31. Peng, L.-M., Ren, C., Dudarev, S.L. and Whelan, M.J. (1996) Debye–Waller factors and absorptive scattering factors of elemental crystals, *Acta Cryst. A*, **52**, 456–470.
32. Peng, L.-M., and Dudarev, S.L. (1993) Tensor theories of high energy electron diffraction and their use in surface crystallography, *Surface Sci.*, **298**, 316–330.
33. Peng, L.-M. and Dudarev, S.L. (1993) Direct determination of crystal and surface structures in THEED, *Ultramicroscopy*, **52**, 312–317.
34. Peng, L.-M. (1995) New developments of electron diffraction theory, In *Advances in Imaging and Electron Physics*, Hawkes, P.W. (Ed.), Vol. 90, Academic Press, London, pp. 205–351.
35. Dudarev, S.L., Peng, L.-M. and Whelan, M.J. (1992) A treatment of RHEED from a rough surface of a crystal by an optical potential method, *Surface Sci.*, **279**, 380– 394.
36. Peng, L.-M. (1997) Anisotropic thermal vibrations and dynamical electron diffraction by crystals, *Acta Cryst. A*, **53**, 663–672.
37. Spence, J.C.H. and Zuo, J.M. (1992) *Electron Microdiffraction*, Plenum Press, New York and London.
38. Zuo, J.M. and Spence, J.C.H. (1991) Automated structure factor refinement from convergent-beam patterns, *Ultramicroscopy*, **35**, 185–196.
39. Ashcroft, N.W. and Mermin, N.D. (1976) *Solid State Physics*, Saunders College, Philadephia.
40. Cowley, J.M. (1992) Scattering factors for the diffraction of electrons by crystalline solids, In *International Tables for Crystallography*, Wilson, A.J.C. (Ed.), Volume C. Kluwer Academic Publishers, Dordrecht/Boston/London.
41. Willis, B.T.M. and Pryor, A.W. (1975) *Thermal Vibrations in Crystallography*, Cambridge University Press, Cambridge.
42. Giacovazzo, C. (1992) *Fundamentals of crystallography*, Oxford University Press, Oxford.
43. Cowley, J.M. (1928) *Diffraction Physics*, 2nd edn., North-Holland, Amsterdam. (1990)
44. Bethe, H. (1928) Theorie der Beugung von Elektronen an Kristallen, *Ann. Physik.*, **87**, 55– 129.
45. Gjønnes, J. (1962) The dynamical potentials in electron diffraction, *Acta Cryst.*, **15**, 703– 707.
46. Hirsch, P.B., Howie, A., Nicholson, R.N., Pashley, D.W. and Whelan, M.J. (1977) *Electron Microscopy of Thin Crystals*, Krieger Publishing Company, Malabar and Florida.
47. Bird, D.M. (1989) Theory of zone axis electron diffraction, *J. Electron Microsc. Tech.*, **13**, 77.
48. Vincent, R., Bird, D.M. and Steeds, J.W. (1984) Structure of augeas determined by convergent-beam electron diffraction – II. Refinement of structural parameters, *Phil. Mag. A*, **50**, 765–786.
49. Bird, D.M. and Saunders, M. (1992) Inversion of convergent-beam electron diffraction patterns, *Acta Cryst. A*, **48**, 555–562.
50. Speer, S., Spence, J.C.H. and Ihrig, E. (1990) On differentiation of the scattering matrix in dynamical transmission electron diffraction, *Acta Cryst. A*, **46**, 763–772.
51. Lewis, A.L., Villagrana, R.F. and Metherall, A.J.F. (1978) A description of electron diffraction from higher order Laue zones, *Acta Cryst. A*, **34**, 138–140.
52. Rous, P.J. (1992) The tensor approximation and surface crystallography by low-energy electron diffraction, *Prog. Surf. Sci.*, **39**, 3–63.
53. *The NAG Fortran Library Manual, Mark 16*. The numerical algorithms group limited, Oxford, 1993.
54. Spence, J.C.H. (1998) *Acta. Cryst. A*.

10

‘Compton microscope effect’?: image of intra-unit-cell atom theoretically observed in compton $B(\mathbf{r})$-function

TEIJI KOBAYASI

College of Medical Sciences, Tohoku University, 2-1 Seiryo-machi, Aoba-ku, Sendai 980-8575, Japan

1. Introduction

In the field of Compton scattering the real space function it B($\mathbf{r}$) for the electron system is defined by the Fourier inversion of the distribution function of electron momentum density (EMD) $\rho(\mathbf{q})$ [1–8]:

$$\rho(\mathbf{q}) = \int B(\mathbf{r}) \exp(-\mathrm{i}\mathbf{q} \cdot \mathbf{r})\, \mathrm{d}^3 r, \tag{1}$$

$$B(\mathbf{r}) = \sum_{\mathbf{q}} \rho(\mathbf{q}) \exp(\mathrm{i}\mathbf{q} \cdot \mathbf{r})/\Omega, \tag{2}$$

where Ω is the crystal volume. In the independent electron model, the EMD function for the spin-degenerating material is given by

$$\rho(\mathbf{q}) = 2 \sum_{n} \sum_{\mathbf{k}} \left| \int \Psi_{n\mathbf{k}}(\mathbf{r}) \exp(-\mathrm{i}\mathbf{q} \cdot \mathbf{r})\, \mathrm{d}^3 r/\sqrt{\Omega} \right|^2, \tag{3}$$

where $\Psi_{n\mathbf{k}}$ is the wave function of an electron with wave vector $\mathbf{k}$ in the nth occupied band.

The $B(\mathbf{r})$-function was originally introduced as a mathematical intermediate in order to attain high accuracy in calculating EMD or Compton profile $J(q_z)$, which is represented under the impulse approximation as [7, 9]

$$J(q_z) = (1/2\pi)^2 \int_{-\infty}^{\infty} \mathrm{d}q_x \int_{-\infty}^{\infty} \mathrm{d}q_y\, \rho(q_x, q_y, q_z). \tag{4}$$

As was pointed out previously [6, 8], in pseudo-potential (PP) approach to these quantities for valence electron systems of semiconducting materials Si and Ge, it is far more favorable to adopt the indirect derivation of EMD via $B(\mathbf{r})$ based on Equation (1), not on Equation (3), both in treatment and in numerical accuracy.

In the course of the PP calculations of these quantities for Si [10] and Ge [11], a characteristic local pattern which reflects position, shape and size of a specific atom in the crystal is observed on the contour map of the valence electron $B(\mathbf{r})$-function. The atom is one of the two atoms in the unit cell of diamond structure. It seems as

Paul G. Mezey and Beverly E. Robertson (eds.), Electron, Spin and Momentum Densities and Chemical Reactivity, 169–178

if the $B(\mathbf{r})$-function works as a *microscope* to detect the structural information of the intra-unit-cell atom in the crystal. The appearance of the atomic pattern has its origin in the core-orthogonalization (CO) taken explicitly into our PP theory.

In the PP theory, the valence electron wave function is composed of two parts. The main part is the pseudo-wave function describing a relatively smooth-varying behavior of the electron. The second part describes a spatially rapid oscillation of the valence electron near the atomic core. This atomic-electron-like behavior is due to the fact that, passing the vicinity of an atom, the valence electron recalls its native outermost atomic orbitals under a relatively stronger atomic potential near the core. Quantum mechanically the situation corresponds to the fact that the valence electronic state should be orthogonal to the inner-core electronic states. The second part describes this CO. The CO terms explicitly contain the information of atomic position and atomic core orbitals.

The purposes of this paper are to discuss the CO effect on the $B(\mathbf{r})$-function and to show that the appearance of the atom-like image can be explained by using the fact that the $B(\mathbf{r})$-function can be described in terms of the autocorrelation function among the electron wave functions over the occupied electronic states. Autocorrelative overlap between the CO terms explicitly containing the atomic information has a possibility to enhance a specific atom on the $B(\mathbf{r})$-function map. The overlap explains why images of the other distant atoms are not pronounced.

2. Method of calculation

In the PP framework, the valence electron wave function $\Psi_{n\mathbf{k}}$ orthogonalized to the inner core electron wave function $\Psi_{c\mathbf{k}}$'s is given by [12]

$$\Psi_{n\mathbf{k}}(\mathbf{r}) = N_{n\mathbf{k}}\left[\Phi_{n\mathbf{k}}(\mathbf{r}) - \sum_{c}\langle\Psi_{c\mathbf{k}} \mid \Phi_{n\mathbf{k}}\rangle\Psi_{c\mathbf{k}}(\mathbf{r})\right], \tag{5}$$

where $\Phi_{n\mathbf{k}}$ is the pseudo-part of the valence electron wave function, $\Psi_{c\mathbf{k}}$ the wave function of the cth core electronic state and $N_{n\mathbf{k}}$ the normalization constant. We assume that $\Psi_{c\mathbf{k}}$ can be well described by the Bloch sum of the ionic core orbitals ϕ_c under the tight-binding-limit approximation as

$$\Psi_{c\mathbf{k}}(\mathbf{r}) = \sum_{m}\sum_{j=1}^{s}\phi_c(\mathbf{r} - \mathbf{R}_m - \mathbf{t}_j)\exp[\mathrm{i}\mathbf{k}\cdot(\mathbf{R}_m + \mathbf{t}_j)]/\sqrt{sN}, \tag{6}$$

where $\mathbf{R}_m$ is the primitive translational vector pointing the mth unit cell in the crystal, $\mathbf{t}_j$ the non-primitive one within the unit cell with s atoms ($s = 2$ for Si and Ge), and N the total number of unit cells. We introduce plane wave expansions of $\Phi_{n\mathbf{k}}$ and $\Psi_{c\mathbf{k}}$:

$$\Phi_{n\mathbf{k}}(\mathbf{r}) = \sum_{\mathbf{G}} C_{n\mathbf{k}}^{\text{pseudo}}(\mathbf{G})\exp[\mathrm{i}(\mathbf{k} + \mathbf{G})\cdot\mathbf{r}]/\sqrt{\Omega}, \tag{7}$$

$$\Psi_{c\mathbf{k}}(\mathbf{r}) = \sum_{\mathbf{G}} b_{c\mathbf{k}}(\mathbf{G})\exp[\mathrm{i}(\mathbf{k} + \mathbf{G})\cdot\mathbf{r}]/\sqrt{\Omega}, \tag{8}$$

where $\mathbf{G}$ is the reciprocal lattice vector, and

$$b_{c\mathbf{k}}(\mathbf{G}) = S(\mathbf{G})\phi_c(\mathbf{k}+\mathbf{G}), \tag{9}$$

$$S(\mathbf{G}) = 1/s\sum_{j=1}^{s}\exp(-\mathrm{i}\mathbf{G}\cdot\mathbf{t}_j), \tag{10}$$

$$\phi_c(\mathbf{q}) = \sqrt{s/\Omega_0}\int\phi_c(\mathbf{r})\exp(-\mathrm{i}\mathbf{q}\cdot\mathbf{r})\mathrm{d}^3r, \tag{11}$$

where Ω_0 is the unit cell volume. Substitution of Equations (7) and (8) into Equation (5) yields

$$\Psi_{n\mathbf{k}}(\mathbf{r}) = \sum_{\mathbf{G}} C_{n\mathbf{k}}(\mathbf{G})\exp[\mathrm{i}(\mathbf{k}+\mathbf{G})\cdot\mathbf{r}]/\sqrt{\Omega}, \tag{12}$$

$$C_{n\mathbf{k}}(\mathbf{G}) = N_{n\mathbf{k}}[C_{n\mathbf{k}}^{\text{pseudo}}(\mathbf{G}) - \sum_{c}\sum_{\mathbf{G}'} b_{c\mathbf{k}}^{*}(\mathbf{G}')b_{c\mathbf{k}}(\mathbf{G})C_{n\mathbf{k}}^{\text{pseudo}}(\mathbf{G}')]. \tag{13}$$

Substituting Equation (12) into Equation (3) and its result into Equation (2), we will obtain the key expression for $B(\mathbf{r})$-function as follows:

$$B(\mathbf{r}) = 2\sum_{n}\sum_{\mathbf{k}}\sum_{\mathbf{G}}|C_{n\mathbf{k}}(\mathbf{G})|^2\exp[\mathrm{i}(\mathbf{k}+\mathbf{G})\cdot\mathbf{r}]/\Omega. \tag{14}$$

Contribution of the CO terms is defined by

$$\Delta B(\mathbf{r}) = B(\mathbf{r}) - B^{\text{pseudo}}(\mathbf{r}), \tag{15}$$

where $B^{\text{pseudo}}(\mathbf{r})$ is the $B(\mathbf{r})$-function of the pseudo-valence electron system described by the pseudo-wave function with no CO terms.

In order to visualize $B(\mathbf{r})$ on a contour map, let us expand it in terms of the cubic harmonics as follows [5, 10, 11, 13, 14]:

$$B(\mathbf{r}) = \sum_{l}\sum_{i} B_{li}(r)K_l^i(\Omega_{\mathbf{r}}), \tag{16}$$

where K_l^i are the l(angular momentum)th order cubic harmonics and i distinguishes the different independent harmonics with the same l. For the diamond structure, we need the harmonics belonging to the Γ_1-representation of the O_h group symmetry. The term of $l = 0$ in Equation (16) describes the spherically symmetric behavior in $B(\mathbf{r})$ and the terms with non-zero l describe its anisotropy. The expansion coefficient functions $B_{li}(r)$ are given by [11]

$$\begin{aligned} B_{li}(r) &= \int B(\mathbf{r})K_l^{i*}(\Omega_{\mathbf{r}})\,\mathrm{d}\Omega_{\mathbf{r}} \\ &= (4\pi/\Omega)i^l 2\sum_{n}\sum_{\mathbf{k}}\sum_{\mathbf{G}}|C_{n\mathbf{k}}(\mathbf{G})|^2 j_l(|\mathbf{k}+\mathbf{G}|r)K_l^{i*}(\Omega_{\mathbf{k}+\mathbf{G}}), \end{aligned} \tag{17}$$

where j_l is the lth order spherical Bessel function. The EMD can be constructed by using the corresponding expansion coefficients given by

$$\rho_{li}(q) = 4\pi(-i)^l \int_0^\infty B_{li}(r) j_l(qr) r^2 \, \mathrm{d}r. \tag{18}$$

3. Numerical calculations

The wave function coefficients $C_{n\mathbf{k}}^{\mathrm{pseudo}}(\mathbf{G})$'s have been solved under the 3L + NL(d) type non-local pps of Heine–Abarenkov form [15]. The potential parameters used for Si are $V_\mathrm{L}(3) = -0.20532$, $V_\mathrm{L}(8) = 0.03548$, $V_\mathrm{L}(11) = 0.07239$, $A_2(\mathrm{Si}^{4+}) = -2.0773$ in units of Ry and $\mathrm{R}_2(\mathrm{Si}^{4+}) = 1.25$ a.u. and, for Ge, $V_\mathrm{L}(3) = -0.24220$, $V_\mathrm{L}(8) = 0.02548$, $V_\mathrm{L}(11) = 0.05264$, $A_2(\mathrm{Ge}^{4+}) = 167.53$ in units of Ry and $\mathrm{R}_2(\mathrm{Ge}^{4+}) = 0.98$ a.u. The Chadi–Cohen's 10-special-point scheme is adopted for making **k**-meshes [16]. It contains 256 **k**-points in the first Brillouin zone. All plane waves with the reciprocal lattice vector **G** satisfying $|\mathbf{k}+\mathbf{G}|^2 - \mathbf{k} \leqq 20(2\pi/a)^2$ are taken into the expansion of $\Phi_{n\mathbf{k}}$ where a is the lattice constant, which are 10.26327 a.u. (Si) and 10.6772 a.u. (Ge). The corresponding vector set includes 137 reciprocal lattice vectors. The core electronic states for Si are originated from the 1s, 2s, 2p ionic states and, for Ge, from the 1s, 2s, 2p, 3s, 3p and 3d states. The Roothaan–Hartree–Fock wave functions are used for the ionic core electron orbitals [17]. Because the core orbitals are highly localized, a set of reciprocal lattice vectors for their Fourier components was forced to include all of the 4621 reciprocal lattice vectors up to the very large shell of the (13,7,7) $(2\pi/a)$ group.

In the cubic harmonics expansion of $B(\mathbf{r})$, a full convergence has been attained by inclusion of $l \leq 22$ in which the first 16 harmonics belonging to the Γ_1-representation are contained [$l = 0, 4, 6, 8, 10, 12$ $(i = 1, 2)$, 14, 16 $(i = 1, 2)$, 18 $(i = 1, 2)$, 20 $(i = 1, 2)$ and 22 $(i = 1, 2)$].

Contour map calculations in the three-dimensional zone of **r** are concentrated on the $(1\bar{1}0)$ plane containing the five fundamental directions of [001], [112], [111], [221] and [110] shown in Figure 1. The cube shown in Figure 1 has 4 times the volume of the unit cell containing two atoms. The intra-unit-cell atoms are, for example, atoms A and B in Figure 1, with the bond length of $(\sqrt{3}/4)a$.

4. Results

Figure 2 represents the contour behavior of $B(\mathbf{r})$ of Si and its variation along the [111] direction of bond. In Figure 2(a) the whole $B(\mathbf{r})$ including the spherically symmetric component is shown and in Figure 2(b) the anisotropic part. The distant parameter r is in units of a. The contour spacings are 0.1 in (a) and 0.01 in (b) in units of $2/\Omega_0$, respectively. Figure 2(c) shows the variation of $B(\mathbf{r})$ along the [111] direction and Figure 2(d) of the anisotropic part. Figure 3 represents the corresponding results for Ge.

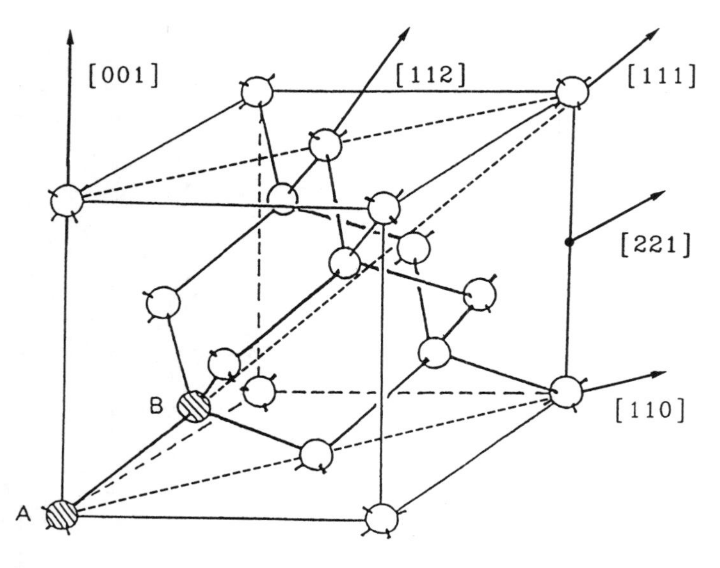

Figure 1. The atomic configuration of the diamond structure and the five directions in the $(1\bar{1}0)$ plane.

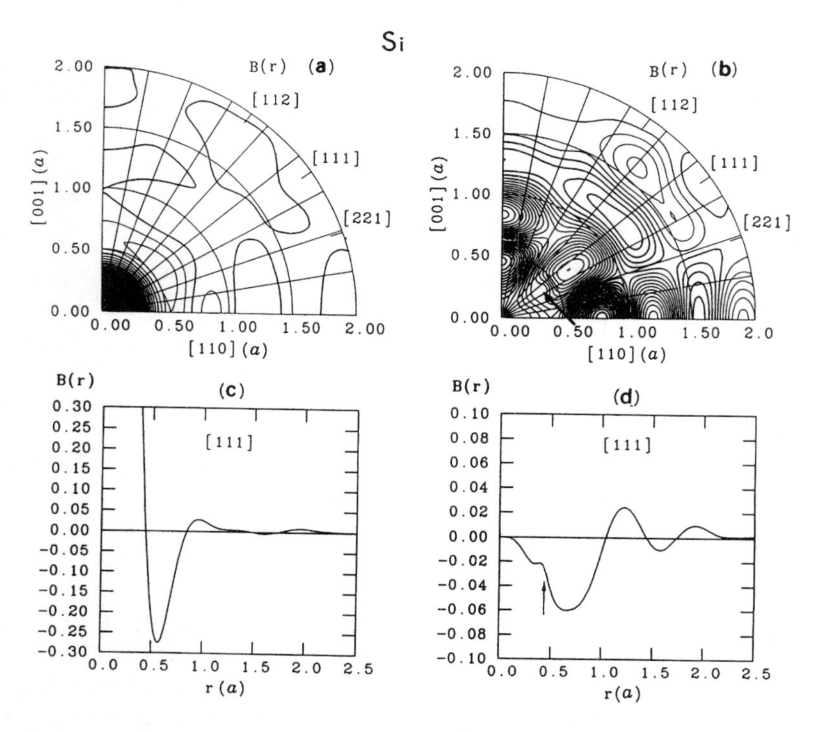

Figure 2. (a) and (b): Contour map of $B(\mathbf{r})$ of Si on the $(1\bar{1}0)$ plane. (a) Total $B(\mathbf{r})$ and (b) its anisotropic part. (c) and (d): Variation along the [111] direction. (c) Total $B(\mathbf{r})$ and (d) its anisotropic part. Arrow indicates a local pattern around the point $(1, 1, 1)a/4$.

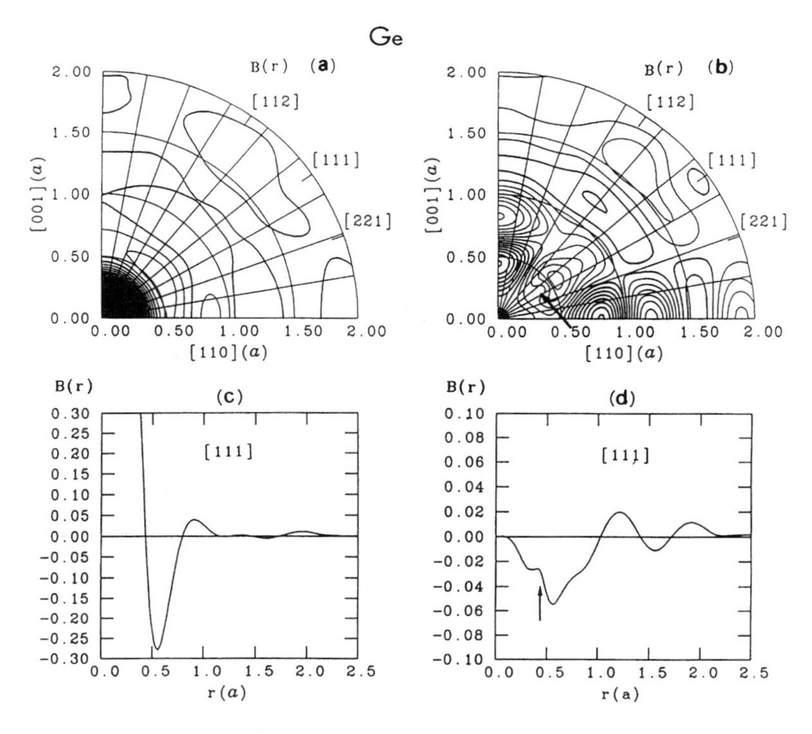

Figure 3. (a) and (b): Contour map of $B(\mathbf{r})$ of Ge on the $(1\bar{1}0)$ plane. (a) Total $B(\mathbf{r})$ and (b) its anisotropic part. (c) and (d): Variation along the [111] direction. (c) Total $B(\mathbf{r})$ and (d) its anisotropic part. Arrow indicates a local pattern around the point $(1, 1, 1)a/4$.

The CO contribution $\Delta B(\mathbf{r})$ of Si is shown in Figure 4. In Figure 4(a) the contour map of the whole $\Delta B(\mathbf{r})$ is drawn and, in Figure 4(b) the anisotropic part. The contour spacing is 0.005. The variation along the [111] direction of the whole $B(\mathbf{r})$ is shown in Figure 4(c) and the anisotropic part is in Figure 4(d). Figure 5 represents the corresponding results for Ge.

Characteristic local pattern is indicated by an arrow in Figures 2–5. The pattern is discussed in the next section.

5. Discussions

As can be seen from Figure 2(a) for Si and Figure 3(a) for Ge, the $B(\mathbf{r})$-function has a large spherically symmetric part around $\mathbf{r} = (0, 0, 0)a$ and it sharply damps outward. Anisotropic behavior of $B(\mathbf{r})$ is well observed in Figures 2(b) and 3(b). We notice a local pattern of contour lines around the point at $\mathbf{r} = (1, 1, 1)a/4$ in the [111] direction. The pattern is arrowed in Figure 2(b), (d) and Figure 3(b), (d). If we put the atom A on $\mathbf{r} = (0, 0, 0)a$, the atom B is on $\mathbf{r} = (1, 1, 1)a/4$ by the bond length $(\sqrt{3}/4)a$ $(= 0.4330a)$ apart. The local pattern is enhanced on the contour maps of

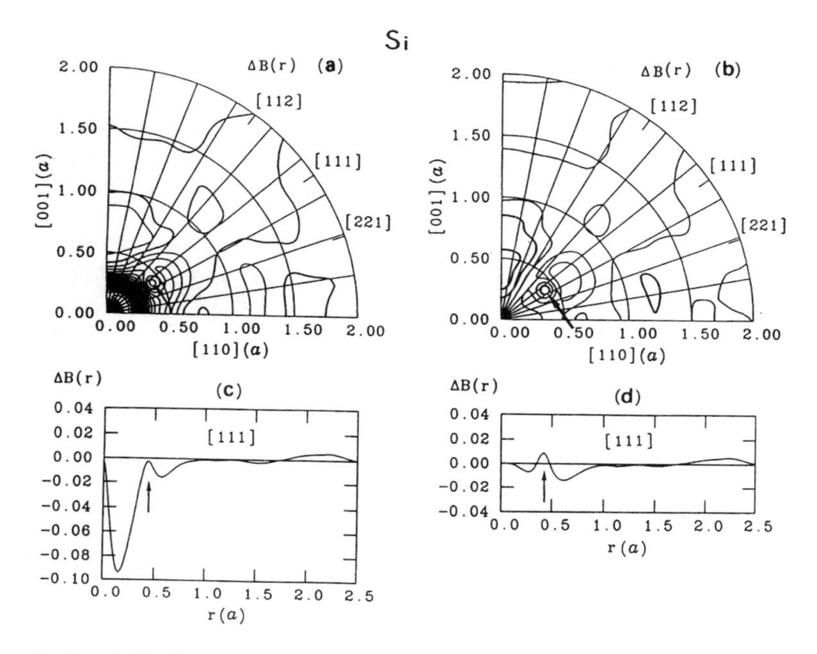

Figure 4. (a) and (b): Contour map of the CO contribution $\Delta B(\mathbf{r})$ of Si. (a) Total $\Delta B(\mathbf{r})$ and (b) its anisotropic part. (c) and (d): Variation along the [111] direction. (c) Total $\Delta B(\mathbf{r})$ and (d) its anisotropic part. The local pattern in Figure 2 becomes clear as a contour-circle centered on the point $(1, 1, 1)a/4$.

the CO contribution $\Delta B(\mathbf{r})$. As the arrow indicates in Figure 4(a), (b) and Figure 5(a), (b), the pattern appears clearly as a small circle centered on the point $(1, 1, 1)a/4$. In Figures 4(d) and 5(d) it shows a sharp peak at $r = 0.433a$. The radius of the circle is roughly equal to $0.1a$, which is nearly equal to the physical core radius R_c. Examples of R_c are $0.41\text{Å} = 0.076a$ for Si^{4+} and $0.53\text{Å} = 0.094a$ for Ge^{4+} [18]. From these observations it can be concluded that we observe a kind of image of one of the intra-unit-cell atoms as the weak but characteristic local pattern in $\Delta B(\mathbf{r})$ or in $B(\mathbf{r})$. The position, size and shape of the atom are well reproduced quantitatively in $\Delta B(\mathbf{r})$. It seems as if the Compton scattering had a microscope effect for detecting a local structure through a process of $J(q_z)$'s $\rightarrow \rho(\mathbf{q}) \rightarrow B(\mathbf{r}),\ \Delta B(\mathbf{r})$.

The microscope effect can be explained by using the fact that the $B(\mathbf{r})$-function is equivalently described in terms of the autocorrelation function of the valence electron wave functions as follows [7]:

$$B(\mathbf{r}) = 2 \sum_{n} \sum_{\mathbf{k}} \int \Psi^{*}_{n\mathbf{k}}(\mathbf{r}')\Psi_{n\mathbf{k}}(\mathbf{r}' + \mathbf{r})\, \mathrm{d}^3 r'/\Omega. \tag{19}$$

For the sake of simplicity, we consider an example of a one-dimensional periodic system of length L with N atoms with one core electronic state per atom. The inter-atom space is a. The pseudo-valence electron is assumed to be in a single plane wave

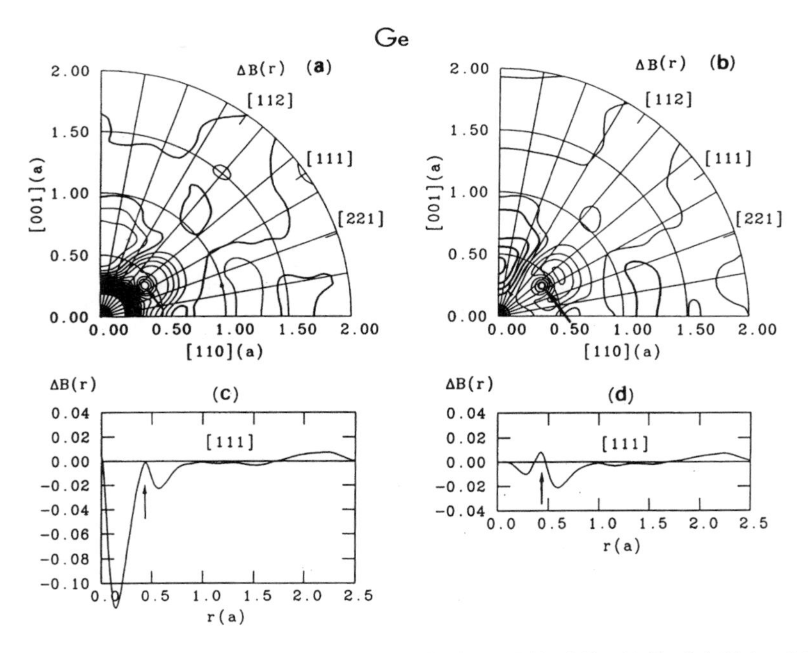

Figure 5. (a) and (b): Contour map of the CO contribution $\Delta B(\mathbf{r})$ of Ge. (a) Total $\Delta B(\mathbf{r})$ and (b) its anisotropic part. (c) and (d): Variation along the [111] direction. (c) Total $\Delta B(\mathbf{r})$ and (d) its anisotropic part. The local pattern in Figure 3 becomes clear as a contour-circle centered on the point $(1, 1, 1)a/4$.

state. The orthogonalized wave function is

$$\Psi_k(x) = N_k\left[\exp(\mathrm{i}kx)/\sqrt{L} - \phi^*(k)\sum_m \exp(\mathrm{i}kR_m)\phi(x - R_m)/\sqrt{N}\right], \quad (20)$$

where ϕ is a core orbital function satisfying

$$\int \phi^*(x - R_m)\phi(x - R_n)\,\mathrm{d}x = \delta_{R_m,R_n}. \quad (21)$$

Here, $\mathbf{R}_m = ma$ is the mth atom position, and

$$\phi(k) = \int \phi(x)\exp(-\mathrm{i}kx)\,\mathrm{d}x/\sqrt{a}, \quad (22)$$

$$N_k = 1/[1 - |\phi(k)|^2]^{1/2} \quad \text{(normalization constant)}. \quad (23)$$

The x-integration is taken over the length L.

Substituting Equation (20) into the $B(x)$ of the one-dimensional system, we obtain

$$
\begin{aligned}
B(x) &= 2\sum_k \int \Psi_k^*(x')\Psi_k(x'+x)\,\mathrm{d}x'/L \\
&= 2\sum_k N_k^2 \exp(\mathrm{i}kx)[1-2|\phi(k)|^2]/L \\
&\quad + 2\sum_k N_k^2|\phi(k)|^2 \sum_n \exp(\mathrm{i}kR_n)\delta_{x,R_n}/L.
\end{aligned} \tag{24}
$$

At $x = 0$, $B(0)$ is equal to the uniform density of electrons. The first term of the right hand side makes a bulk peak around $x = 0$. It sharply damps outside, because the k-integration over the occupied states is similar in structure to the following damping oscillation function:

$$
\int_{-G}^{G} \exp(\mathrm{i}kx)\,\mathrm{d}k = 2\sin(Gx)/x. \tag{25}
$$

The second term in the right hand side of Equation (24) comes from the autocorrelation containing the CO terms and is of higher order contribution. At the atom position $x = R_n$, it gives the contribution of $2\sum_k N_k^2|\phi(k)|^2\exp(\mathrm{i}kR_n)/L$ to $B(x)$. Due to the k-integration, it damps rapidly as x increases. The largest value of the second term at $x = 0$ is absorbed into the value of the first term at $x = 0$ to reproduce the uniform density of electrons $[= 2\sum_k^{\text{occup}}/L = B(0)]$. As a result, the CO contribution is marked only on the nearest neighbor atom, reflecting its shape and size through the core orbital ϕ. These facts explain why the atom-like pattern appears and why it is limited on the atom sites close to the origin. If the unit cell contains s atoms, the δ-function in Equation (24) is replaced by $1/s\sum_{i=1}^{s}\sum_{j=1}^{s}\exp[\mathrm{i}k(t_i - t_j)]\delta_{x,R_n+t_i-t_j}$. In this case, therefore, the most pronounced atom-like image can be observed at the position of the shortest distance among the intra-unit-cell atom–atom distances.

Experimentally, the EMD function $\rho(\mathbf{q})$ can be reconstructed from a set of Compton profiles $J(q_z)$'s, and $B(\mathbf{r})$ from the EMD. However, $\Delta B(\mathbf{r})$ is not a direct experimental product. By combining the experimental $B(\mathbf{r})$ with theoretical $B^{\text{pseudo}}(\mathbf{r})$, we need to derive a semiexperimental $\Delta B(\mathbf{r})$. Since the atomic image is very weak, many problems must be cleared in experimental resolution, in reconstruction (for example, selection of a set of directions and range of q_z's), in various deconvolution procedures and so on. First of all, high resolution experiments are desirable.

The effect can be applied, for example, to estimate a bond length or atomic spacing, to observe valence electron spin distribution around a specific atom and to derive information of the nearest neighbor atom distribution in a disordered system such as amorphous, under an expansion of the theory.

References

1. Pattison, P. and Williams, B. (1976) Fermi surface parameters from fourier analysis of Compton profiles, *Solid State Commun.*, **20**, 585–588.

2. Mueller, F.M. (1977) Anisotropic momentum densities from Compton profiles: silicon, *Phys. Rev.*, **B15**, 3039–3044.
3. Shülke, W. (1977) The one-dimensional Fourier transform of Compton profiles, *Phys. Stat. Sol.(b)*, **82**, 229–235.
4. Kramer, B., Krusius, P., Schröder, W. and Schülke, W. (1977) Fourier-transformed Compton profiles: a sensitive probe for the microstructure of semiconductors, *Phys. Rev. Lett.*, **38**, 1227–1230.
5. Mijnarends, P.E. (1977) Reconstruction of three-dimensional distribution, In *Compton Scattering*, Williams, B. (Ed.), McGraw-Hill, New York, pp. 323–345.
6. Nara, H., Shindo, K. and Kobayasi, T. (1979) Pseudopotential approach to anisotropies of Compton-profiles of Si and Ge, *J. Phys. Soc. Jpn.*, **46**, 77–83.
7. Cooper, M.J. (1985) Compton scattering and electron momentum determination, *Rep. Prog. Phys.*, **48**, 415–481.
8. Kobayasi, T. (1994) Fourier inversion formalism for the calculation of angular correlation of positron annihilation radiation of semiconductors, *Bull. Coll. Med. Sci. Tohoku Univ.*, **3**, 11–22.
9. Berggren, K.F., Manninen, S., Paakkari, T. *et al.* (1977) Solids, In *Compton Scattering*, Williams, B. (Ed.), McGraw-Hill, New York, pp. 139–208.
10. Kobayasi, T. (1996) Core-orthogonalization effect on the Compton profiles of valence electrons in Si, *Bull. Coll. Med. Sci. Tohoku Univ.*, **5**, 149–164.
11. Kobayasi, T., Nara, H., Timms, D.N. and Cooper, M.J. (1995) Core-orthogonalization effects on the momentum density distribution and the Compton profile of valence electrons in semiconductors, *Bull. Coll. Med. Sci. Tohoku Univ.*, **4**, 93–104.
12. Callaway, J. (1974) *Quantum Theory of the Solid State (Part A)*, Academic Press, New York.
13. Mueller, F.M. and Priestley, M.G. (1966) Inversion of cubic de Haas–van Alphen data, with an application to palladium, *Phys. Rev.*, **148**, 638–643.
14. Nara, H., Kobayasi, T., Takegahara, K., Cooper, M.J. and Timms, D.N. (1994) Optimal number of directions in reconstructing 3D momentum densities from Compton profiles of semiconductors, *Computational Materials Sci.*, **2**, 366–374.
15. Kobayasi, T. and Nara, H. (1993) Properties of nonlocal pseudopotentials of Si and Ge optimized under full interdependence among potential parameters, *Bull. Coll. Med. Sci. Tohoku Univ.*, **2**, 7–16.
16. Chadi, D.J. and Cohen, M.L. (1973) Special points in the Brillouin zone, *Phys. Rev.*, **B8**, 5747–5753.
17. Clementi, E. (1965) Tables of atomic functions, in Supplement to the paper '*Ab initio* computations in atoms and molecules', *IBM J. Res. Develop.*, **9**, 2–19.
18. Pauling, L. (1960) *The Nature of the Chemical Bond and the Structure of Molecules and Crystals*, 3rd edn, Cornel Univ. Press, Ithaca.

11

New light on electron correlation in simple metals: inelastic X-ray scattering results vs. current theoretical treatment

A. KAPROLAT, K. HÖPPNER, CH. STERNEMANN and W. SCHÜLKE

University of Dortmund, Institute of Physics, Otto Hahn Straße 4, D-44221 Dortmund, Germany

1. Introduction

In this contribution we will deal with electron–electron correlation in solids and how to learn about these by means of inelastic X-ray scattering both in the regime of small and large momentum transfer. We will compare the predictions of simple models (free electron gas, jellium model) and more sophisticated ones (calculations using the self-energy influenced spectral weight function) to experimental results. In a last step, lattice effects will be included in the theoretical treatment.

In Section 2 we will present a short overview of the theoretical approach widely used to describe inelastic X-ray scattering results at low momentum transfer ($q \leq$ 1 a.u.), leading to the dynamical structure factor $S(\mathbf{q}, \omega)$. Section 3 will confront these theoretical descriptions to experimental measurements of $S(\mathbf{q}, \omega)$ for Li and liquid Al. In Section 4 we will extend the theoretical treatment to large momentum transfers, leading to the so-called Compton profile $J(p_z)$ of the solid using the impulse approximation. Section 5 then again confronts this approach to high- and ultra-high resolved Compton measurements on Li.

2. Inelastic X-ray scattering at low momentum transfer

The principal geometry of an inelastic X-ray scattering experiment is outlined in Figure 1. The experimental outcome of this kind of experiment, the so-called double differential cross section $\mathrm{d}^2\sigma/\mathrm{d}\omega_2\,\mathrm{d}\Omega$ is defined as the flux of X-ray photons scattered into an energy interval $[\hbar\omega_2, \hbar\omega_2 + \hbar\,\mathrm{d}\omega_2]$ and a certain interval of solid angle $[\Omega, \Omega + \mathrm{d}\Omega]$, defined by the special experimental arrangement. During the scattering process, the photon wave vector changes from $\mathbf{k}_1$ to $\mathbf{k}_2$ and its energy from $\hbar\omega_1$ to $\hbar\omega_2$. If one neglects resonant and magnetic contributions and calculates up to the first-order perturbation theory, this quantity is directly proportional to the dynamical structure factor $S(\mathbf{q}, \omega)$, where $\hbar\mathbf{q} = \hbar\mathbf{k}_2 - \hbar\mathbf{k}_1$ and $\hbar\omega = \hbar\omega_2 - \hbar\omega_1$ are momentum and energy transferred to the scattering system, respectively. $S(\mathbf{q}, \omega)$ can be expressed in terms of the Fourier transform in time of the two-particle correlation function [1]:

$$\frac{\mathrm{d}^2\sigma}{\mathrm{d}\omega\,\mathrm{d}\Omega} \sim S(\mathbf{q}, \omega) \sim \int \mathrm{e}^{-\mathrm{i}\omega t} \sum_{j,j'} \left\langle \mathrm{e}^{\mathbf{iqr}_{j'}(0)} \mathrm{e}^{-\mathbf{iqr}_{j}(t)} \right\rangle \mathrm{d}t, \tag{1}$$

Paul G. Mezey and Beverly E. Robertson (eds.), Electron, Spin and Momentum Densities and Chemical Reactivity, 179–193

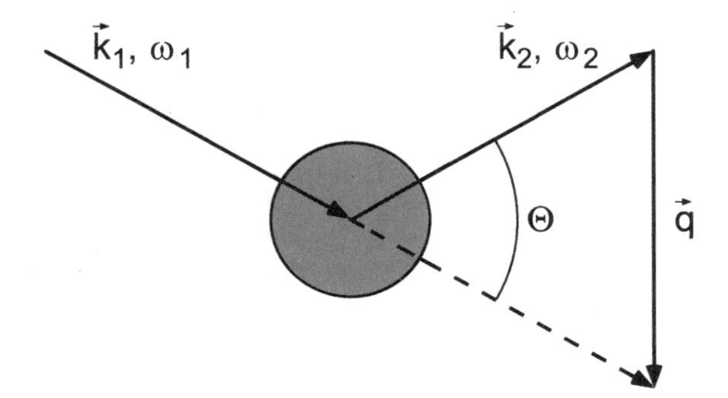

Figure 1. Principal geometry.

$r_{j'}(0)$ and $r_j(t)$ being the position in space of electron j' at time $t = 0$ and of electron j at time t. One can clearly see that $S(\mathbf{q}, \omega)$ contains in a somewhat integral way information about all possible electron–electron correlations, as it connects the scattering phases of electrons at different points in space and time. Via the fluctuation dissipation theorem, $S(\mathbf{q}, \omega)$ is often expressed in terms of the dielectric response function $\varepsilon^{-1}(\mathbf{q}, \omega)$ which gives the modification of an external applied electric potential $\Phi_{\mathrm{ext}}(\mathbf{q}, \omega)$ by the electron gas of the solid:

$$S(\mathbf{q}, \omega) \sim \mathrm{Im}\, \varepsilon^{-1}(\mathbf{q}, \omega), \tag{2}$$

$$\varepsilon^{-1}(\mathbf{q}, \omega) = \frac{\Phi_{\mathrm{ext}}(\mathbf{q}, \omega) + \Phi_{\mathrm{ind}}(\mathbf{q}, \omega)}{\Phi_{\mathrm{ext}}(\mathbf{q}, \omega)} = 1 - \frac{4\pi e^2}{q^2} \chi(\mathbf{q}, \omega). \tag{3}$$

$\varepsilon^{-1}(\mathbf{q}, \omega)$ is connected to the polarizability $\chi(\mathbf{q}, \omega)$ in the way shown in Equation (3). Let us turn to the most simple model case, the jellium model, in which the electrons are assumed to be free particles, moving embedded in a positive uniform charge background, obeying only Pauli's principle but not showing any electron–electron interactions. For this case of a non-interacting electron gas, the dielectric response function is given by

$$\frac{1}{\varepsilon(\mathbf{q}, \omega)} = 1 + \frac{4\pi e^2}{V q^2} \sum_{\mathbf{k}} \frac{f(E_{\mathbf{k}}) - f(E_{\mathbf{k+q}})}{\hbar\omega + E_{\mathbf{k}} - E_{\mathbf{k+q}} + \mathrm{i}O^+}, \tag{4}$$

revealing the density of states of possible electron–hole pairs.

Interaction of the electrons in the framework of the self-consistent field approximation is accounted for by considering the induced density fluctuations as a response of independent particles to $\Phi_{\mathrm{ext}} + \Phi_{\mathrm{int}}$ via Poissons equation [2]. This means, physically, that collective excitations of the electrons can occur, taken into account via a chain of electron–hole excitations. These collective excitations show up in $S(\mathbf{q}, \omega)$ as a distinct energy loss feature. Figure 2 shows the shape of the real and imaginary parts of the dielectric function in RPA ($\varepsilon_{\mathrm{r}}(\mathbf{q}, \omega)$, $\varepsilon_{\mathrm{i}}(\mathbf{q}, \omega)$) and the resulting dielectric response

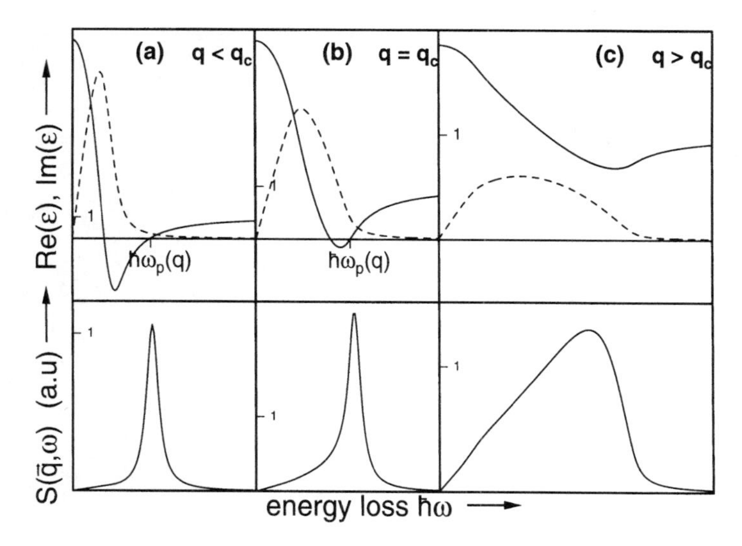

Figure 2. Dielectric function in RPA.

function [3]:

$$\operatorname{Im} \varepsilon^{-1}(\mathbf{q}, \omega) = \frac{-\varepsilon_{\mathrm{i}}(\mathbf{q}, \omega)}{\varepsilon_{\mathrm{r}}^2(\mathbf{q}, \omega) + \varepsilon_{\mathrm{i}}^2(\mathbf{q}, \omega)} \sim S(\mathbf{q}, \omega). \tag{5}$$

One can clearly see, that for small q, a strong peak in $S(\mathbf{q}, \omega)$ dominates, where ε_{r} and ε_{i} are close to zero, thus indicating the independent collective correlation of the electrons. For increasing q, ε_{i} gets broader and $\mathrm{S}(\mathbf{q}, \omega)$ reveals the spectrum of possible electron–hole excitations.

The next step to include electron–electron correlation more precisely historically was the introduction of the (somewhat misleading) so-called local- field correction factor $g(q)$, accounting for statically screening of the Coulomb interaction by modifying the polarizability [4]:

$$\chi(\mathbf{q}, \omega) = \frac{\chi^0(\mathbf{q}, \omega)}{1 - (4\pi e^2/q^2)(1 - g(q))\chi^0(\mathbf{q}, \omega)}. \tag{6}$$

There exist quite a lot of different approaches to calculate the shape of $g(q)$ with largely varying results [4–8] (Figure 3).

As will be shown in Section 3, inelastic X-ray scattering experiments can help to decide which theoretical approach is appropriate. One must keep in mind that this static correction is far from an appropriate description of electron correlations. A more accurate way is to account for dynamical screening by writing $\chi(\mathbf{q}, \omega)$ in terms of the one-particle Greens function $G(\mathbf{p}, \varepsilon)$ corrected for many-particle effects by a

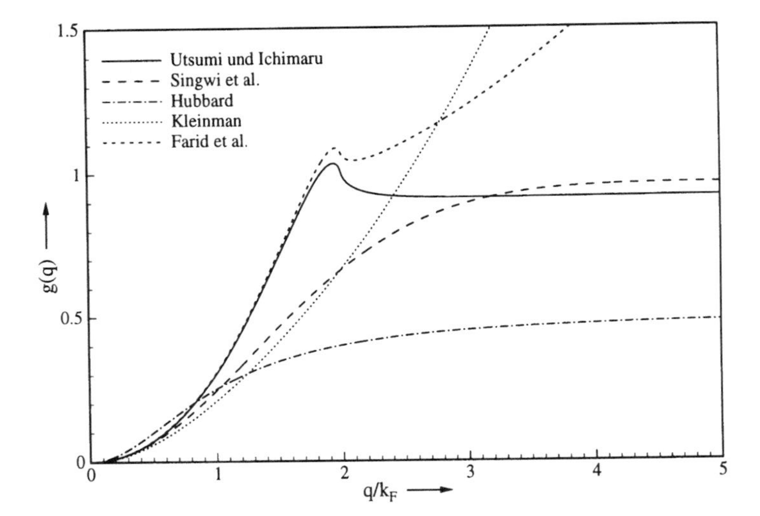

Figure 3. Local-field correction factor calculations.

vertex function Λ [9]:

$$\chi^{\mathrm{SC}}(\mathbf{q}, \omega) = 2 \int \frac{\mathrm{d}\varepsilon}{2\pi \mathrm{i}} \int \frac{\mathrm{d}^3 p}{(2\pi)^3} \underbrace{G(\mathbf{p}, \varepsilon) G(\mathbf{p} + \mathbf{q}, \varepsilon + \omega)}_{\text{one particle Greens function}} \underbrace{\Lambda(\mathbf{p}, \varepsilon, \mathbf{q}, \omega)}_{\text{vertex function}}. \quad (7)$$

This expression usually is rewritten as a geometrical series in an irreducible particle–hole interaction:

$$\begin{aligned} \chi^{\mathrm{SC}}(\mathbf{q}, \omega) = 2 \int \frac{\mathrm{d}\varepsilon}{2\pi \mathrm{i}} \int \frac{\mathrm{d}^3 p}{(2\pi)^3} \Bigg[& G(\mathbf{p}, \varepsilon) G(\mathbf{p} + \mathbf{q}, \varepsilon + \omega) \\ & + G(\mathbf{p}, \varepsilon) G(\mathbf{p} + \mathbf{q}, \varepsilon + \omega) \int \frac{\mathrm{d}\varepsilon'}{2\pi \mathrm{i}} \int \frac{\mathrm{d}^3 p'}{(2\pi)^3} \underbrace{I(\mathbf{p}, \varepsilon, \mathbf{p}', \varepsilon', \mathbf{q}, \omega)}_{\text{irred. part.–hole interaction}} \\ & \times G(\mathbf{p}', \varepsilon') G(\mathbf{p}' + \mathbf{q}, \varepsilon' + \omega) + \cdots \Bigg], \end{aligned} \quad (8)$$

which of course is rather complicated to compute. Equation (8) can nevertheless be used to get an expression for χ^{SC} that accounts for self-energy effects in the so-called off-shell way, by again replacing the full particle–hole interaction by the statically screened Coulomb interaction $(4\pi e^2/q^2)g(q)$ and obtaining [10]:

$$\chi^{\mathrm{SC}}(\mathbf{q}, \omega) = \frac{\chi^{\mathrm{SE}}(\mathbf{q}, \omega)}{1 + (4\pi e^2/q^2) g(q) \chi^{\mathrm{SE}}(\mathbf{q}, \omega)}, \quad (9)$$

with

$$\chi^{\mathrm{SE}}(\mathbf{q}, \omega) = 2 \int \frac{\mathrm{d}\varepsilon}{2\pi \mathrm{i}} \int \frac{\mathrm{d}^3 p}{(2\pi)^3} G(\mathbf{p}, \varepsilon) G(\mathbf{p}, \varepsilon) G(\mathbf{p}+\mathbf{q}, \varepsilon+\omega). \tag{10}$$

Using the connection between $G(\mathbf{p}, \varepsilon)$ and the spectral density function $A(\mathbf{p}, \varepsilon)$,

$$G(\mathbf{p}, \varepsilon) = \frac{1}{2\pi} \int_{-\infty}^{\infty} \frac{A(\mathbf{p}, E)\,\mathrm{d}E}{\varepsilon - E - \mathrm{i}O^+} \tag{11}$$

leads to

$$\operatorname{Im} \chi^{\mathrm{SE}}(\mathbf{q}, \omega) = -\int \frac{\mathrm{d}^3 p}{(2\pi)^3} \int_{-\omega+E_f}^{E_f} \frac{\mathrm{d}E}{2\pi} A(\mathbf{p}, E) A(\mathbf{p}+\mathbf{q}, E+\omega), \tag{12}$$

that is, the convolution of the spectral density function A for ground state and excited state. $A(\mathbf{p}, E)$ expresses the probability for the system to be in a state with energy E above ground state right after injection of an electron (analogous for injection of holes).

The improvement compared to the representation of Equation (6) is, that self-energy effects are included via the influence of the self-energy Σ on A:

$$A(\mathbf{p}, E) = \frac{1}{\pi} \frac{-\operatorname{Im} \Sigma(\mathbf{p}, E)}{\left(E - \varepsilon_p^0 - \operatorname{Re} \Sigma(\mathbf{p}, E)\right)^2 + \left(\operatorname{Im} \Sigma(\mathbf{p}, E)\right)^2}. \tag{13}$$

Figure 4 shows the result of a calculation of $A(\mathbf{p}, E)$ by Lundquist [11]. From Equation (13) it is clear that prominent structures in $A(\mathbf{p}, E)$ arise, when $\operatorname{Re} \Sigma$ approaches the value $E - \varepsilon_p^0$, ε_p^0 being the energy of a free particle, while $\operatorname{Im} \Sigma$ is nearly zero. So, for large p, A will exhibit a peak following the quasi-particle dispersion and broadened by approximately one half of the plasmon energy, whereas for small p two prominent peaks are to be seen: one is the quasi-particle excitation, the other is attributed to the excitation of a so-called plasmaron, the latter holding approximately one-third of the total spectral weight.

In a last step of this section, the on-shell approximation shall be applied to Equation (12) by replacing

$$\Sigma(\mathbf{p}, E) \to \Sigma\left(\mathbf{p}, \frac{p^2}{2m}\right)$$

and setting the real part of $\Sigma(\mathbf{p}, E)$ equal to zero, its imaginary part to Γ_p, the inverse lifetime of the excited particle–hole pair. This leads to the following expression for χ:

$$\operatorname{Im} \chi(\mathbf{q}, \omega) = -\int \frac{\mathrm{d}^3 p}{(2\pi)^3} \times \frac{1}{\pi^2} \frac{(\Gamma_{\mathbf{p}} + \Gamma_{\mathbf{p}+\mathbf{q}})}{\left(\omega - (\mathbf{p}\cdot\mathbf{q}/m) - (q^2/m)\right)^2 + \left(\Gamma_{\mathbf{p}} + \Gamma_{\mathbf{p}+\mathbf{q}}\right)^2}. \tag{14}$$

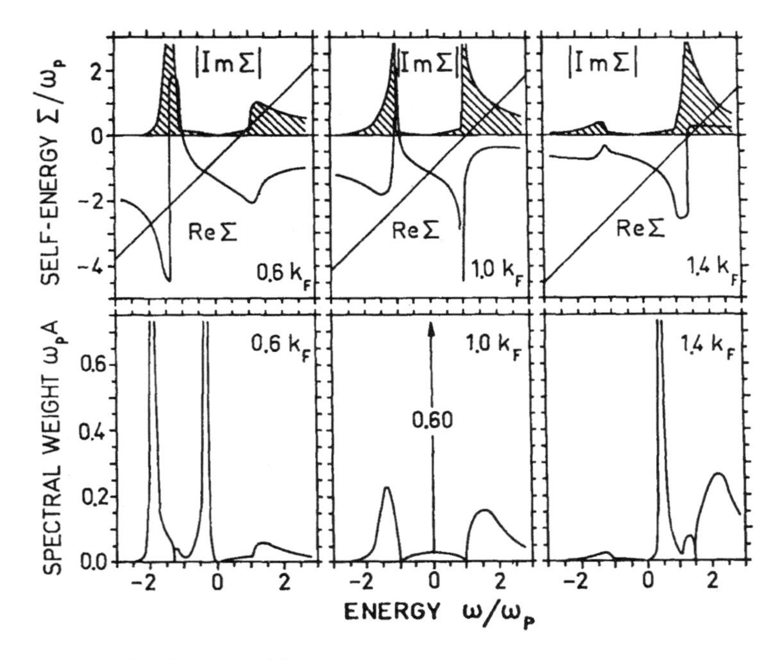

Figure 4. $A(\mathbf{p}, E)$ after Lundquist [11].

This is nothing but the imaginary part of the Lindhard-polarizability χ^0, but with finite O^+ that now is given by the imaginary part of the self-energy, revealing the lifetime of the excited state.

To sum up this section, electron–electron correlations can to a first approximation, be included into the response function by introducing a static local-field correction factor modifying the Lindhard or RPA dielectric function, accounting only for statically screened Coulomb interaction. Including self- energy effects *on-shell* leads to the inclusion of lifetime effects as well. The more sophisticated way to use the one-particle Greens function together with the irreducible particle–hole interaction turns out to be too complicated to compute. Accounting again only for static Coulomb interaction leads to the *off-shell* representation of χ as the convolution of the spectral density function for ground and excited state, which in turn contains the self-energy. It should be mentioned that the latter approach should, in principle, be physically more significant when compared to accounting only for statically screened Coulomb interaction.

3. Inelastic X-ray scattering experiments on lithium and liquid aluminum

The typical experimental setup (here the experiment established at beamline G3/ HASYLAB [12] is shown) is outlined in Figure 5. The white synchrotron radiation is monochromatized by a double crystal monochromator using the Ge (311) reflection

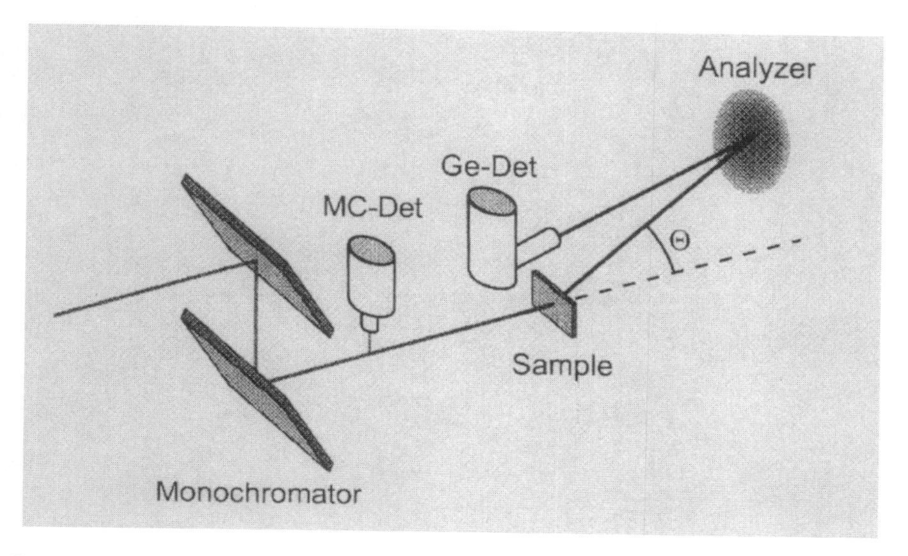

Figure 5. Experimental setup.

in (+/−) setting to a primary energy of $\approx$ 8keV. This primary radiation is scattered under a certain angle Θ that defines Ω and therefore the momentum transfer $\mathbf{q}$ and energy analyzed using a spherically bent analyzer in nearly backscattering geometry. Measurement of a spectrum is performed in inverse geometry, that is, by varying the primary energy and keeping the analyzer at a constant angle.

A simple metal like lithium or aluminum should best reveal the properties of the jellium model. To be sure, all long range order influence has been switched off, we measured $S(\mathbf{q}, \omega)$ of liquid Al ($T = 1000\,\mathrm{K}$). Figure 6 shows the result of a measurement for $|\mathbf{q}| = 1.5\,\mathrm{a.u.}$ together with theoretical calculations.

The solid points show the experimental result, the long dashed line the calculation of a $g(q)$-modified Lindhard response function according to Equation (6), using $g(q)$ after Utsumi and Ichimaru [5]. The solid line gives the result of a calculation that also takes into account self-energy effects *on-shell*, that is, introducing the lifetime of the involved states into the calculation according to Equation (14). One can clearly see that the latter reproduces the experimental result quite nicely.

Quite surprisingly, the *off-shell* method of accounting for self-energy effects using the spectral density function $A(\mathbf{p}, E)$ according to Equation (12) given by the short dashed curve in Figure 6 is far off the experimental values, although this method should be of greater physical significance. We believe that a possible reason for this failure is the neglect of the vertex correction function together with a cancellation of *off-shell* self-energy effects and the vertex correction.

To make this point clear, one has to look at the involved effects using diagram techniques. Self-energy effects, that is the deformation of the electron cloud around a single electron which then reacts back on this electron, can be divided into two

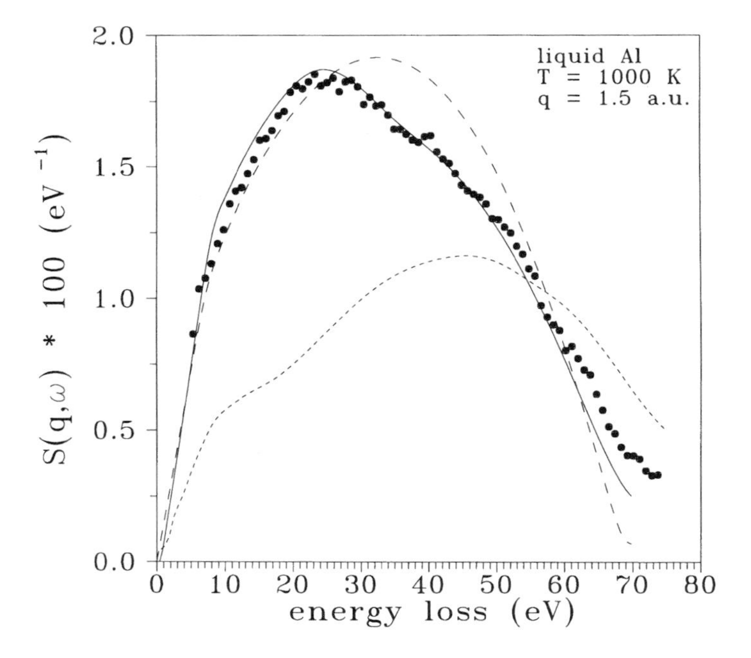

Figure 6. $S(\mathbf{q}, \omega)$ experiment on Al vs. theoretical calculations, for details see text.

principally different groups. The first type of process is described by the creation of a hole and a plasmon during the interaction process (Figure 7, upper left). In this case, the lifetime of the intermediate state is not restricted and therefore on-shell. The other type of self-energy process involves the emission of a plasmon, which reacts back on the emitting electron (Figure 7, upper right).For this kind of process, the lifetime of the intermediate process must be smaller than the lifetime of the original electron–hole pair and therefore off-shell. The vertex correction on the other hand consists of processes in which electron and hole exchange a plasmon directly (Figure 7, lower part). One clearly sees that off-shell self-energy processes and dynamical vertex cancel because for the first one a plasmon acts on an electron, for the latter, the plasmon acts on a hole so that they interfere destructively. Neglect of vertex correction while accounting for off-shell self-energy processes must then lead to a considerable error. As dynamical vertex and off-shell self-energy effects cancel, taking into account a statically screened Coulomb interaction via the local-field factor together with the self-energy on-shell will be appropriate to describe the experimental results as can be seen from Figure 6.

As was mentioned in Section 2, there exists a variety of different theoretical approaches to calculate the local field factor $g(q)$. Following Farid *et al.* [7], the behavior of $g(q)$ for large q is connected to the size z of the step in the occupation number function $n(k)$ for $k = k_F$, k_F being the Fermi-momentum (see Figure 8). This on-shell self energy:

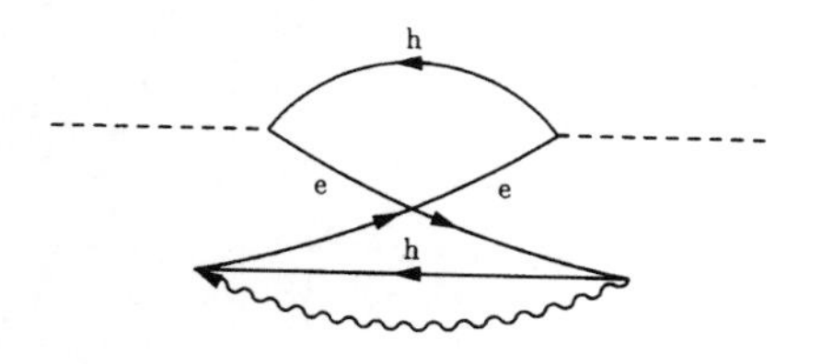

off-shell self energy:

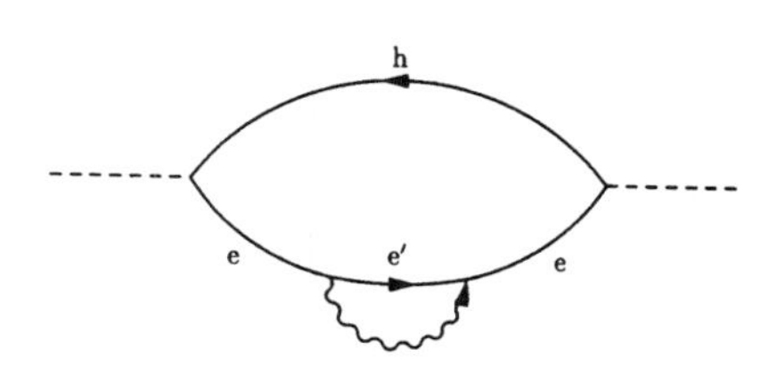

dynamical vertex:

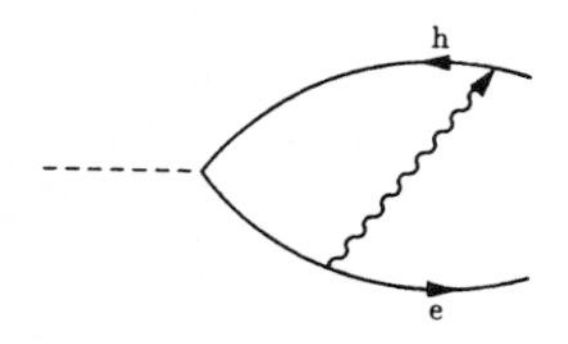

dynamical vertex:

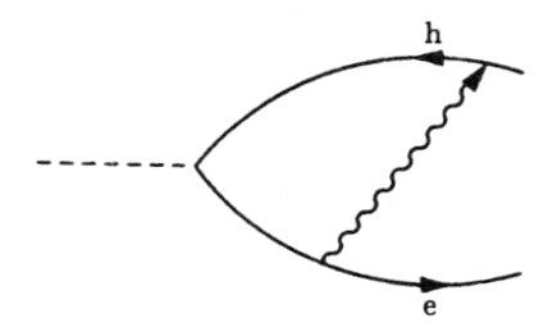

Figure 7. Cancellation of off-shell self-energy processes and dynamical vertex.

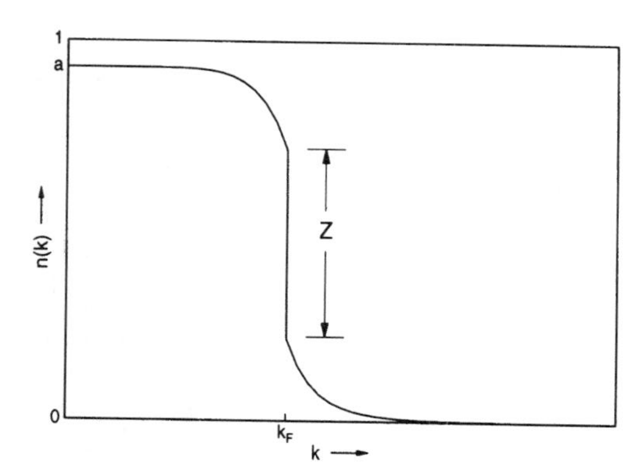

Figure 8. Occupation number function for correlated electrons.

quantity z is strongly determined by electron–electron correlation, so experimental evidence about the exact behavior of $g(q)$ would be highly desirable.

Fortunately, inelastic X-ray scattering can provide a means of determining semi-empirically $g(q)$ from $S(\mathbf{q}, \omega)$ measurements of simple metals. The application of $g(q)$ to χ^0 according to Equation (6) tends to shift χ^0 on the energy loss scale to lower energy losses.

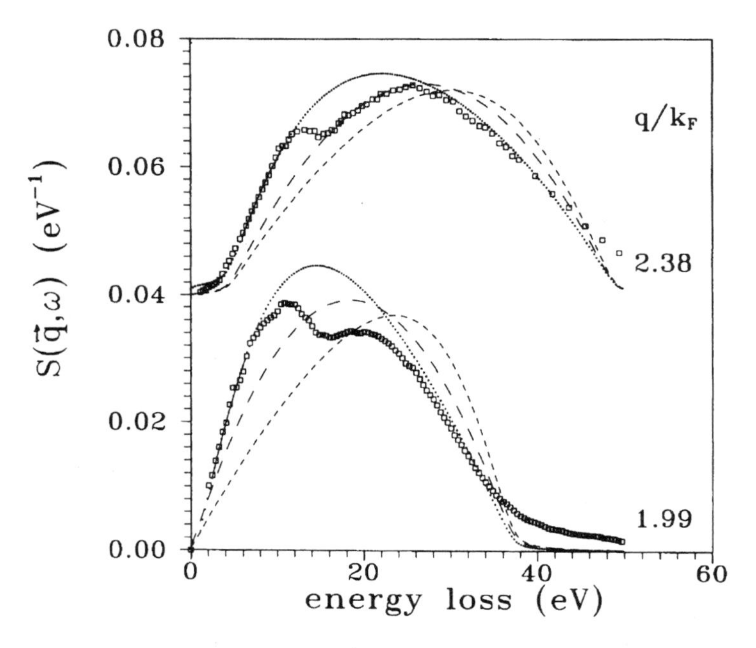

Figure 9. $S(\mathbf{q}, \omega)$ and local-field corrected of χ^0 Li.

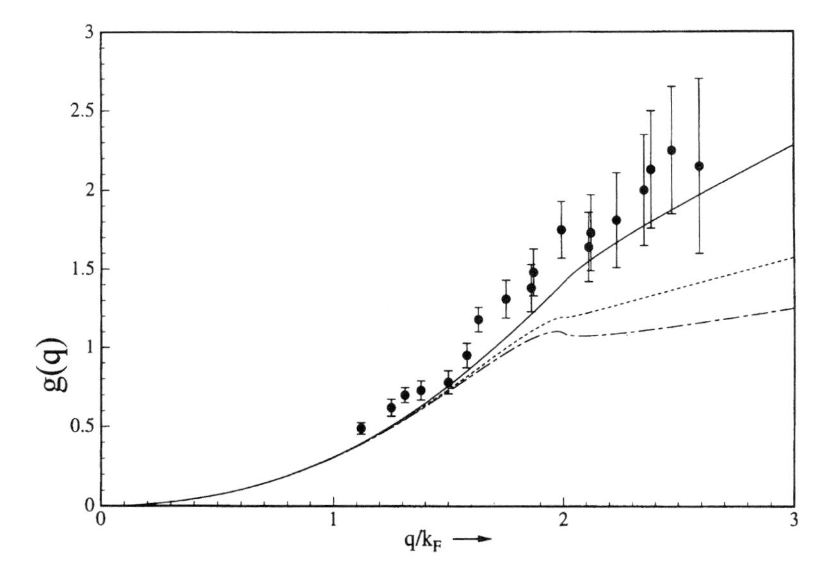

Figure 10. $g(q)$ for different z [7] compared to the experimental findings.

This can be seen from Figure 9 showing experimental $S(\mathbf{q}, \omega)$ (points) of Li for two different values of q together with (a) the pure Lindhard response function χ^0 calculated for the Li parameters (short dashed line), and (b) a $g(q)$-modified calculation according to Equation (6) using the $g(q)$-value of Utsumi and Ichimaru [5] (long dashed line). It can be seen how the application of this local field correction shifts χ^0 toward the correct position but it is obvious that this correction is not sufficient to reproduce the experimental findings. Using now the $g(q)$-value as a parameter, one can determine $g(q)$ for different q simply by fitting the local field corrected χ^0 to the experimental data. This procedure is described in more detail elsewhere [13]. Figure 9 shows the obtained local field corrected χ using the optimum $g(q)$ values in small points, showing remarkable agreement with the experimental data.

To determine the behavior of $g(q)$ for large q, we performed measurements of $S(\mathbf{q}, \omega)$ of Li for $1.1\,\text{a.u.} < q < 2.6\,\text{a.u.}$ and performed for each spectrum a fit of the $g(q)$-modified χ^0 to the experimental data. Figure 10 shows the result of this semi-empirical determination of $g(q)$ together with the shape of the local-field correction factor after Farid *et al.* [7] calculated for different values of z: solid line ($z = 0.1$), dashed line ($z = 0.5$) and dash-dotted line ($z = 0.7$). One clearly sees that the curve for the surprisingly small value of $z = 0.1$ fits our experimental findings best.

4. Inelastic X-ray scattering with high momentum transfer

If the geometrical parameters of an inelastic scattering experiment are set up in a way that large momenta are transferred to the sample (that is, a large scattering angle is chosen) and the amount of energy transfer is large compared to characteristic energies of the valence electrons, the so-called impulse approximation [14] can be applied when calculating the double differential cross section. One assumes that the scattering process happens so fast due to the large energy transfer, that the electrons do not rearrange during the process, the interaction potential remains constant and therefore the information about the scattering system is restricted to ground state information. Detailed calculation leads to the connection of the double differential cross section to the so-called Compton profile $J(p_z)$ via

$$\frac{\mathrm{d}^2\sigma}{\mathrm{d}\omega\,\mathrm{d}\Omega} \sim \underbrace{\int \mathrm{d}\mathbf{p}\,\rho(\mathbf{p})\delta\left(\frac{\hbar^2 q^2}{2m} - \frac{\hbar\mathbf{q}\cdot\mathbf{p}}{m} - \hbar\omega\right)}_{\text{Compton profile}\, J(\mathbf{p}\cdot\mathbf{q}) = J(p_z)}, \tag{15}$$

the Compton profile being the projection of the momentum space density $\rho(\mathbf{p})$ of the electrons on the direction of the momentum transfer $\mathbf{q}$.

So, by measuring $J(p_z)$ for a variety of different $\mathbf{q}$-directions, one can, in principle, reconstruct in three-dimension the momentum space density.

Application of the formalism of the impulse approximation to the double differential cross section in terms of the dielectric response (Equation 12), that is, using free-electron-like final states $E = |\mathbf{p}+\mathbf{q}|^2/2m$ in the calculation of $A(\mathbf{p}+\mathbf{q}, E+\hbar\omega)$

yields

$$\operatorname{Im} \chi^{SE}(\mathbf{q}, \omega) = \frac{1}{\pi} \int \frac{\mathrm{d}^3 p}{(2\pi)^3} \rho(\mathbf{p}) \times \frac{\operatorname{Im} \Sigma(\mathbf{p}+\mathbf{q}, (p^2/2m)+\omega)}{(\omega - (\mathbf{p}\cdot\mathbf{q}/m) + q^2/2m)^2 + (\operatorname{Im} \Sigma(\mathbf{p}+\mathbf{q}, (p^2/2m)+\omega))^2}, \tag{16}$$

which is nothing but the Compton profile in the jellium model, convoluted with $A(\mathbf{p}+\mathbf{q}, E+\hbar\omega)$. It should be noted that this creates a principal limitation for the momentum space resolution that can be achieved in ultra-high resolved Compton experiments due to the finite width of $A(\mathbf{p}+\mathbf{q}, E+\hbar\omega)$ (see Section 2). One can connect the information contained in Compton profiles to the occupation number function $n(\mathbf{k})$ [15, 16] via the reciprocal form factor $B(z)$ [17–19] which is the Fourier transform of a Compton profile

$$B(z) = \frac{1}{2\pi} \int J(p_z)\, \mathrm{e}^{\mathrm{i}p_z z}\, \mathrm{d}p_z, \tag{17}$$

$n(\mathbf{k})$ for metals is then given as a series in $B(\mathbf{R})$, $\mathbf{R}$ being a lattice translation vector:

$$\sum_{\nu} n_{\nu}(\mathbf{k}) = \sum_{\mathbf{R}} B(\mathbf{R})\, \mathrm{e}^{-\mathrm{i}\mathbf{R}\cdot\mathbf{k}}; \quad \nu\text{: band index.} \tag{18}$$

Figure 11 shows the influence of correlation and lattice effects on the shape of $n(\mathbf{k})$ for the case of lithium. The short dashed line shows $n(\mathbf{k})$ according to the jellium model with no electron–electron interaction included. Inclusion of correlation effects can be described using a model-$n(\mathbf{k})$:

$$n(k) = \begin{cases} (1-a) - (1/2)(1-a-z)(k/k_{\mathrm{F}})^8, & \text{for } k < k_F, \\ 1/2(1-a-z)(k_{\mathrm{F}}/k)^8, & \text{for } k < k_F, \end{cases} \tag{19}$$

where a is determined by the normalization condition

$$4\pi \int_0^{\infty} n(k) k^2\, \mathrm{d}k \frac{4\pi}{3} k_{\mathrm{F}}^3. \tag{20}$$

This yields the long dashed curve for a step of $z = 0.7$. In triangles, circles and squares, GWA-calculations by Kubo [20] for different directions are given. One can see a dependence of $n(\mathbf{k})$ of the direction of $\mathbf{k}$ and, as was seen from IXSS experiments already, that the calculated step tends to be smaller than 0.3. Given as a solid line is a model-$n(\mathbf{k})$ for $z = 0.1$ for comparison.

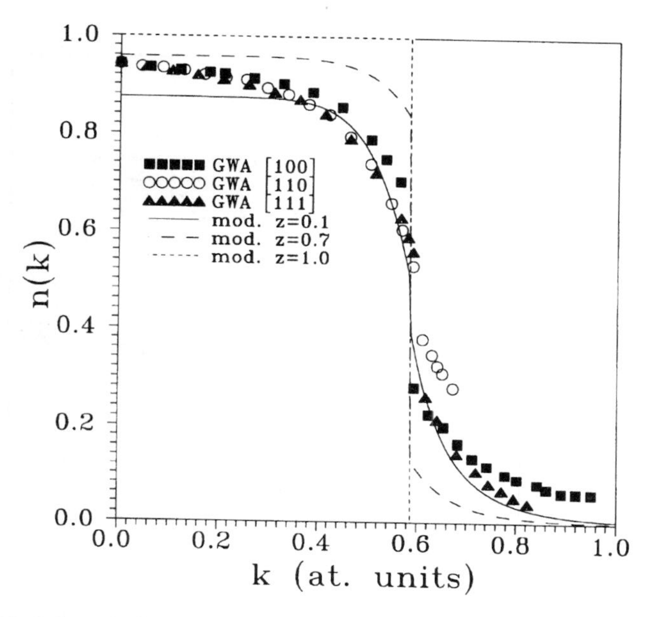

Figure 11. Calculations of $n(\mathbf{k})$.

5. Compton scattering from lithium

To provide experimental information about $n(\mathbf{k})$ from Compton profile measurements, we performed Compton measurements on Li using an experimental setup described elsewhere [21]. The momentum space resolution obtained was $\Delta P_z = 0.12$ a.u., 11 different directions of $\mathbf{q}$ were measured. From the obtained profiles we calculated their Fourier transforms and took from these the $B(\mathbf{R})$-values, given by triangles in Figure 12.

For comparison, the results of the GWA-calculation [20] are plotted in squares. We fitted the model-$n(k)$-based $B(r)$ according to Equations (19) and (18) using the step as fit parameter and got a remarkable agreement of the so calculated $B(r)$-function with the experimental values $B(\mathbf{R})$ again for the small value $z = 0.1$.

To get more direct experimental information about a correlation induced smearing of the step z in $n(\mathbf{k})$, we performed ultra-high resolved Compton profile measurements, using the standard IXSS as in Figure 5, choosing a large scattering angle. These experiments are of a preliminary nature and up to now suffer from poor statistics. We achieved an energy resolution of $\Delta E < 2$ eV yielding a momentum space resolution of $\Delta p_z < 0.02$ a.u. Figure 13 shows a typical raw spectrum.

A smearing of $n(\mathbf{k})$ at $k = k_F$ should influence the Compton profile in the following way: at the Fermi-break, the Compton profile should change slope abruptly, changing from the narrow valence electron profile into the much broader profile of the core electrons, if the electrons are modelled as being free without correlation. As

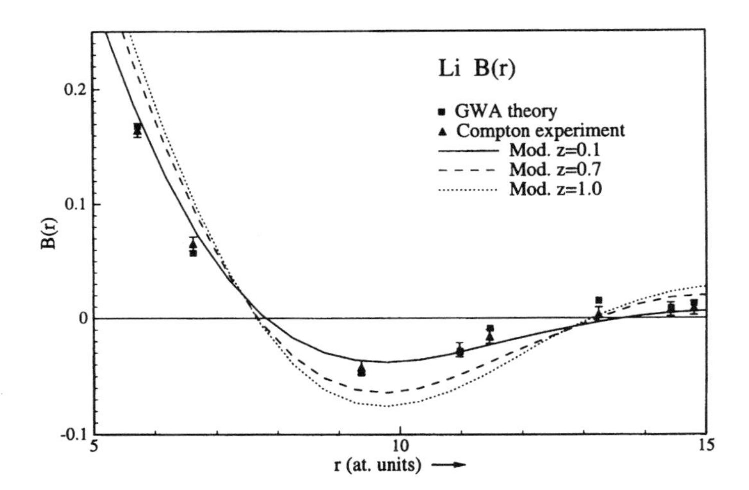

Figure 12. $B(r)$-calculations, $B(\mathbf{R})$ from experiment (for Li).

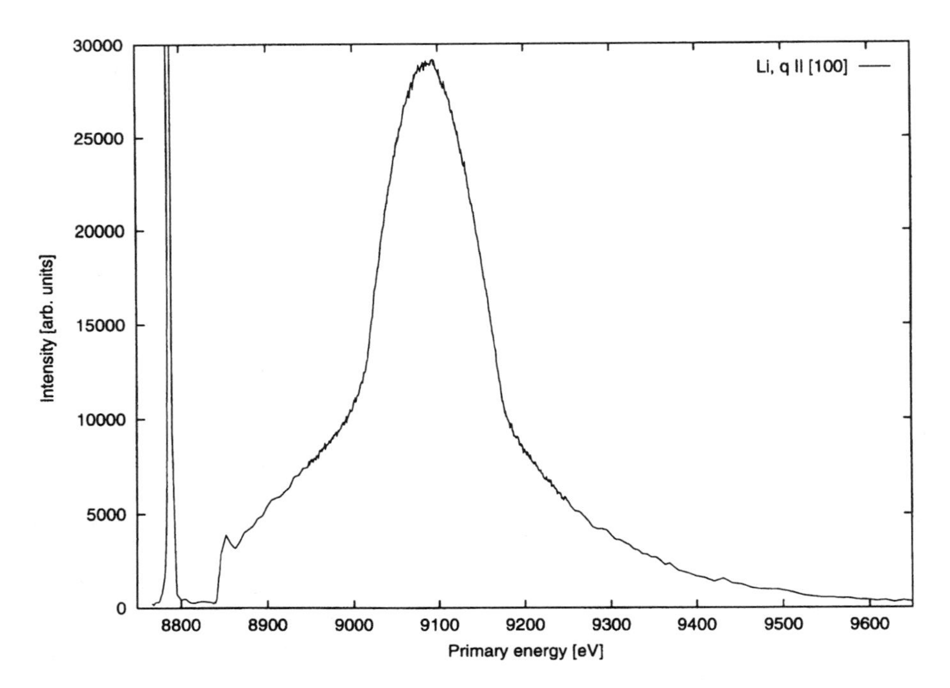

Figure 13. Typical raw spectrum of ultra-high resolved Compton experiments on Li.

correlation effects begin to smear out the abrupt change in $n(\mathbf{k})$, also this transition in slope from valence to core profile becomes less distinct, which can be seen directly from the second derivative of the Compton profile.

We find the second derivative of the Compton profile for $\mathbf{q} \parallel (111), (100)$ to be broadened beyond the experimental resolution with an additional $\Delta E \approx 5$eV which is due to the convolution with $A(\mathbf{k}+\mathbf{q}, E+\hbar\omega)$, as described in the previous section. For $\mathbf{q} \parallel (110)$, we find an additional broadening of the second derivative of the order of some eV which we ascribe to lattice effects on the electron correlation, predicted by the GWA-calculation.

6. Conclusion

To conclude, we have demonstrated how inelastic X-ray scattering experiments, both for small and large momentum transfer, can provide information about electron–electron correlation and lattice effects on correlation.

We have shown for the case of Li that the step in the occupation number function is surprisingly small: $z \approx 0.1$ and provided semi-empirically obtained values for the local-field correction factor. For the case of Al, we showed the additional cancellation of self-energy and vertex correction.

Furthermore, we showed for the first time the principal possibility of obtaining the correlation induced smearing of the occupation number function from ultra-high resolved Compton spectra and presented the first test experiments on Li.

References

1. van Hove, L. (1954) *Phys. Rev.*, **95**, 249.
2. Pines, D. and Nozieres, P. (1966) *The Theory of Quantum Liquids*, Vol. 1, Benjamin, New York, Amsterdam.
3. Lindhard, J. (1954) *Matematisk-Fysiske Meddelelser*, **28**, 8.
4. Hubbard, J. (1958) *Proc. Roy. Soc. A*, **243**, 336.
5. Utsumi, K., Ichimaru, S. (1980) *Phys. Rev. B*, **22**, 5203.
6. Kleinman, L. (1967) *Phys. Rev.*, **160**, 585.
7. Farid, B., Heine, V., Engel, G.E. and Robertson, I.J. (1993) *Phys. Rev. B*, **48**, 11602.
8. Singwi, K.S., Tosi, M.P. and Land, R.H. (1970) *Phys. Rev. B*, **1**, 1044.
9. Awa, K., Yasukara, H. and Asahi, T. (1981) *Solid State Comm.*, **38**, 1285.
10. Green, F., Neihon, D. and Szymanski, J. (1987) *Phys. Rev. B*, **35**, 124.
11. Lundquist, B.I. (1968) *Phys. kond. Materie*, **7**, 117.
12. Berthold, A., Mourikis, S., Schmitz, J.R., Schülke, W. and Schulte-Schrepping, H. (1992) *Nucl. Instrum. Meth. A*, **317**, 373.
13. Schülke, W., Höppner, K. and Kaprolat, A. (1996) *Phys. Rev. B*, **54**, 17464.
14. Eisenberger, P. and Platzman, P.M. *Phys. Rev. A*, **2**, 415.
15. Schülke, W. (1977) *Phys. Stat. Sol. B*, **80**, K67.
16. Schülke, W. (1978) *Jap. J. Appl. Phys.*, **17**, 332.
17. Schülke, W. (1977) *Phys. Stat. Sol. B*, **82**, 229.
18. Pattison, P., Weyrich, W. and Williams, B.G. (1977) *Solid State Comm.*, **21**, 967.
19. Weyrich, W., Pattison, P. and Williams, B.G. (1979) *Chem. Phys.*, **41**, 271.
20. Kubo, Y. (1997) *J. Phys. Soc. Jpn.*, **66**, 8.
21. Schülke, W., Stutz, G., Wohlert, G. and Kaprolat, A. (1996) *Phys. Rev. B*, **45**, 14381.

12

The measurement of spectral momentum densities of solids by electron momentum spectroscopy

MAARTEN VOS and ERICH WEIGOLD

Research School of Physical Sciences and Engineering, Institute of Advanced Studies, Australian National University, Canberra ACT 0200, Australia

1. Introduction

Electron momentum spectroscopy (EMS) or (e, 2e) spectroscopy is based on kinematically complete measurements of high energy electron impact ionisation of a suitable target material, which may be atoms or molecules in the gaseous state or condensed matter. Electrons of well-defined energy and momentum E_0 and $\mathbf{k}_0$ are directed onto the target material, and the energies and momenta of the two emitted (scattered and ejected) electrons are measured [1–3]. The kinematics are chosen to be that for free electron–electron collisions (allowing for the relatively small binding energy of the target electron [1]), or as it is commonly expressed, Bethe-ridge kinematics. In order to ensure a clean knockout of the struck electron the kinematics is also chosen to involve a large momentum transfer $\mathbf{K}$ from the incident electron to the ejected electron. These are the conditions for clean binary (e, 2e) collisions or EMS, where for an N electron target the $N-1$ electrons not observed can, to a good approximation, be treated as 'spectators'.

In general, one energy and angle analyser, denoted by f, detects emitted electrons that are faster than those detected in the other analyser, denoted by s. (Due to the indistinguishability of electrons it does not matter which is the 'scattered' electron or indeed whether their energies are equal, as is the case in 'symmetric' kinematics.) To ensure that the two detected electrons come from the same event, fast timing techniques are used [1, 2].

For each pair of detected electrons the binding energy ω and ion recoil momentum $\mathbf{p}$ are recorded. In a clean knockout, the recoil momentum $\mathbf{p} = -\mathbf{k}$, where $\mathbf{k}$ is the momentum of the bound electron when it is struck. Thus from energy and momentum conservation

$$\omega = E_0 - E_{\mathrm{f}} - E_{\mathrm{s}} \tag{1}$$

and

$$\mathbf{k} = \mathbf{k}_{\mathrm{f}} + \mathbf{k}_{\mathrm{s}} - \mathbf{k}_0. \tag{2}$$

At high enough energies the free electrons can be described by plane waves, and the differential cross section is given by [2]

$$\sigma(k, \omega) = C f_{\mathrm{ee}} \sum_{\mathrm{av}} |\langle f | a_k | 0 \rangle|^2 \delta(\omega - E_0 + E_{\mathrm{f}} + E_{\mathrm{s}}), \tag{3}$$

Paul G. Mezey and Beverly E. Robertson (eds.), Electron, Spin and Momentum Densities and Chemical Reactivity, 195–208

where C is a constant depending on the energies of the electrons, f_{ee} is the electron–electron collision factor which is also essentially constant in EMS kinematics [2], $|f\rangle$ and $|0\rangle$ are the electronic final ionic and initial (usually ground) states, and the operator $a_{\mathbf{k}}$ annihilates an electron of energy momentum $\mathbf{k}$ in the initial many-body target state $|0\rangle$. $\sum_{av}$ indicates a sum over final-state and average over initial-state degeneracies. For a non-oriented molecular or atomic target this means that differential cross section must be averaged over the solid angle $\hat{\mathbf{k}}$, i.e. spherically averaged. In addition if vibrational states are not resolved, the average over vibrations can be very well approximated by taking the initial and final states to those corresponding to the initial-state equilibrium positions [2].

We can consider EMS to be a direct probe for the energy–momentum spectral density function

$$A(\mathbf{k}, \omega) = \sum_{av} |\langle f|a_{\mathbf{k}}|0\rangle|^2 \delta(\omega - E_0 + E_f + E_s). \tag{4}$$

For atoms and molecules it is usual to make the weak coupling expansion for the one electron target-ion overlap amplitude [1, 2]:

$$\langle f|a_{\mathbf{k}}|0\rangle = \langle f|i\rangle\langle i|a_{\mathbf{k}}|0\rangle, \tag{5}$$

where $|i\rangle$ is a one-hole state formed by annihilating an electron from the 'orbital' i in the target state. One can define the experimental orbital $\psi_i(\mathbf{k})$ (the Dyson orbital) by

$$\psi_i(\mathbf{k}) \equiv \langle i|a_{\mathbf{k}}|0\rangle. \tag{6}$$

The differential cross section is then proportional to the spectroscopic factor $S_f^{(i)}$ (or pole strength):

$$S_f^{(i)} = |\langle f|i\rangle|^2, \tag{7}$$

which is the probability that the final ion state $|f\rangle$ contains the one-hole state $|i\rangle$. Thus the spectroscopic factors $S_f^{(i)}$ give the intensities of the transitions (main and satellite lines) for a given manifold of ion states belonging to orbital i.

For atomic hydrogen, $\psi_i(\mathbf{k}) \equiv \psi_{1s}(\mathbf{k})$, the wave function for the 1s ground state, and $S_f^{(i)} = 1$, since there are no electron–electron correlations. Figure 1 shows the measurements of $|\psi_{1s}(\mathbf{k})|^2$ obtained by Lohmann and Weigold [4]. Momentum is given in atomic units, as in the rest of this work. The results are independent of energy and in excellent agreement with the momentum density given by the absolute square of the Schrödinger momentum space wave function (solid curve).

Electron correlations show up in two ways in the measured cross sections. If the initial target state is well described by the independent particle Hartree–Fock approximation, the experimental orbital (6) is the Hartree–Fock orbital. Correlations in the ion can then lead to many transitions for ionisation from this orbital, rather than the expected single transition, the intensities of the lines being proportional to the spectroscopic factors $S_f^{(i)}$.

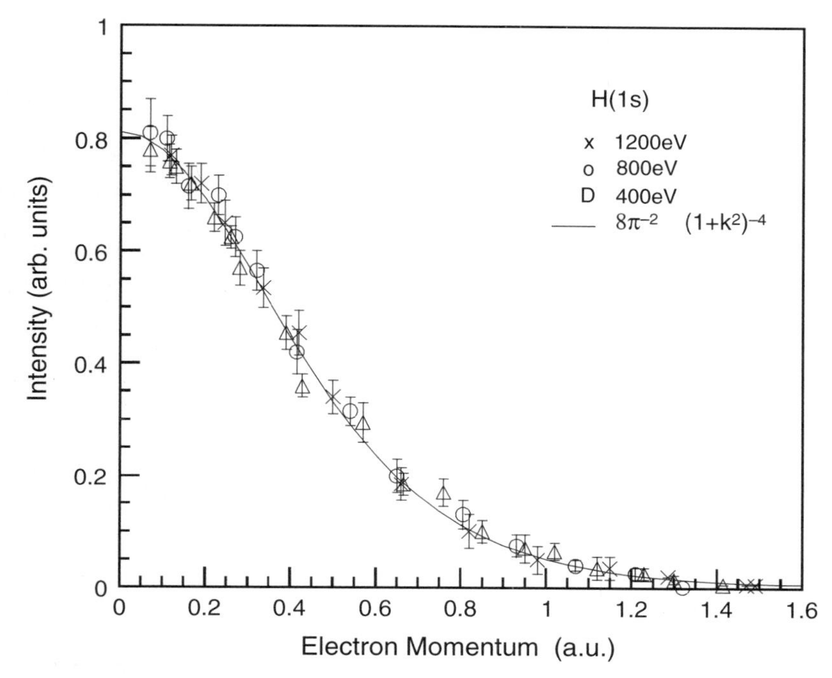

Figure 1. The electron momentum density for atomic hydrogen measured by EMS for the indicated energies compared with the square of Schrödinger wave function (solid curve) [4].

The simple linear molecule ethyne C_2H_2 is such a case. Figure 2 shows the electron binding energies measured by EMS and the momentum densities for the corresponding peaks in the spectrum. There are four peaks below 25 eV corresponding to the four valence orbitals of ethyne (the $1\pi_u$, $3\sigma_g$, $2\sigma_u$, and $2\sigma_g$ orbitals). The binding energy spectrum shows additional structure at high binding energies due to electron correlations. The momentum density of this structure has the same shape as that calculated with the independent particle wave function for the inner valence $2\sigma_g$ orbital. It has, however, only 38% of the strength of that orbital, the main $2\sigma_g$ transition at 23.6 eV having nearly all of the remaining strength. The independent particle orbital momentum densities agree very well with the measured ones. The spectroscopic factor for the observed $2\sigma_u$ transition is also a little smaller than unity, so some of the strength of this orbital may contribute at higher binding energies.

Secondly, correlations in the initial state can lead to experimental orbital momentum densities significantly different from the calculated Hartree–Fock ones. Figure 3 shows such a case for the outermost orbital of water, showing how electron–electron correlations enhance the density at low momentum. Since low momentum components correspond in the main to large r components in coordinate space, the importance of correlations to the chemically interesting long range part of the wave function is evident.

In solids, as in atoms and molecules, the spectral density $A(\mathbf{k}, \omega)$ contains much more information than simply the band peak position, i.e. the band dispersion. The

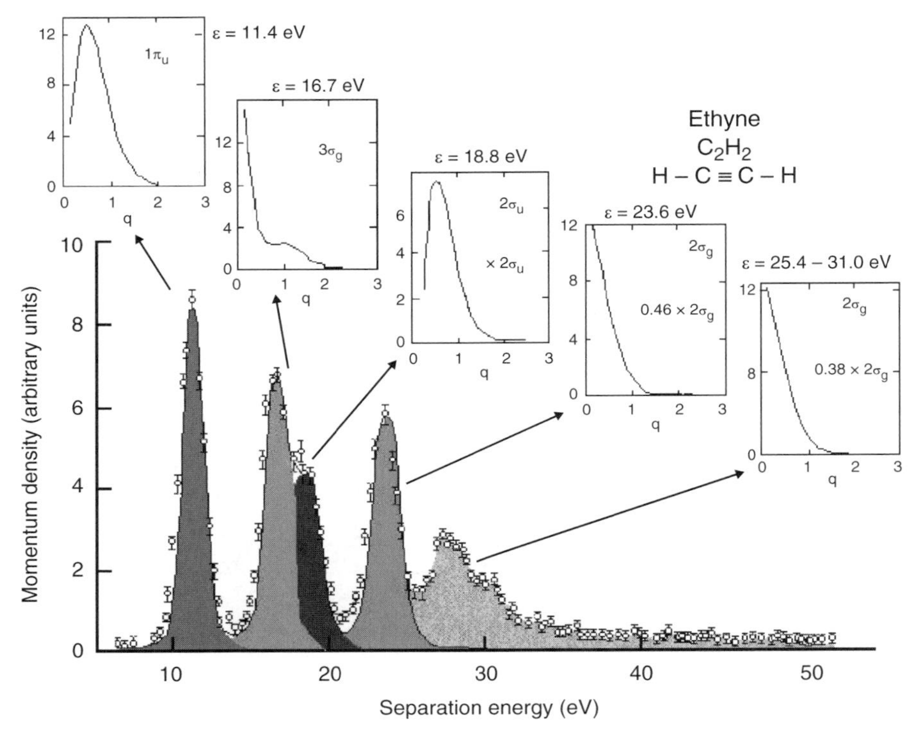

Figure 2. The binding energy spectrum for valence electrons of ethyne and the corresponding measured and calculated self-consistent-field independent particle orbital momentum densities [5].

width in energy of the main quasi-particle peak in $A(\mathbf{k}, \omega)$ gives the quasi-particle lifetime at momentum **k** and peak energy ω. The magnitude of $A(\mathbf{k}, \omega)$ is the probability of the particle having momentum **k** and energy ω. Due to correlations the spectral density function may contain additional satellite structures [9], which should be observable by EMS.

It is important to note that since the momentum **k** measured in EMS is the real electron momentum, EMS of solids does not require the object to be a single crystal (as for instance in ARPES). Thus using EMS one can obtain the full spectral density function $A(\mathbf{k}, \omega)$ for amorphous and polycrystalline targets as well as for single crystals. During the last two decades considerable progress has been made in the application of EMS to the study of the electronic structures of atoms and molecules [2]. Although the first EMS measurement was on a solid target [10], technical problems severely limited its application to solids until the recent development of the Flinders high resolution multiparameter spectrometer [11]. The severest limitation was the poor energy and momentum resolution and the low count rates in the earlier spectrometers. For the study of solids it is essential to use very high energy beams in order to reduce multiple scattering in the target.

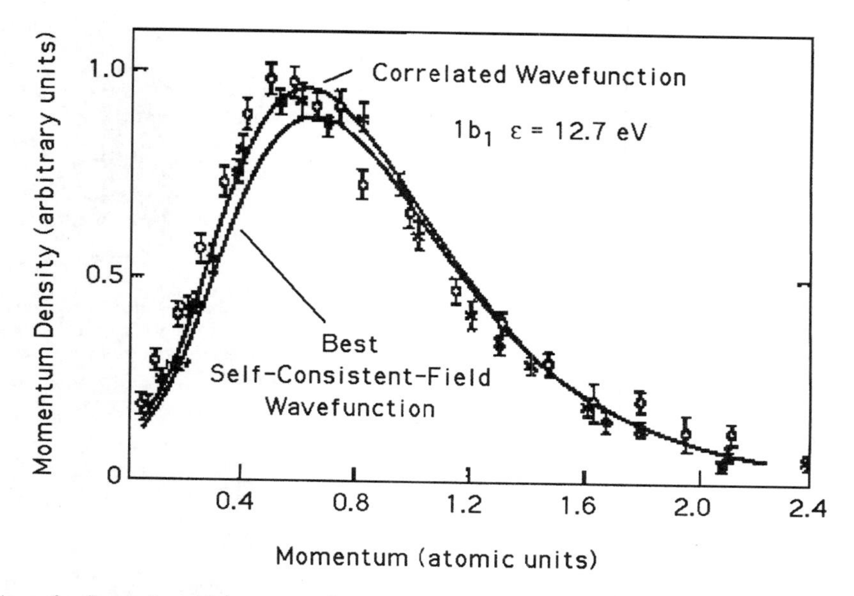

Figure 3. Comparison of the measured momentum distributions of the outermost valence orbital for water [6–8] with spherically averaged orbital densities from Hartree–Fock limit and correlated wave functions [6].

For high energy electrons it is more difficult to obtain the sort of energy resolution ($\lesssim$ 1 eV) required for valence band studies. In addition the cross section decreases as the energy is increased and this leads to low coincidence count rates, and also to poor momentum resolution due to the need for large detector solid angles to help increase the count rate. Canney *et al.* [12] achieved an energy resolution of $\lesssim$ 0.9 eV and coincidence count rates of the order of 200 counts/minute, compared to the earliest measurements where the energy resolution was $\gtrsim$ 100 eV and coincidence count rates of the order of 0.1 counts/minute. We will now briefly describe the Flinders spectrometer and associated techniques before discussing some recent examples of EMS applied to amorphous, polycrystalline, and single crystal materials.

2. Experimental details

The kinematics of the Flinders spectrometer is shown in Figure 4. The plane formed by $\mathbf{k}_0$ and the mean direction of $\mathbf{k}_\mathrm{f}$ is the laboratory z–x plane (horizontal) and the y-direction is the vertical. The coincidence spectrometer can record events simultaneously over the predetermined azimuthal angular range ($-18^\circ < \phi_\mathrm{f} < 18^\circ$, $180^\circ - 7^\circ < \phi_s < 180^\circ + 7^\circ$). This means that the cross section as a function of electron momentum $\mathbf{k}$ is sampled simultaneously over a range of k_y with k_x and k_y fixed and essentially zero for polar angles $\theta_\mathrm{f} = 14^\circ$ and $\theta_\mathrm{s} = 76^\circ$ and $E_\mathrm{f} = 18.8 \pm 0.01$ keV, $E_\mathrm{s} = 1.2 \pm 0.02$ keV and $E_0 = 20$ keV $+ \omega$. θ_s can be varied over a small range about 76° so that one can sample events with $k_x \neq 0$ and

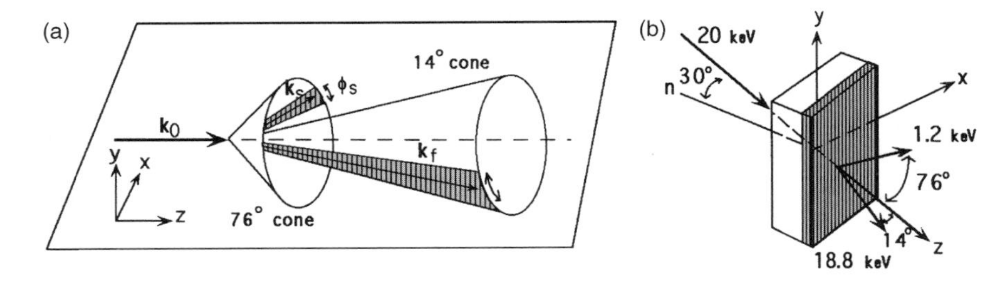

Figure 4. Kinematics of the solid-state EMS spectrometer [11]. (a) The polar angles made by $\mathbf{k}_f$ and $\mathbf{k}_s$ with respect to the incident (z) direction are $\theta_f = 14^\circ$ and $\theta_s = 76^\circ$. In (b) is shown a typical sample membrane relative to the electron trajectories. The surface sensitivity is largely determined by the escape depth of the 1.2 keV electrons (~ 2 nm) and is indicated by the shaded area.

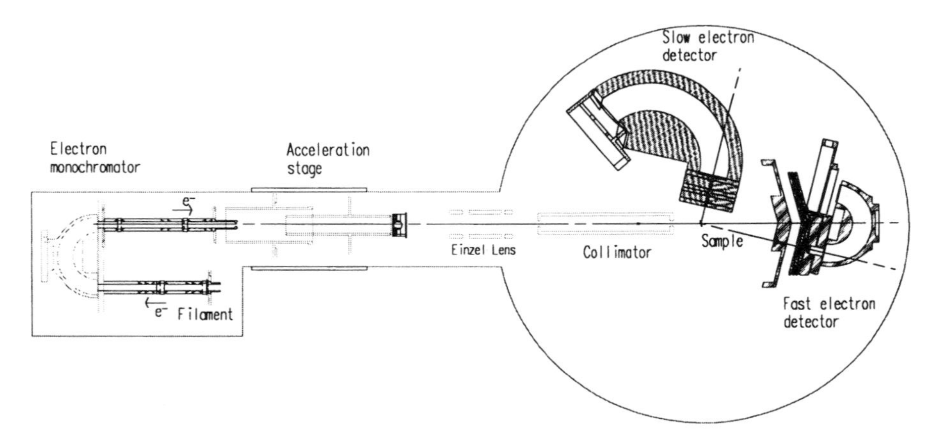

Figure 5. A horizontal schematic cut through the EMS spectrometer showing the monochromated and collimated incident beam and the hemispherical (fast) and toroidal (slow) energy and angle dispersive analysers as well as the retarding lens systems.

$k_z \neq 0$. Under normal operating conditions the azimuthal angular range is restricted to $\phi_f = \pm 10^\circ$ and $\phi_s = \pi \pm 6^\circ$, which restricts k_y to the range ± 2.5 au.

A horizontal cut through the spectrometer is shown schematically in Figure 5. It shows the electron gun with its monochromator and accelerator stages, the sample position, and the slow (toroidal) and fast (hemispherically) electron electrostatic analysers with their retarding lens stacks. The mean pass kinetic energies of the electrons through the analysers are 200 eV for the toroidal (slow) analyser and 100 eV for the hemispherical (fast) analyser, the dispersion in energy being in the radial direction in the plane shown in the figure. The dispersion in angle is in the direction perpendicular to that plane. On the exit planes of each analyser, a stack of chevron mounted microchannel plates, followed by a Gear-type resistive two-dimensional position sensitive anode, provides the fast timing signal as well as the four position determining signals [11]. Each analyser is carefully calibrated so that from the arrival position on the anode the energy and angle (i.e. momentum) of each detected electron can be determined. For each coincident pair of detected electrons, the binding energy ω and bound electron momentums **k** (Equations (1) and (2)) can then be obtained and recorded. The arrival position allows one to also infer the trajectory of the detected electron through the analyser, and in this way transit time variations can be corrected for in the timing spectrum, leading to significant improvement in the signal to noise ratio. The whole experiment and data acquisition is under complete computer control. The data reduction techniques are outlined by Vos *et al.* [13].

For transmission EMS measurements the requirements for target sample preparation are severe. First of all the target has to be a very thin (10–20 nm) free-standing membrane with a diameter of at least 0.3 mm. The composition has to be well known, and the exit surface has to be clean, due to the surface sensitivity resulting from the small mean free path of the slow ejected electron (Figure 1). Therefore much ingenuity and time has gone into developing suitable sample preparation techniques.

The target preparation and characterisation facilities are shown schematically in Figure 6. Details of typical sample preparation are given in Fang *et al.* [14]. Samples

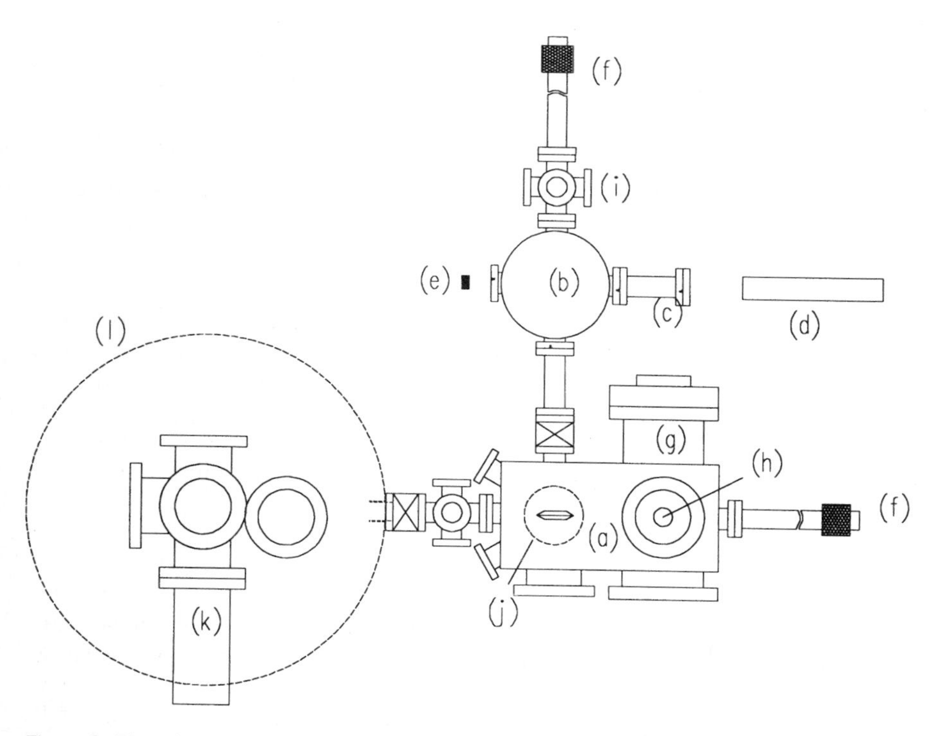

Figure 6. Plan of the target preparation facilities consisting of UHV preparation chamber (a), (reactive) ion etching chamber (b), ion etching gun (c), laser (d), photon detector (e), transfer arms (f), Auger system for surface analysis (g), sample manipulator and annealing facility (h), load lock and optical microscope for viewing sample (i), evaporator (j), transmission diffractometer (k), and vacuum tank for main spectrometer (l).

can be thinned by cleaving and/or electrochemical and chemical etching to a preliminary stage. They can then be further thinned by either reactive ion (plasma) etching or etching by the ion beam facility. A laser interferometer can be used for monitoring the thickness during the ion etching.

The target preparation and characterisation facility consists of two vacuum chambers connected in series with the main collision chamber. The chamber furthest from the collision chamber has a dual function. It serves as a chamber where reactive etch gases can be used and where the etch gas pressure can be maintained in the Torr range. The intermediate buffer chamber (low 10^{-10} Torr range) has an Auger system for characterising the surface and an annealing stage. Targets can also be prepared in this chamber by evaporation onto one surface of a thin free-standing film. Argon ion sputtering can also take place in this chamber. Ultra-high vacuum (UHV) conditions are maintained in the preparation chambers, the target samples being inserted using a load lock system. Similarly, it is possible to transfer the sample rapidly from one chamber to another under UHV conditions.

A transmission electron diffraction facility, mounted on top of the EMS spectrometer can be used to further characterise the sample. The target is mounted on a manipulator in the main spectrometer chamber. This manipulator allows rotation about the vertical (y-axis) and the surface normal, as well as movement in the x-, y- and z-directions. The sample can be aligned along a crystal direction using the diffraction set-up, and then transferred to the measurement stage retaining this alignment. Thus momentum densities can be determined along chosen crystallographic directions.

With these facilities a range of high quality targets can be produced. Samples already successfully fabricated include graphite and silicon single crystals, free-standing films of copper, aluminium oxide, silicon and silicon oxide. A considerable number of amorphous or polycrystalline targets have been made by evaporating sample material onto a thin amorphous carbon substrate. Samples studied in this way include aluminium, fullerene, silicon, germanium and copper. A 30 Å overlayer of the material is generally sufficient to attenuate the signal from the carbon backing by several orders of magnitude. In some cases such as copper, where 'islands' are formed on evaporation onto the backing, the characteristic carbon energy–momentum density traces underlie the sample data. However, since the carbon energy–momentum densities have been accurately measured, they can be subtracted from the data.

3. Solid state results

One of the attractive aspects of EMS is that it allows for the study of atoms, molecules and solids in a unified framework. In atoms and molecules we have discrete orbitals, whereas in solids there is a momentum distribution continuously varying with energy. The transition from molecules to solids is well illustrated in the case of a C_{60} film. The interaction between the different C_{60} molecules is small, and is expected to fall within the resolution of the present spectrometer (≈ 1 eV). We expect thus that the EMS measurements can be described by the theory of isolated molecules. In Figure 7 we show an energy spectrum of C_{60}. Non-zero intensity is found for binding energies much larger than expected for a C_{60} molecule. This is attributed (at least for a large

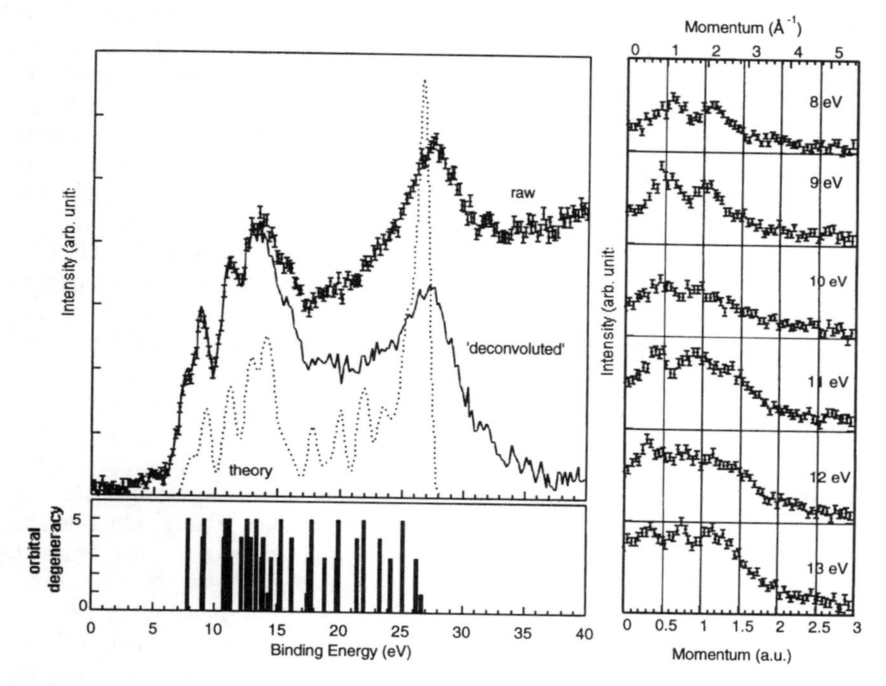

Figure 7. EMS results for C_{60}. In the left top panel we show the energy spectrum as obtained by integrating over a momentum range from 0 to 1.7 au. Raw data (error bars) are shown as well as these data approximately corrected for multiple energy loss events (solid line). In the lower left panel we show the calculated energy positions of the different levels plus their degeneracy (excluding spin degeneracy). The theoretical spectrum (dotted line) is obtained by summing the momentum distribution as described in the text. The right panel shows the experimentally obtained momentum distributions for binding energies as indicated.

part) to multiple scattering. One of the electrons involved in the (e, 2e) event suffered energy loss due to additional scattering. Thus in spite of the large energies used in these experiments and the small thickness of the film (5–10 nm), the thickness of the film cannot be made small enough so that multiple scattering becomes negligible. This is an important difference with the gas-phase experiments, where multiple scattering is negligible. In the figure the additional energy loss processes are corrected for in an approximate deconvolution procedure.

The calculated position of the 120 occupied orbitals, and their degeneracy is indicated below the experimental data. The energy position of the highest occupied molecular orbital is aligned with the edge of the experimental spectrum. At small binding energies there are still experimental indications for the discrete nature of the orbitals. At larger binding energies this is washed out, presumably due to increased lifetime broadening. From the calculated orbitals a theoretical spectrum was derived, using a broadening that is due to experimental resolution only. This reproduces the main structures well, except in the inner valence region, where the measured spectrum is much more spread out in energy than given by the self-consistent field calculation.

As in ethyne (Figure 2) this is probably due to electron–electron correlations in the inner valence region.

These large molecules can be considered as a small cluster as well, i.e as a small solid. In the left panel of Figure 6 we show the measured momentum densities for small binding energies. With decreasing binding energy the peaks move slowly to larger momentum values. In a true solid the spacing of the levels becomes infinitely small, and the momentum densities peak at well-defined values at each energy. The dependence of peak position on binding energy is referred to as 'dispersion' and it is this relation between binding energy and momentum that is usually presented as the result of calculations of the electronic structure of materials. Thus the development of the orbital momentum density with binding energy shows the signature of an emerging solid. More details of the transition from molecular to solid state behaviour can be found in [15]. A complete description of the C_{60} results is given in [16].

The outer most levels in C_{60} are due to 'π orbitals'. These are formed by 2p electrons which have their orbitals oriented along the radius of the molecule. The different environment inside and outside the spherical molecule causes the double-peaked structure in the momentum densities. In graphite the π band is formed by 2p orbitals oriented perpendicular to the sheets of carbon atoms. Using single-crystal graphite films we have a unique opportunity to study the effects of the orientation of these 2p orbitals in detail.

In the left panel of Figure 8 we show the band structure calculation of graphite in the repeated zone scheme, together with a drawing of the top half of the first Brillouin zone. The band structure is for the Γ–M direction. As the dispersion is very small along the c-axis we would find a similar result if we add a constant p_c component to the line along which we calculate the dispersion [17]. The main difference is that the splitting of the $\sigma 1$ and π band, caused by the fact that the unit cell comprises two layers, disappears at the Brillouin zone boundary (i.e. if the plot would correspond to the A–L direction).

We show the experimentally obtained spectral momentum density for the electrons with their momentum directed along the graphitic planes (i.e. along Γ–M with $p_c = 0$) in the central panel of Figure 8. We observe only one continuous structure. From Γ–M, in the first Brillouin zone, the measured intensity is at binding energies corresponding to the σ_1 band. In the second Brillouin zone the experimentally observed intensity is along the σ_2 band. There is no indication of any intensity related to the π band or the σ_3 band. For the π band this is due to the orientation of the $2p_c$ electrons. Their wave functions have a nodal plane at the layer formed by the carbon honeycomb structure. This causes a nodal plane in the momentum representation of these orbitals for $p_c = 0$, and hence they are not observed under these conditions. The absence of the σ_3 band is not accidental either, it can be shown from symmetry arguments that its intensity should be equal to zero under these conditions [18].

By changing the scattering parameters we can 'tune in' to electrons with a well-defined, constant momentum value along the c-axis. In the present case $p_c = 0.25$ au i.e. the resulting measurement was on the boundary of the Brillouin zone, along the A–L direction. Now two structures are visible, both the $\sigma_{1,2}$ and the π band. The dispersion of the $\sigma_{1,2}$ band has not changed noticeably from that found for the Γ–M

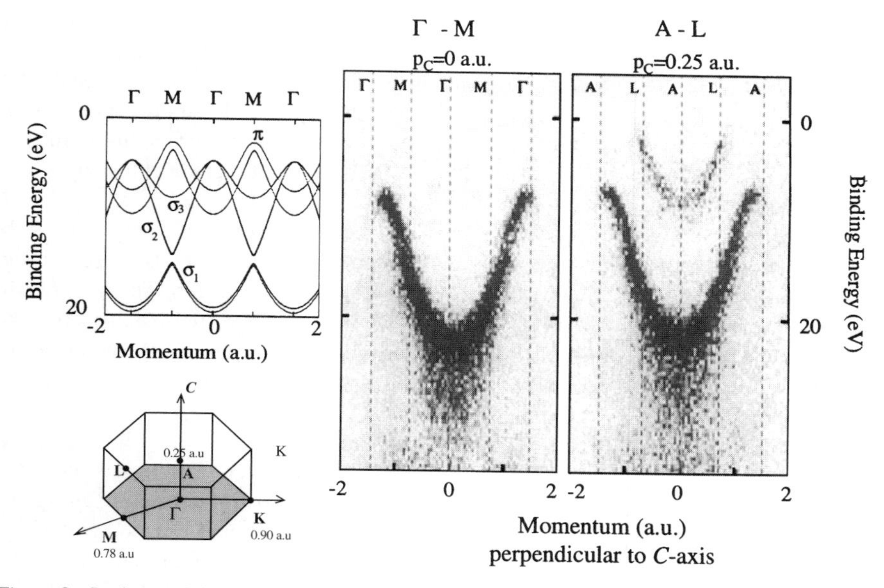

Figure 8. In the top left panel we show the band structure in the repeated zone scheme for the Γ–M direction. In the lower left panel we show the top half of the graphite Brillouin zone. The measurement presented in the central and right panel are for the Γ–M and A–L directions. Darker shading corresponds to larger intensities. Note that the π band is visible in the latter but absent in the first.

measurement. The σ_3 band is still absent. At the boundary of the first Brillouin zone the π band has not yet reached its maximum intensity. It is expected to be reached for p_c values around 0.75 au [17].

A comparison of the band structure diagram and these two measurements shows that experimentally the main measured intensity is constrained to a few of the bands present. In the first Brillouin zone the σ_1 band is found to be occupied, in the second zone σ_2. No sign of σ_3 or the π band is found for the Γ–M measurement. For the A–L measurement the same bands as for the Γ–M measurement contribute but in addition the π band is observed, mainly in the first Brillouin zone. These experiments are a beautiful, direct observation of the nodal plane of the π electrons in momentum space.

Thus, in addition to the dispersion itself, we get information about which band is occupied in which Brillouin zone. This is a consequence of the fact that EMS measures real momentum, and not, like for example angle-resolved photoemission, crystal momentum.

According to theory the measured intensity should be directly proportional to the momentum density. Thus not only the peak position is meaningful, but also the area under the peak can be directly interpreted as the momentum density. To what extent is this confirmed by the experiment? There are two main difficulties in verifying this claim.

One is multiple scattering. Both elastic and inelastic scattering contribute to a smooth background on which the clean events are superimposed. Assumptions on the shape of the background make large differences to the area attributed to a peak. As the transport of keV electrons through solids is quite well understood it is possible to simulate these multiple scattering effects, using Monte Carlo procedures. In this way it is possible to compare the experiment with theory that includes these multiple scattering effects.

The other problem is the theory. In a solid all electrons interact strongly with each other due to the long range nature of the Coulomb field. This leads to screening which reduces the effective range of the electron–electron interaction. The final result is that, even for simple metals, which appear at first sight as free-electron metals (as far as dispersion is concerned), the full many-body calculations of the spectral-momentum density (the spectral function) give considerable intensity away from the single-particle branch [9].

Let us illustrate these effects for the case of aluminium [19]. In Figure 9 (left panel) we show the momentum density near the Fermi level. Additional inelastic scattering events shift intensity away from the Fermi level, to higher binding energies. So at the Fermi level the only observed background is due to elastic scattering. The dotted line is the result of a linear muffin tin orbital (LMTO) band structure calculation, in the local density approximation. The theoretical results are broadened by lifetime broadening as determined empirically by Levinson *et al.* [20], but this is only important for energies away from the Fermi level. The LMTO calculation has peaks that coincide approximately with the experimental peak positions, but much less intensity away from these peaks. Inclusion of multiple scattering effects, using Monte Carlo simulations for the incoming and outgoing electron trajectories improves the agreement considerably [21]. Thus the intensity in between the two main peaks seems to be a result of elastic scattering.

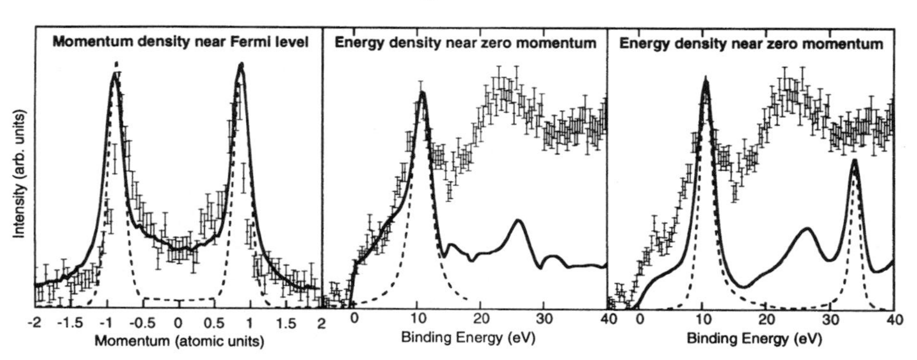

Figure 9. The measured momentum density of an aluminium film. In the left panel we show the measured momentum density near the Fermi level (error bars), the result of the LMTO calculations (dashed line) and the result of these calculations in combination with Monte Carlo simulations taking into account the effects of multiple scattering (full line). In the central panel we show in a similar way the energy spectrum near zero momentum. In the right panel we again show the energy spectrum, but now the theory is that of an electron gas, taking approximately into account the effects of electron–electron correlation (dashed) and this electron gas theory plus Monte Carlo simulations (solid line).

In the central panel we show a measured energy spectrum near zero momentum compared with the LMTO calculation and the LMTO calculation plus simulation of multiple scattering events.

The same normalisation of theory to the experiment is used as in the momentum density plot. Clearly the Monte Carlo simulation compares better with the experiment than the LMTO calculation by itself, but at high binding energies there is still a significant amount of intensity missing in the theory.

Replacing the LMTO theory with a calculation that neglects the effect of the crystal lattice, but takes, within the random phase approximation, electron–electron correlation into account does not lead to a perfect fit either. This is shown in the right panel. This theory includes lifetime broadening, as this is a consequence of electron–electron correlations. The obtained lifetime broadening is somewhat smaller than that obtained experimentally by Levinson *et al.* As a result the 'fit' at low binding energy is not as good as in the LMTO case. It results however in more intensity at large binding energy, due to satellites corresponding to a coupled hole–plasmon final state. However, this intensity is concentrated in a narrow energy range, whereas the experiment shows excess intensity over a broad range of energies. For more details see [22].

4. Conclusions

Electron momentum spectroscopy gives direct information about the binding energy of electrons *and* their distribution in momentum space. As it can be applied to atoms, molecules and solids it gives a very unified picture of the electronic structure of matter. For solids multiple scattering is a complicating factor. It can be reduced significantly by going to larger energies of the incoming and outgoing particles. For this purpose a new high energy spectrometer is under construction at the ANU. With this spectrometer we plan to achieve results that are able to test not only the one-electron type theories like band structure calculations, but true many-body calculations as well.

Acknowledgements

This work was done as part of a Special Research Centre at the Flinders University of South Australia, funded by the Australian Research Council. It could not have been successful without the contribution of all people involved in this centre.

References

1. McCarthy, I.E. and Weigold, E. (1988) *Rep. Prog. Phys.*, **51**, 299–392.
2. McCarthy, I.E. and Weigold, E. (1991) *Rep. Prog. Phys.*, **54**, 789–879.
3. Dennison, J.R. and Ritter, A.L. (1996) *J. Electr. Spectrosc.*, **77**, 99–142.
4. Lohmann, B. and Weigold, E. (1981) *Phys. Lett.*, **86A**, 139.
5. Weigold, E., Zhao, K. and von Niessen, W. (1991) *J. Chem. Phys.*, **94**, 3468–3479.
6. Bawagan, A.O., Brion, C.E., Davidson, E.R. and Feller, D. (1987) *Chem. Phys.*, **113**, 19.
7. Cambi, R., Ciullo, G., Sgamellotti, A., Brion, C.E., Cook, J.P.D., McCarthy, I.E. and Weigold, E. (1984) *Chem. Phys.*, **91**, 373–381.
8. Pascual, R. and Weigold, E. (1988) (unpublished).

9. Lundqvist, B.I. (1968) *Phys. Kondens. Mat.*, **7**, 117.
10. Camilloni, R., Giardini Guidoni, A., Tiribelli, R. and Stefani, G. (1972) *Phys. Rev. Lett.*, **29**, 618.
11. Storer, P.J., Caprari, R.S., Clark, S.A.C., Vos, M. and Weigold, E. (1994) *Rev. Sci. Instr.*, 2214.
12. Canney, S.A., Brunger, M.J., McCarthy, I.E., Storer, P.J., Utteridge, S., Vos, M. and Weigold, E. (1997) *J. Elect. Spectr. Rel. Phenom.*, **83**, 65–76.
13. Vos, M., Caprari, R.S., Storer, P., McCarthy, I.E. and Weigold, E. (1996) *Can. J. Phys.*, **74**, 829–836.
14. Fang, Z., Guo, X., Utteridge, S., Canney, S.A., McCarthy, I.E., Vos, M. and Weigold, E. (1997) *Rev. Sci. Instr.*, **68**, 4396.
15. Vos, M. and McCarthy, I.E. (1995) *Rev. Mod. Phys.*, **67**, 713.
16. Vos, M., Canney, S.A., McCarthy, I.E., Utteridge, S., Michalewicz, M.T. and Weigold, E. (1997) *Phys. Rev. B*, **56**, 1309.
17. Kheifets, A.S. and Vos, M. (1995) *J. Phys.: Condens. Matter*, **7**, 3895.
18. Harthoorn, R. and Mijnarends, P.E. (1978) *J. Phys. F*, **8**, 1147.
19. Canney, S.A., Vos, M., Kheifets, A.S., McCarthy, I.E. and Weigold, E. (1997) *J. Phys.: Condens. Matter*, **9**, 1931.
20. Levinson, H.J., Greuter, F. and Plummer, E.W. (1983) *Phys. Rev. B*, **27**, 727.
21. Vos, M. and Bottema, M. (1996) *Phys. Rev. B*, **54**, 5946.
22. Canney S., Kheifets, A., Vos, M. and Weigold, E. (1998) *Proc. 7th Int. Conf. Electron. Spectrosc.*, Chiba, Japan, *J. Elect. Spectr. Rel. Phenom.*, **88–91**, 247.

13

Accurate structure factor determination using 100 keV synchrotron radiation

T. LIPPMANN[1], D. WAASMEIER[2], A. KIRFEL[3] and J.R. SCHNEIDER[1]

[1]*HASYLAB/DESY, Notkestr. 85, D-22603 Hamburg, Germany*
[2]*Mineralogisches Institut, Universität Würzburg, Am Hubland, D-97074 Würzburg, Germany*
[3]*Mineralogisch-Petrologisches Institut, Universität Bonn, Poppelsdorfer Schloß, D-53115 Bonn, Germany*

The accuracy of experimentally determined structure factors is limited by various error sources, which may be introduced by the experimental method itself or during the data reduction stage. A reduction of those errors is expected by the use of high-energy synchrotron radiation ($E(I_0) \geq 100$ keV) as primary beam source, because absorption and extinction corrections are negligible in most practical cases.

Conventional diffractometers for structure factor determination purposes are operated in the 'low energy' regime up to 40 keV, whereas instruments at high-energy beamlines were constructed for different purposes (e.g. diffuse scattering experiments). Thus, there are principal differences in design and equipment, and our first investigations were focused on tests of the instrumental prerequisites.

Scattering on the Triple-Axis-Diffractometer [1, 2] at the HASYLAB high-energy beamline BW5 is performed in the horizontal plane using an Eulerian cradle as sample stage and a germanium solid-state detector. The beam is monochromatized by a single-crystal monochromator (e.g. Si 111, FWHM: 5.8″), focused by various slit systems (Huber, Risø) and iron collimators and monitorized by a scintillation counter. The instrument is controlled by a μ-VAX computer via CAMAC.

In order to provide all necessary measurement routines for a structure factor data collection the Four-Circle software package DIF4 [3] was adapted to the BW5 control software SPECTRA ON_LINE [4]. The program allows for fast reflection search (i.e. rotation photograph, 'peak hunting', cone scan), reflection centering, automatic reflection indexing, and setting up and refinement of the orientation matrix. Flexible data collection, either using reflection lists or *hkl*-files, includes on-line data reduction and graphical presentation and processing of a scan during the data acquisition.

Test measurements were performed on Cuprite, Cu_2O, for various reasons: it has a well-known structure (cubic, space group Pn$\bar{3}$m, $a_0 = 4.2696$, $Z = 2$) and has been examined by conventional X-ray single crystal diffractometry [5, 6] and by synchrotron radiation in the 'low energy' regime [7]. As a consequence of the

Paul G. Mezey and Beverly E. Robertson (eds.), Electron, Spin and Momentum Densities and Chemical Reactivity, 209–212

special positions of Cu (4b) and O (2a) all reflections can be separated into four parity groups:

Group	Dominated by	Intensity
eee	Cu, O	Strong
ooo	Cu	Strong
ooe	O	Weak
eeo	—	Very weak

e = even, o = odd.

Assuming spherical symmetric charge densities, the fourth group is forbidden. Hence, these reflections provide information about the anisotropic copper displacement parameters and chemical bonding, and a correct determination of these 'forbidden' reflections gives evidence of the data quality.

Experiments were carried out during several runs of DORIS III ($E = 4.5$ GeV). The beam size was $1 \times 1\,\mathrm{mm}^2$, the sample–detector distance was ≈ 1200 mm and the detector aperture $9 \times 9\,\mathrm{mm}^2$. ω stepscans were applied with 1 s per step for the stronger and 3 s per step for the weak reflections. The sample was a sphere of ≈ 190 µm diameter.

The first investigations were focused on some basic topics. In order to check the dynamic range various reflections of all parities were recorded, and an intensity ratio $I(\mathrm{eee})/I(\mathrm{eeo}) \approx 5 \cdot 10^5$ was covered. The strongest reflections were measured using iron absorbers of up to 27 mm thickness. A couple of strong and medium-intensity reflections were investigated several times during various DORIS-runs in order to test the reproducibility. The data were averaged and internal consistencies of 0.5–1.5% were achieved, showing good reproducibilities and stable beam conditions (i.e. beam position) at the beamline BW5. The same tests were carried out on a couple of symmetry-equivalent reflections and good internal consistencies were also found. Conclusively, the instrumental prerequisites for structure factor measurement purposes (i.e. stability, good alignment) were verified.

After these tests a dataset (CU96) was recorded, which consists of

(i) 485 reflections of all parities; $0 < (\sin\Theta)/\lambda \lessapprox 1\,\text{Å}^{-1}$; $h, k, l > 0$; 120 unique;
(ii) 103 ooe ('oxygen') reflections with $\approx 1 < (\sin\Theta)/\lambda \lessapprox 1.4\,\text{Å}^{-1}$; $h, k, l > 0$; 46 unique;
(iii) 122 eeo ('forbidden') reflections with $0 < (\sin\Theta)/\lambda \lessapprox 1\,\text{Å}^{-1}$; $h, k, l > 0$; 23 unique;

Integral intensities were obtained after dead-time corrections, background subtraction and normalization to averaged monitor counts. The Lp correction was applied in the usual way. Since the polarization ratio was not measured at BW5 so far, 90% linear horizontally polarized radiation was assumed for all scans. Calculations show that even a change in the beam polarization of 10% would effect the intensities of the highest order reflections of less than 1.5%.

No absorption corrections were carried out. The correction for TDS was evaluated using TDSCOR [8]. Elastic constants were taken from Hallberg and Hanson [9].

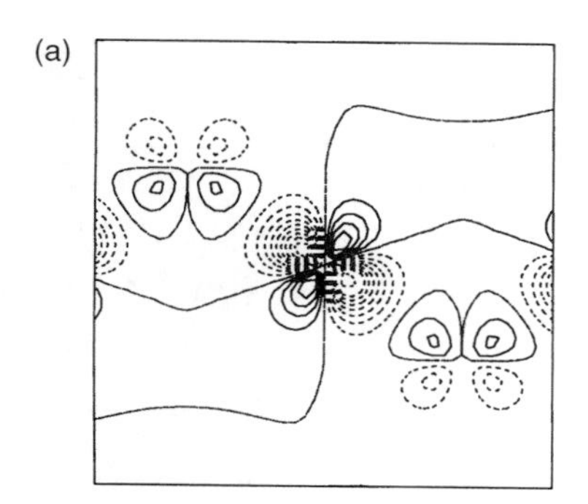

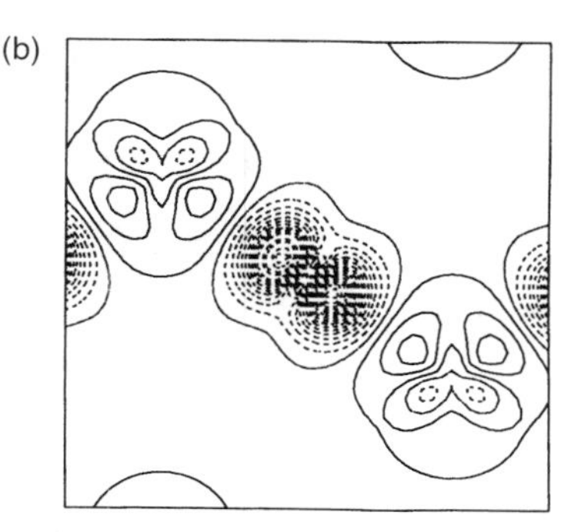

Figure 1. Electron-density maps (5 Å × 5 Å) in the plane (110). Horizontal axes: [001], vertical axes: [110]. (a) Monopoles omitted and (b) monopoles included. Contours at 0.05 e Å^{-3}, negative – broken, positive – full lines.

Averaging of symmetry-equivalent and multiply measured reflections was performed using AVSORT [8]. Fifteen ooe and eeo reflections were considered unobserved ($I < \sigma$) and the remaining data set consists of 128 unique reflections with an internal consistency $R_{\text{int}}(F^2) = 0.0064$.

A second data set (CU97, 1535 reflections of all parities, $0 < (\sin\Theta)/ \lessapprox 1.3$; $h, k, l > 0$ and $h, k, l < 0$) was recorded in continuous scan mode (i.e. the detector was read out *during* the ω-moves). This scan mode accelerated the data acquisition and enhanced the accuracy of the derived integral intensities. Averaging of these data yielded 120 reflections with an internal consistency $R_{\text{int}}(F^2) = 0.0038$.

Thermal parameters of conventional independent-atom refinements using BLFLS [8] were applied as starting values for full multipole refinements, which were performed with VALRAY [10]. Both data sets were successfully refined. The results were compared to those published by Kirfel and Eichhorn [7], and good agreement was found.

Using the refinement parameters, electron density maps were calculated. Figure 1 shows an example derived from CU96. According to all refinement results the intensities of the strong reflections were too low. Therefore, they were omitted for the electron density studies. The problem is currently under study. Additionally, low-temperature measurements are planned for the near future.

Acknowledgements

We thank K. Eichhorn for providing the source code of DIF4, Th. Kracht for help with the adaptation of the software to SPECTRA ON_LINE, and H.-G. Krane for providing the cuprite sample.

References

1. Schneider, J.R. (1995) *Cond. Matter News*, **4**, 11–19.
2. Bouchard, R. *et al.*, submitted to *J. Synchrotron Radiation*.

3. Eichhorn, K., DIF4 User's Guide, HASYLAB 1991.
4. Kracht, Th., SPECTRA, Technische Notiz, DESY/HASYLAB 94-01, 1994.
5. Lewis, J., Schwarzenbach, D. and Flack, H.D. (1982) *Acta Cryst.*, **A38**, 733–739.
6. Restori, R., Schwarzenbach, D. (1986) *Acta Cryst.*, **B42**, 201–208.
7. Kirfel, A. and Eichhorn, K. (1990) *Acta Cryst.*, **A46**, 271–284.
8. Eichhorn, K., Crystal program package, HASYLAB 1990.
9. Hallberg, J. and Hanson, R.C. (1970) *Phys. Status Solidi*, **42**, 305–310.
10. Stewart, R.F. and Spackman, M.A., VALRAY User's Manual, Pittsburgh, PA, USA, 1981.

14

Charge density data from CCD detectors

A. ALAN PINKERTON

Department of Chemistry, University of Toledo, Toledo, Ohio 43606, USA

1. Introduction

The advent of CCD detectors for X-ray diffraction experiments has raised the possibility of obtaining charge density data sets in a much reduced time compared to that required with traditional point detectors. This opens the door to many more studies and, in particular, comparative studies. In addition, the length of data collection no longer scales with the size of the problem, thus the size of tractable studies has certainly increased but the limit remains unknown. Before embracing this new technology, it is necessary to evaluate the quality of the data obtained and the possible new sources of error. The details of the work summarized below has either been published or submitted for publication elsewhere [1–3].

As area detectors (other than multiwire systems) are not energy discriminating devices, a potential source of error lies in the contamination of the data with harmonics of the assumed wavelength of the primary beam. The importance of this effect has been estimated for molybdenum Kα radiation using a graphite monochromator [1].

The quality of the intensity data obtainable has been assessed from an experiment on oxalic acid obtained at 100 K with a CCD detector. In this experiment the contamination of $\lambda/2$ to the measured intensities was eliminated by appropriate choice of the generator voltage. Various criteria for judging the quality of the data are discussed below [2].

The advantages of measuring at very low temperatures are well established [4]. Because of the increased speed of data collection, it now becomes feasible to consider the use of liquid helium as a cryogen. A prototype open-flow helium cooling device using mainly off-the-shelf components has been developed [3].

2. $\lambda/2$ contamination

The characteristic radiation employed for a typical diffraction experiment is commonly obtained by using a crystal monochromator. Harmonics of λ contributing to the primary beam leaving the X-ray source also contribute to the 'monochromatic' beam arriving at the sample. Limiting ourselves to Mo-Kα radiation, only $\lambda/2$ is important under normal operating conditions (50 kV). For any reflection $2h2k2l$ due to λ, there will be a contribution due to $\lambda/2$ to the intensity of the reflection hkl if this harmonic is present in the primary beam. When using a scintillation counter to detect the scattered radiation, this can be removed by energy discrimination. With

Paul G. Mezey and Beverly E. Robertson (eds.), Electron, Spin and Momentum Densities and Chemical Reactivity, 213–223

CCD detectors (and also image plates), this contamination is a potential source of systematic errors because these devices cannot discriminate with respect to energy.

There are four approaches to remove or account for this effect: (i) primary radiation free of any $\lambda/2$ component can be produced; (ii) a Si or Ge monochromator may be used; (iii) the $\lambda/2$ component of the scattered radiation can be determined; (iv) an independent determination of the amount of $\lambda/2$ scattering may be carried out.

$\lambda/2$ free radiation

A primary beam that is free of $\lambda/2$ radiation is produced when the accelerating voltage is reduced below the threshold required for $\lambda/2$ generation. This is given by [5]

$$\lambda_{\min} = \frac{hc}{eV} = 12{,}398\frac{1}{V}.$$

For Mo-K$\bar{\alpha}$ radiation, $\lambda/2 = 0.3554$ Å which is produced at a threshold voltage of 34.9 kV. Operating an X-ray source at this potential also results in a drastic reduction in intensity of the desired characteristic radiation [5].

$$I(\text{K}\alpha) = k(V - V_0)^{1.63}.$$

Thus, reducing the potential from 50.0 to 34.9 kV will reduce the intensity of the desired radiation (Mo-K$\bar{\alpha}$) by 68%. Although some of this loss in intensity may be recovered by increasing the filament current, this is not a convenient solution for most experiments. For samples which are strongly diffracting, e.g. minerals, this is a reasonable approach.

Use of Si or Ge monochromators

The amplitude of F_{222} for Si or Ge is close to zero, therefore the contribution of $\lambda/2$ to the 111 reflection is zero. Hence, the 111 reflection from a Si or Ge monochromator is used to obtain $\lambda/2$ free radiation. However, these monochromators also drastically reduce the intensity of the primary beam compared to the graphite monochromators found in most commercial diffractometers.

Experimental determination

A structure factor obtained from an experiment where there is $\lambda/2$ contamination may be written as

$$F'_{hkl} = F_{hkl} + kF_{2h2k2l}.$$

Thus, we may obtain a best value for k from a modification of the normal least squares procedure. As $k \ll 1$, F'_{hkl} will only significantly differ from F_{hkl} when F_{hkl} is small and F_{2h2k2l} is large, i.e. the reflections carrying the most information about k are those very reflections that are poorly observed. Hence, this is not a reliable method.

Independent determination of the $\lambda/2$ contribution to the scattering

A method for correcting intensities from film data was proposed by Guinier [6] where two films were used. These were separated by a metal filter designed to absorb radiation λ and let the more penetrating $\lambda/2$ radiation through. Subtraction of the intensities on the second film from those on the first gave intensities free from $\lambda/2$ contamination.

A method to measure the $\lambda/2$ contribution to the observed intensity at any reciprocal lattice point was proposed by Rees [7] for neutron data and is equally applicable to X-rays. By comparing the intensity of strong reflections with at least one index odd with that of pure $\lambda/2$ reflections (which appear at reciprocal lattice nodes) we obtain a direct measure of the two components. With an area detector, this information is always contained in the measurement but is normally ignored if the reflections have been correctly indexed. This information may be extracted using existing software by defining a new unit cell in which all the axes have been doubled. Now all of the original reflections in the data set have even indices; all of those with any odd index are pure $\lambda/2$ reflections. From the comparison of appropriate pairs of reflections, the ratio of the intensities of the two components of the primary beam may be estimated. As this is a property of the primary beam, it is best performed as a separate experiment designed to optimize the integration of the half integral reflections under standard operating conditions. It is, of course, a function of the accelerating voltage but not of the filament current. The value should be stable with time except for effects due to changes in the absorption of the window of the X-ray source. The difference between diffractometers is small but not negligible (see below).

Using three spherical crystals – the standard ylide crystal provided by Siemens Analytical Instrumentation, ruby and ammonium hydrogen tartrate (Enraf-Nonius standard crystal) – such an experiment has been carried out using two SMART CCD diffractometers. Before integration [8], all of the cell axes were multiplied by 2. Duplicate measurements were then averaged, and all odd reflections with values of $F^2 > 15$ esd's were compared with the reflection with double the indices to obtain the best value of k for the expression $F^2_{hkl} = kF^2_{2h2k2l}$. The average values of k obtained for the two diffractometers were 0.0014(2) and 0.00106(5).

These values were used to correct the intensities for nine representative data sets, organic crystals, organometallics and minerals, and the data compared with respect to systematic absences and space group assignment (i) with no corrections to the data, (ii) with an absorption correction (SADABS [9]), (iii) with only $\lambda/2$ correction, and (iv) with both absorption and $\lambda/2$ correction. Analogously four different refinements per sample were carried out based on F^2.

The magnitudes of the corrections varied quite widely over the nine data sets, the average correction being less than one esd, however the maximum correction was 47.4σ. In all cases there was an improvement in the number of 'observed' systematic absences and, hence, space group assignment.

The effects of the $\lambda/2$ correction on the final refinements of these routine data sets was negligible. There was no significant change in the final agreement factors and the changes in the geometrical parameters were all smaller than one esd. However, although it has been suggested that this is also the case for charge density data sets, it has yet to be rigorously demonstrated.

3. Oxalic acid

Experiment

An extensive data set on oxalic acid dihydrate was obtained at 100.0(1) K using a Siemens SMART Platform diffractometer and graphite monochromated Mo-Kα radiation. $\lambda/2$ contamination was eliminated by running the generator at 35 kV and 50 mA. Intensity data were collected using 0.3° omega scans with a detector distance of 3 cm. Maximum redundancy in the data was obtained by using four phi settings (0, 90°, 180°, 270°) for each of five detector positions (−35°, −65°, −95°, −109° and −50°) in 2θ. The 60 s frames were measured for the three lowest resolution detector positions, and 120 s frames for the two others. For the first four detector positions, 600 frames were measured for each phi setting and 500 for the final detector position.

Data reduction

The unit cell (Table 1) and orientation matrix were determined from the XYZ centroids of 8192 reflections with $I > 20\sigma(I)$. The intensities (SAINT [8]) were corrected for beam inhomogeneity and decay, and the esd's adjusted using SADABS [9]. An absorption correction was applied ($T_{\min}$ 0.949, $T_{\max}$ 0.983) and symmetry and multiply measured reflections averaged with SORTAV [10].

Of 46,135 reflections measured (29,973 with $I > 2\sigma(I)$), only 156 reflections were missing to $\sin\theta/\lambda = 1.34$ Å^{-1}; 5102 reflections were unique of which 2681 had been measured more than nine times (symmetry equivalents plus multiple measurements). The merging R values were $R1 = 0.037$ and $R2 = 0.024$ for 4809 accepted means. Examination of the reflection statistics (Table 2) with respect to $F_o^2/\sigma(F_o^2)(Q)$ and $\sin\theta/\lambda(S)$ indicates the usual trends and suggests that the data should be adequate for a charge density study.

Table 1. Comparison of unit cell parameters.

This Work	IUCr Study
$a = 6.1024(1)$ (Å)	$\bar{a} = 6.102(6)$
$b = 3.4973(1)$ (Å)	$\bar{b} = 3.501(7)$
$c = 11.9586(2)$ (Å)	$\bar{c} = 11.964(17)$
$\beta = 105.771(1)$ (°)	$\bar{\beta} = 105.80(5)$

Table 2. Reflection statistics.

	R1	R2	R*w*	N_{terms}	N_{means}
(a) With respect to intensity					
$Q < -2.0$	0.0527	0.0610	0.0758	8	4
$-2.0 < Q < -1.0$	0.4186	0.4591	0.4343	156	41
$-1.0 < Q < 0.0$	0.9073	0.9195	0.8695	3157	443
$0.0 < Q < 1.0$	0.8215	0.8215	0.8745	8342	1033
$1.0 < Q < 2.0$	0.3585	0.3668	0.4385	4621	566
$2.0 < Q < 3.0$	0.2105	0.2358	0.2622	3449	404
$3.0 < Q < 4.0$	0.1380	0.1622	0.1732	2522	292
$4.0 < Q < 6.0$	0.0963	0.1137	0.1194	4499	468
$6.0 < Q < 8.0$	0.0710	0.0856	0.0866	3551	333
$8.0 < Q < 10.0$	0.0543	0.0666	0.0656	2553	230
$10.0 < Q < 20.0$	0.0346	0.0417	0.0418	7804	634
$20.0 < Q < 50.0$	0.0219	0.0288	0.0252	5232	382
$50.0 < Q$	0.0144	0.0173	0.0161	547	48
(b) With respect to resolution					
$S < 0.500$	0.0192	0.0235	0.0227	3891	313
$0.500 < S < 0.600$	0.0241	0.0245	0.0304	3262	214
$0.600 < S < 0.650$	0.0286	0.0265	0.0370	1900	129
$0.650 < S < 0.700$	0.0327	0.0295	0.0431	2126	155
$0.700 < S < 0.750$	0.0385	0.0345	0.0483	2790	181
$0.750 < S < 0.800$	0.0467	0.0391	0.0571	3343	207
$0.800 < S < 0.850$	0.0479	0.0378	0.0575	2998	213
$0.850 < S < 0.900$	0.0533	0.0423	0.0608	3682	270
$0.900 < S < 0.950$	0.0657	0.0502	0.0730	3872	289
$0.950 < S < 1.000$	0.0772	0.0577	0.0854	3343	298
$1.000 < S < 1.050$	0.1026	0.0750	0.1140	3418	352
$1.050 < S < 1.100$	0.1119	0.0778	0.1197	3169	386
$1.100 < S < 1.150$	0.1081	0.0823	0.1253	2296	338
$1.150 < S < 1.200$	0.1449	0.1041	0.1642	2393	415
$1.200 < S < 1.250$	0.1334	0.1054	0.1520	2347	480
$1.250 < S < 1.300$	0.1525	0.1165	0.1821	1253	396
$1.300 < S < 1.350$	0.1433	0.1139	0.1776	358	156

$Q = F_{\circ}^2/\sigma(F_{\circ}^2)$, $S = \sin\theta/\lambda$, $\mathrm{R1} = \sum|Y - \bar{Y}|/\sum|Y|$, $\mathrm{R2} = N\left(\sum(Y - \bar{Y})^2/\sum Y^2\right)$, $Rw = N\left(\sum w((Y - \bar{Y})/\sigma(Y))^2/\sum w(Y/\sigma(Y))^2\right)$ where $Y = F_{\circ}^2$.

Refinements

Starting coordinates were taken from Stevens and Coppens (hereafter SC) [11] and all refinements were carried out on F^2 using the XD suite of programs [12]. Four different refinements were carried out using statistical weights throughout and the results are summarized in Table 3. Refinement **I** is an independent atom refinement; **II** is a high angle refinement ($1.00 < \sin\theta/\lambda < 1.34\,\text{Å}^{-1}$) with the hydrogen atoms fixed at the neutron positions [13] with isotropic thermal parameters fixed at the values obtained from **I**; **III** is a kappa refinement to assign atomic charges [14] with hydrogen parameters fixed as in **II**; a complete atom centered multipole refinement [15] was carried out in **IV** with hydrogen atoms treated as in **II** with one atom directed dipole and quadrupole population varied. All other atoms were refined as previously [11] up

Table 3. Summary of least squares refinements.

	Refinement			
	I	**II**[a]	**III**[a]	**IV**[a]
$\sin\theta/\lambda$ range (Å^{-1})	0.00–1.20	1.00–1.33	0.00–1.33	0.00–1.20
N_{obs}	3860	2895	5166	3860
N_{V}	50	37	50	108
Scale factor	1.812(2)	1.87(2)	1.813(4)	1.812(4)
$R(F^2)^b$	0.0540	0.1170	0.0457	0.0532
$Rw(F^2)^b$	0.0769	0.1630	0.0777	0.0299
$R(F)(I > 2\sigma(I))$	0.0281	0.0423	0.0282	0.0190
GOF	1.01	0.84	0.86	0.73

[a]Hydrogen coordinates fixed at neutron positions with isotropic thermal parameters from **I**.
[b]All reflections.

to the hexadecapole level with mirror symmetry imposed in the plane of the oxalic acid molecule and in the bisecting plane perpendicular to the plane of the water molecule. Isotropic extinction was included in all four refinements, however, the value obtained from refinement **I** was held constant for the high angle refinement. For purposes of comparison, particularly of the multipole refinements, an identical set of refinements was carried out using the SC data [11] and the results compared. As the SC data only extend to $\sin\theta/\lambda = 1.2$ Å^{-1}, the current data was also limited to this resolution for refinement **IV**.

Positional parameters of the non-hydrogen atoms obtained from refinements **I** and **II** are in good agreement with those of SC (1980) or Dam, Harkema and Feil (hereafter DHF) [16] from X-ray data as well as those from neutron data [13, 17].

The values for the thermal displacement parameters fall in the same range as reported in the IUCr study [18], however, all refinements gave values that were systematically smaller than those obtained by SC [11]. In contrast, the values are quite similar to the values reported by DHF [16]. It is tempting to suggest that the thermal parameters are too small due to the presence of TDS contamination in the intensity data, however, it is more likely that the experimental temperatures were not identical. The agreement of thermal parameters from refinement **II** with those obtained from one neutron data set [17] is quite satisfying, however, agreement with a second neutron study [13] reported for the same temperature differed by 15%, again suggesting a problem with temperature calibration across these experiments.

Charge densities

The atomic charges estimated from a kappa refinement are given in Table 4 as well as those obtained from the SC data [11]. The main difference is the positive charge obtained for the C atom from the CCD data (in agreement with chemical intuition) compared to a small negative one obtained using the point detector data.

Difference density maps from the current data and the SC [11] data after high order refinements are shown in Figure 1(a) and (b). The main features of the maps agree

Table 4. Comparison of the results of the kappa refinements.

	This work		SC	
	q	κ	q	κ
O(1)	−0.42(3)	0.976(3)	−0.18(3)	0.986(3)
O(2)	−0.51(2)	0.972(3)	−0.32(3)	0.973(3)
O(3)	−0.63(3)	0.960(3)	−0.44(4)	0.986(4)
C(1)	0.27(3)	1.032(5)	−0.06(5)	0.990(7)
H(1)	0.49(2)	1.49(6)	0.29(3)	1.20(3)
H(2)	0.41(2)	1.34(3)	0.35(2)	1.22(3)
H(3)	0.41(2)	1.34(3)	0.35(2)	1.22(3)

with respect to the bonding and lone pair regions, however, there are much deeper negative regions close to the nuclear positions for the current data. In contrast, a similar map reported by DHF [16] also has similar negative regions.

Comparison of the model maps (Figure 1(c) and (d)) again shows qualitative agreement. All regions agree to within one contour level (0.05 e $Å^{-3}$) except for the lone pair regions for O(2) and O(3) which are significantly sharpened compared to those obtained with the SC data.

Examination of the multipole populations gives no indication of the discrepancy observed in the model maps, all populations from parallel refinements agreeing to within two esd's (Table 5). The one striking exception is the monopole population (P_r) for carbon. This must be a simple difference in the partitioning of the charge density between atom centers in the model as there is no discernible difference in the model maps around the carbon position.

It has been suggested that the integrated intensities of weak reflections are overestimated by the current integration algorithm (SAINT, [8]). The ratio between F_0^2 for the current data and that of SC [11], scaled from the multipole refinements, has been calculated for all common reflections (Figure 2). The agreement for the strong reflections is good, however, either the CCD detector has indeed overestimated the weak intensities or else they are underestimated by the scintillation counter.

4. Helium cooling

Most previous attempts to obtain X-ray diffraction data at very low temperatures ($<$ 80 K) have used custom built systems with closed cycle helium refrigerators mounted on large, robust four circle diffractometers. In order to remove the inherent disadvantages of these systems – cost, single application, absorption and scattering of the windows – we have built an open flow system from mainly off-the-shelf components which uses liquid helium as the cryogen. This is not the first open flow helium system [19, 20] but is the first that is mainly off-the-shelf and is mountable on any diffractometer. It is based on an ADP Helitran ESR cryostat with modifications to the nozzle assembly and to the direction of the gas flow. The lowest temperature is estimated to be $<$30 K. At the current price for liquid helium in

Figure 1. Difference map after high order refinement – data cutoff at $\sin\theta/\lambda = 0.9$ Å^{-1} for reflections with $F^2 > 2\sigma(F^2)$; (a) this work; (b) SC. Model map after multipole refinement – data cutoff at $\sin\sigma/\lambda = 0.9$ Å^{-1} for reflections with $F^2 > 2\sigma(F^2)$; (c) this work; (d) SC.

the US, with the increase in measuring speed available with a CCD detector, the economics per data set are comparable with those using liquid nitrogen on a conventional diffractometer.

5. Conclusion

Clearly the CCD data is adequate for a charge density analysis. It is also possible that the weak data, contrary to popular wisdom, is actually better than that from point

Table 5. Comparison of refined multipole parameters.

	C(1)		O(1)		O(2)		O(3)	
	This work	SC	This work	SC	This work	SC	This work	SC
κ'	0.981(3)	0.974(4)	0.974(3)	0.987(2)	1.000(5)	0.978(2)	1.014(6)	0.999(3)
P_v	3.87(4)	4.31(3)	6.22(3)	6.06(2)	6.47(3)	6.27(2)	5.83(5)	5.77(3)
P_{+11}	0.026(14)	0.063(8)	−0.015(9)	−0.028(4)	−0.051(10)	−0.045(6)	−0.100(9)	−0.082(5)
P_{-11}	−0.015(11)	0.018(6)	0.001(7)	−0.055(5)	0.010(8)	0.010(4)	−0.021(8)	0.011(5)
P_{20}	−0.264(12)	−0.255(7)	−0.032(10)	−0.008(6)	−0.078(10)	−0.062(7)	−0.015(8)	−0.037(5)
P_{+22}	0.072(12)	0.069(7)	−0.045(9)	−0.040(4)	−0.003(9)	−0.039(6)	−0.022(8)	−0.016(5)
P_{-22}	−0.044(11)	−0.031(6)	0.027(9)	0.022(5)	0.040(8)	0.003(5)	−0.013(8)	−0.006(5)
P_{+31}	0.031(13)	0.017(7)	0.012(10)	0.019(4)	−0.014(10)	0.005(5)	0.124(8)	0.090(4)
P_{-31}	0.009(11)	−0.013(6)	−0.028(9)	−0.039(4)	−0.015(9)	−0.012(4)	0.020(8)	0.024(4)
P_{+33}	0.287(14)	0.301(7)	0.079(9)	0.076(4)	0.057(9)	0.034(4)	0.013(8)	−0.001(4)
P_{-33}	0.030(16)	0.067(7)	0.012(10)	−0.011(4)	0.016(8)	−0.007(4)	0.006(8)	−0.006(4)
P_{40}	0.038(16)	0.026(8)	0.010(13)	0.017(6)	−0.001(13)	−0.006(6)	−0.019(11)	−0.002(5)
P_{+42}	0.009(17)	0.016(8)	0.021(12)	0.014(5)	−0.003(11)	−0.013(5)	0.062(10)	0.071(5)
P_{-42}	−0.018(17)	−0.010(8)	0.007(11)	−0.005(6)	−0.010(11)	−0.000(5)	−0.001(10)	0.018(5)
P_{+44}	−0.061(20)	−0.031(9)	−0.001(10)	0.028(5)	−0.008(10)	−0.007(5)	−0.010(10)	0.012(5)
P_{-44}	−0.051(15)	−0.007(8)	0.032(10)	0.029(4)	0.010(10)	−0.002(4)	−0.003(10)	−0.009(5)

	H(1)		H(2)		H(3)*	
	This work	SC	This work	SC	This work	SC
κ	1.00	1.00	1.00	1.00	1.00	1.00
P_v	0.75(2)	0.87(1)	0.93(2)	0.86(1)	0.93(2)	0.86(1)
P_{10}	0.197(21)	0.253(12)	0.392(20)	0.384(11)	0.392(20)	0.384(11)
P_{20}	0.330(31)	0.330(20)	0.275(22)	0.243(13)	0.275(22)	0.243(13)

* H(3) populations are constrained equal to H(2).

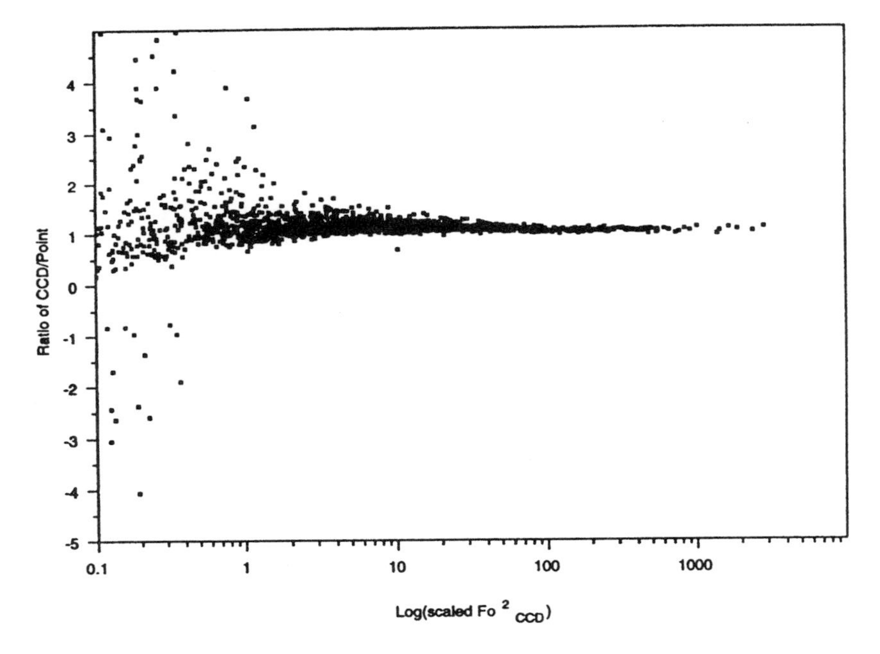

Figure 2. Ratio between scaled F^2 for CCD and point detector data with respect to log F^2.

detectors. Following informal discussions at the 1997 ACA meeting[1] it seems possible that even better data will be obtained by using narrower frames (e.g. 0.1°).

Acknowledgements

We thank Professor E.D. Stevens for providing the data for the 1980 study, the College of Arts and Sciences of the University of Toledo and the Ohio Board of Regents for generous financial support of the X-ray diffraction facility and thank the Office of Naval Research for funding this work (Contract Nos. N00014-95-1-0013 and N00014-95-1-1252).

References

1. Kirschbaum, K., Martin, A. and Pinkerton, A.A. (1997) *J. Appl. Cryst.*, **30**, 514–516.
2. Martin, A. and Pinkerton, A.A. (1998) *Acta Cryst.*, **B54**, 471–478.
3. Hardie, M.J., Kirschbaum, K., Martin, A. and Pinkerton, A.A. (1998) *J. Appl. Cryst.*, **31**, 815–817.
4. Larsen, F.K. (1995) *Acta Cryst.*, **B51**, 468–482.
5. *International Tables for X-Ray Crystallography*, Vol. C, Kluwer Academic Publishers, Dordrecht, 1995.

[1]Contributors to the discussion were the author, Drs. W. Clegg, P. Coppens, C.S. Frampton, T. Koritsanszky, F.K. Larsen, P.M. Mallinson, A. Martin, M.R. Pressprich and E.D. Stevens.

6. Guinier, A. (1963) *X-Ray Diffraction in Crystals, Imperfect Crystals, and Amorphous Bodies*, W.H. Freeman and Co., pp. 175.
7. Rees, B. (1977) *Israel J. Chem.*, **16**, 154–158.
8. Siemens Analytical Instrumentation (1996) SAINT, a program to integrate and reduce raw crystallographic area detector data.
9. Sheldrick, G.M. (1996) SADABS, University of Göttingen, Germany, to be published.
10. Blessing, R.H. (1987) *Cryst. Rev.*, **1**, 3–57.
11. Stevens, E.D. and Coppens, P. (1980) *Acta Cryst.*, **B36**, 1864–1876.
12. Koritsanszky, T., Howard, S., Mallison, P.R., Su, Z., Richter, T. and Hansen, N.K. (1995) XD, a program for refinement and display of charge densities.
13. Feld, R.H., Brown, P.J. and Lehmann, M.S. (1978) *Acta Cryst.*, **A34**, S28; Feld, R.H. (1980) Ph.D. thesis, Philipps-Universität Marburg/Lahn.
14. Coppens, P., Guru Row, T.N., Leung, P., Stevens, E.D., Becker, P.J. and Yang, Y.W. (1979) *Acta Cryst.*, **A35**, 63–72.
15. Hansen, N.K. and Coppens, P. (1978) *Acta. Cryst.*, **A34**, 909–921.
16. Dam, J., Harkema, S. and Feil, D. (1984) *Acta Cryst.*, **B39**, 760–768.
17. Koetzle, T.F and McMullan, R., unpublished results.
18. Coppens, P. (1984) *Acta Cryst.*, **A40**, 184–195.
19. Greubel, K.H., Gmelin, E., Moser, H., Mensing, C. and Walz, L. (1990) *Cryogenics*, **30** (Suppl.), 457–462.
20. Teng, T., Schildkamp, W., Dolmer, P. and Moffat, K. (1994) *J. Appl. Cryst.*, **27**, 133–139.

15

Recent studies in magnetisation densities

E. LELIÈVRE-BERNA

Institut Laue Langevin, BP 156, 38042 Grenoble cedex 9, France(E-mail: lelievre@ill.fr)

1. Introduction

The polarised neutron diffraction (PND) technique [1, 2] applies to single crystals which are magnetically ordered in a ferro- or ferrimagnetic phase under an applied magnetic field. Assuming a good knowledge of the nuclear structure factors N (i.e. the Fourier components of the density of atomic nuclei in the unit cell), the dependence of the elastic scattering cross-section on the initial neutron polarisation gives access to magnetic structure factors M (i.e. the Fourier components of the magnetisation density). In practice, one measures the 'flipping ratio' R between the intensities observed for $(+)$ and $(-)$ initial polarisation states at the peak of each Bragg reflection. The experimental flipping ratios are easily corrected for some instrumental imperfections, and one has to take into account extinction which may occur in the scattering process [3]. As shown by the study of Ce_3Al_{11}, it is even possible to determine the magnetisation density of a twined crystal [4].

2. Methods of analysis

After a simple Fourier inversion of a set of magnetic structure factors M_{hkl}, one can retrieve the magnetisation density. A much better result, e.g. the most probable density map, can be obtained using the Maximum Entropy (MaxEnt) method. It takes into account the lack and the uncertainty of the information: not all the Bragg reflections are accessible on the instrument, and all the values contained in the error bars are satisfactory and have to be considered. However, as this method extracts all the information contained in the data, it is important to keep in mind that it may show spurious small details associated to a low accuracy and/or a specific lack of information located in Q-space.

Another way consists in comparing the measurements to a model. One can express the atomic form factor in the dipolar approximation at low $\sin\theta/\lambda$, or the magnetisation density in a multipolar expansion. In the case of molecular magnets, one generally fits a Hartree–Fock magnetic wave function constructed from standard Slater orbitals at each magnetic site. A more versatile modelling is obtained by expanding the spin density in a superposition of aspherical densities (series expansion in real spherical harmonics).

Paul G. Mezey and Beverly E. Robertson (eds.), Electron, Spin and Momentum Densities and Chemical Reactivity, 225–233

3. Molecular compounds

3d transition metal complexes and organic free radicals form groups of molecular compounds to which PND has been successfully applied [5, 6]. The results provide a basis to understand the antiferro- and ferromagnetic spin-coupling mechanism of molecular-based magnets. It also gives the opportunity to study 2p electron magnetism and to test new theories in fundamental physics with model systems.

Magnetism in molecular compounds is firstly characterised by the spin delocalisation which is required for the existence of exchange interaction. These effects are clearly shown in binuclear compounds such as the Mn(II)Cu(II) heterobimetallics where the two metals are antiferromagnetically coupled [7–9]. One can see in the oxamato-bridged Mn(II)Cu(II) chain that the Cu(II) magnetic orbital is more delocalised toward the nitrogen atom than the oxygen atoms of the copper basal plane (Figure 1). It explains the more pronounced antiferromagnetic exchange interaction observed for the oxamido-bridged Mn(II)Cu(II) pair.

Some confirmatory evidence from PND experiments for spin and charge transfer through hydrogen bonding have been reported in $Ni(NH_3)_4(NO_2)_2$ and $[Co(NH_3)_5(OH_2)][Cr(CN)_6]$ [10–12]. It was suggested that it could provide an explanation for the antiferromagnetic exchange interaction in the complex $Ni(NH_3)_4(NO_2)_2$. The spin density of the $[Cr(CN)_6]$ ion is distinctly different from that in the salt $Cs_2KCr(CN)_6$ where strong features of covalence and spin polarisation are observed [13]: the local density functional calculations are in broad agreement with the experimental results, giving considerable spin transfer to the cation and large spin polarisation effects, with possibly a strong involvement of protons in the hydrogen bonds. In $Cs_2KFe(CN)_6$, 60% of the total moment of the $Fe(CN)_6$ units comes from

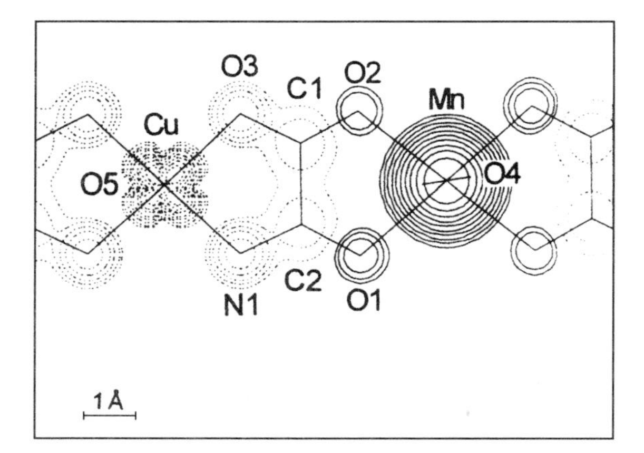

Figure 1. Induced spin density map for $MnCu(pba)(H_2O)_3 \cdot 2H_2O$ at 10 K under 5 T in projection along the perpendicular to the basal plane. Solid and dashed lines are used respectively for negative and positive spin densities. Contour steps are $5\ m\mu_B/Å^2$. The spin delocalisation is more pronounced toward the N atom than the O atoms.

the orbital contribution [14]. The magnetisation density is dominated by the orbital angular momentum, and its distribution shows almost cylindrical symmetry about the ligand direction closest to the applied field. Metal–water bonding interactions have also been studied in a $CsMo(SO_4)_2 \cdot 12D_2O$ single crystal [15]. Most of the spin is concentrated in the Mo(III) t_{2g} orbitals, but there is significant spin transfer to the ligand. The presence of metal–water bonding normal to the plane of the water molecule is shown, as well as the absence of significant in-plane metal–ligand spin transfer which implies that metal–ligand interaction is highly anisotropic.

Several $[TCNE]^{\cdot -}$ (TCNE = tetracyanoethylene) based ferromagnets and ferrimagnets have been successfully synthesised. Among these, $[Fe(C_5Me_5)_2]^{\cdot +}[TCNE]^{\cdot -}$, a solvent soluble salt, is a ferromagnet with an ordering temperature of 4.8 K [16–18]. The MaxEnt reconstruction of the spin density of $[TCNE]^{\cdot -}[Bu_4N]^{\cdot +}$ and the corresponding density functional theory calculations are consistent and show that the excess α-spin on the $[TCNE]^{\cdot -}$ is distributed across the radical. It demonstrates that $[TCNE]^{\cdot -}$ has the possibility for strong magnetic interactions with neighbouring spin sites [19, 20]. Using a non-uniform atomic orbital model, the MaxEnt reconstruction of the spin density shows features contained in the data, but not in the model. There is a significant off-centring, and the nitrogen spin populations are inequal (Figure 2).

Nitronyl and imino nitroxide free radicals are also among the most versatile spin carriers which are widely used in the design of molecular magnets. Their delocalised unpaired electrons make them convenient building blocks and ideal magnetic bridges between magnetic metals, to achieve new compounds with particular

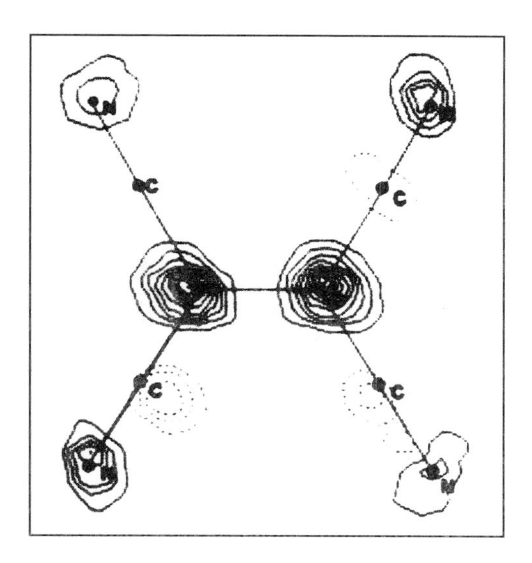

Figure 2. $[TCNE]^{\cdot -}[Bu_4N]^{\cdot +}$ spin density obtained by MaxEnt reconstruction using an atomic orbital model, and subsequent projection onto the molecular plane of $[TCNE]^{\cdot -}$. Positive contour steps are $50\,m\mu_B/Å^2$ and negative contours are dashed (step $10\,m\mu_B/Å^2$). A significant off-centring is present.

magnetic properties. It has been shown that NITϕ (2-phenyl-4,4,4,5,-tetramethyl-4,5-dihydro-1,H-imidazol-3-oxide-1-oxyl), Cu(hfac)$_2$NIT-Me (catena-(m-1,3-2,4,4,5,5-penta-methyl-4,5-dihydro-1, H-imidazol-3-oxide-1-oxyl)bis(hexafluroacetylaceto-nato) copper(II)), p-NPNN (β-para-nitrophenyl nitronyl nitroxide) and m-NPIM (2-(3-nitrophenyl)-4,4,5,5-tetramethyl-4,5-dihydro-1H-imidazol-1-oxyl) exhibit ferromagnetic coupling [21–26], and that $CuCl_2(NIT\phi)_2$ (bis-(2-phenyl-4,4,5,5,-tetramethyl-4,5-dihydro-1, H-imidazol-3-oxide-1-oxyl)copper(II)-chloride) and m-NPIN (meta-nitrophenyl imino nitroxide) exhibit antiferromagnetic interactions [23–26]. To the knowledge of the author, the rather large negative contribution to the spin density on the carbon atom of the O–N–C–N–O fragment is observed systematically on all the nitronyl nitroxides which have been investigated up to now. In $(Cu(hfac)_2NIT\text{-}Me)_n$, 1 and $CuCl_2(NITPh)_2$,2, one can also see that coordination of a nitroxide to a Cu(II) ion results in spin density transfer from the bound oxygen atom to the nitrogen. This effect is more pronounced when the Cu(II) spin density is coupled antiferromagnetically to the spin of the unpaired electron of the O–N–C–N–O fragment (Figure 3).

In all cases, the oversimplified theory formulated by Kahn and Briat [27, 28] allows to guess the sign of magnetic coupling: the exchange interaction between electrons residing on two orbitals is ferromagnetic when those orbitals are orthogonal, and usually antiferromagnetic otherwise. In fact, the understanding of the interaction requires detailed knowledge of the orbital structure. Unrestricted Hartree–Fock calculations might predict [29] a good estimate of T_c, but give generally a wrong spin population. Compared to UHF, the density functional theory does better in predicting the spin polarisation effect [30], but a restricted open-shell Hartree–Fock calculation with configurational interaction corrections is now considered.

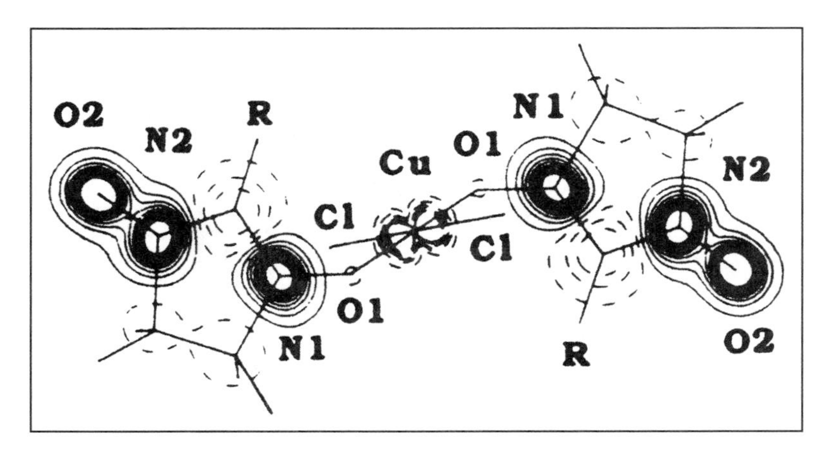

Figure 3. $CuCl_2(NITPh)_2$,2: projection of the spin density along the π^* direction of the nitroxide. Contours $5 \pm n(10)\ m\mu_B/Å^2$. Coordination of a nitroxide to a copper(II) ion results in spin density transfer from the bound O atom to the N atom.

4. Itinerant magnetism

Polarised neutron diffraction technique is also very useful for studying lanthanide- and actinide-transition metal intermetallics. The itinerant-electron theory and the model in which localised magnetic moments are assumed, have been the subject for discussion for many years. The itinerant approach, which is associated to d-electron systems, and mainly to 3d, is based on the strong overlap of the charge distributions of atoms. The magnetism is of spin origin and is governed by the crystal field and in a lower manner by the exchange interaction. It may lead to small magnetically ordered moments and to high ordering temperatures.

Previous experiments have shown that the internal magnetisation density of Ni is fitted well by a 3d-like density superposed on a uniform negative 'diffuse' magnetisation which has been ascribed by band calculations to negative polarisation of the 4s conduction band. This reverse polarisation only occurs when there is long range order, and cannot be induced either by an applied field or by the short range fluctuating local fields which are present in the paramagnetic state [31]. The internal magnetisation density of Ni is well reconstructed by the MaxEnt method with positive values only, and it reveals clearly the magnetic anisotropy of e_g and t_{2g} types. But allowing for negative contribution, the corresponding form factors give contribution to the innermost reflections and vary in an oscillatory manner contrary to what was expected [32]. In fact, effect of a presence of the negative contribution is of the order of the sensitivity of the MaxEnt program, and further investigations are needed at high $\sin\theta/\lambda$.

Recently, it has been shown that 3d elements might also exhibit magnetic instabilities. In the hexagonal Laves phase compound $TiFe_2$, the fact that a site with no ordered moment persists down to low temperatures suggests that the Fe moment is near to instability [33] as it is observed [34–37] for Mn in YMn_2 or $TbMn_2$. Binary cubic alloys might also exhibit high temperature strongly magnetic materials like $ZrFe_2$, in which a strong 3d–4d hybridisation is reported [38].

5. Localised magnetism

Contrary to itinerant magnetism, the localised approach treats electrons as strongly associated with a particular atom. The 4f-magnetism is of spin and orbital origin and governed by the spin–orbit interaction and a lower exchange interaction. The ordering temperatures are usually low and the complex magnetic configurations are promoted indirectly via conduction electrons.

Among lanthanides, cerium and samarium exhibit particular behaviours compared to other rare-earths. For example, in $CeFe_2$, the 4f electrons form energy bands, and the Ce atoms carry a magnetic moment oppositely polarised to the Fe moments [39] as it is the case in $ZrFe_2$. The magnetic form factor measured in the mixed-valence $^{154}Sm^{11}B_6$ is in agreement with a purely Sm^{2+} form factor [40]. This result is surprising owing to the anomalies in the Q-dependence of the form factor and to the strong different magnetic densities expected for the two valence states Sm^{2+} and Sm^{3+}.

6. Actinides: localisation vs itinerency

Apart from d- and 4f-based magnetic systems, the physical properties of actinides can be classified to be intermediate between the lanthanides and d-electron metals. 5f-electron states form bands whose width lies in between those of d- and 4f-electron states. On the other hand, the spin–orbit interaction increases as a function of atomic number and is the largest for actinides. Therefore, one can see direct similarity between the light actinides, up to plutonium, and the transition metals on one side, and the heavy actinides and 4f elements on the other side. In general, the presence or absence of magnetic order in actinides depends on the shortest distance between 5f atoms (Hill limit).

In neptunium intermetallics, the critical Hill spacing is about 3.2 Å, and we should anticipate strong 5f hybridisation leading to non-magnetic behaviour in $NpCo_2$ where $d_{Np-Np} \approx 3.05$ Å. However, hybridisation may also occur with the unpaired 3d electrons of magnetic Cobalt atoms. Many previous experiments have failed to reveal any direct evidence for long-range ordering of the Np moments. A recent work has shown unambiguously the development of long-range order below 13 K, and the PND experiment [41] gives the proof of hybridisation between 5f and 3d electrons, and shows a decrease of the orbital magnetic moment compared to the spin 5f moment. This result is in agreement with single-electron band-structure calculations which predict the reduction of the orbital moments of the actinides 5f electrons.

Concerning induced orbital moments of U-based intermetallic compounds, many PND experiments have been performed and have shown that the ratio μ_L/μ_S can be used as a measure of the hybridisation [42–44] (in the light actinides, orbital and spin moments are oppositely directed and the neutron magnetic form factors are highly sensitive to the ratio μ_L/μ_S). Indeed, this ratio is reduced as compared to the free ion expectations (Figure 4).

Recently, it has been shown that this hybridisation may be anisotropic. In URhAl, there is a strong moment on RhI atoms, but no moment on RhII atoms [45]. This anisotropy is at the origin of the bulk anisotropy. This hybridisation also plays a significative role in the appearance of magnetic anisotropy in U_3X_4-type actinides. In U_3Bi_4 and U_3Sb_4, the different magnetic moments determined on the two U sites are due to a low local symmetry environment [46]. We also observed a strong reduction of their magnitudes compared to the free ion value.

7. Conclusion

The PND technique is a very sensitive method and certainly the most powerful tool for determining magnetisation densities. It reveals unambiguously the spin delocalisation, the polarisation sign, the density shape and the effects of magnetic interactions. It may also separate precisely the spin and orbital contributions, and combined with well adapted data treatment analyses, it gives access to precise quantitative results. Compared to X-ray diffraction technique, the PND technique gives a more direct way of investigating the chemical bonding involved in molecules containing unpaired electrons: only electrons of the outer valence shells are considered. Furthermore,

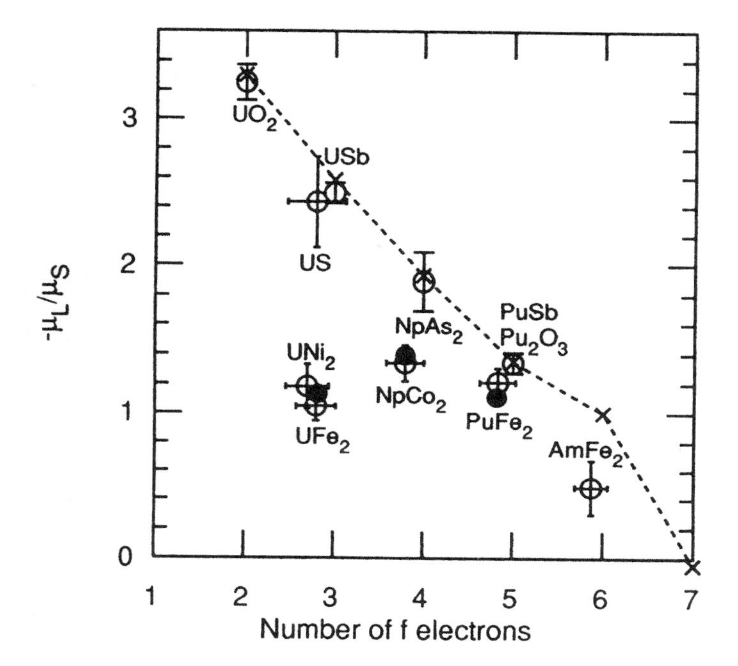

Figure 4. Dependence of the ratio $-\mu_L/\mu_S$ on the number of 5f electrons for light actinide compounds: × free ion values, ○ experimental values, • form band calculations. The hybridisation between 5f and 3d electrons leads to the reduction of the 5f orbital moments (metallic covalency).

contrary to magnetic resonance, PND corresponds to observations concerning the whole space of the unit cell.

References

1. Nathans, R., Shull, C.G., Shirane G. *et al.* (1959) The use of polarized neutrons in determining the magnetic scattering by iron and nickel, *J. Phys. Chem. Solids*, **10**, 138–146.
2. Squires, G.L. (1978) *Introduction to the Theory of Thermal Neutron Scattering*, University Press, Cambridge.
3. Schweizer, J. and Tasset, F. (1980) Polarised neutron study of the RCo_5 intermetallic compounds: I. The cobalt magnetisation in YCo_5, *J. Phys. F: Met. Phys.*, **10**, 2799–2818.
4. Muñoz, A., Betallan, F., Boucherle, J.X. *et al.* (1995) Magnetization density in Ce_3Al_{11}, *J. Phys.: Cond. Mat.*, **7**(46), 8821–8831.
5. Gillon, B. and Schweizer, J. (1989) Study of chemical bonding in molecules: the interest of polarised neutron diffraction, *J. Mol. Phys, Chem. Bio.*, **3**, 111–147.
6. Schweizer, J. (1996) Spin densities in magnetic molecular compounds, *Physica B*, (in press).
7. Baron, V., Gillon, B., Kahn, O. *et al.* (1993) Spin density in a bimetallic magnetic chain $MnCu(pba)(H_2O)_3 \cdot 2H_2O$: a polarised neutron diffraction study, *Mol. Cryst. Liqu. Cryst.*, **233**, 247–256.
8. Baron, V., Gillon, B., Plantevin, O. *et al.* (1996) Spin-density maps for an oxamido-bridged Mn(II)Cu(II) binuclear compound. Polarized neutron diffraction and theoretical studies, *J. Am. Chem. Soc.*, **118**, 11822–11830.

9. Baron, V., Gillon, B., Cousson, A. *et al.* (1997) Spin density maps for the ferrimagnetic chain compound $MnCu(pba)(H_2O)_3 \cdot 2H_2O$ (pba = 1,2-propylenebis(oxamato)): polarized neutron diffraction and theoritical studies, *J. Am. Chem. Soc.*, **119**, 3500–3506.
10. Figgis, B.N., Reynolds, P.A. and Mason, R.J. (1983) Covalent bonding in *trans*-tetraammine-dinitronickel(II) studied by polarised neutron diffraction, *J. Am. Chem. Soc.*, **105**, 440–443.
11. Figgis, B.N. and Kucharski, E.S. (1993) Spin transfer through hydrogen bonding in $[Co(NH_3)_5(OH_2)][Cr(CN)_6]$? *Z. Naturforsch.*, **A48**, 123–126.
12. Figgis, B.N., Kucharski, E.S. and Vrtis, M. (1993) Spin and charge transfer through hydrogen bonding in $[Co(NH_3)_5(OH_2)][Cr(CN)_6]$, *J. Am. Chem. Soc.*, **115**, 176–181.
13. Figgis, B.N., Forsyth, J.B. and Reynolds, P.A. (1987) Spin density of the Hexacyanochromate(III) ion measured by polarized neutron diffraction, *Inorg. Chem.*, **26**, 101–105.
14. Day, P., Delfs C.D., Figgis B.N. *et al.* (1993) Polarized neutron diffraction from $Cs_2KFe(CN)_6$. The orbital moment and its anisotropy, *Mol. Phys.*, **78**, 769–780.
15. Best, S.P., Figgis, B.N., Forsyth, J.B. *et al.* (1995) Spin distribution and bonding in $[Mo(OD_2)_6]^{3+}$, *Inorg. Chem.*, **34**, 4605–4610.
16. Miller, J.S. and Epstein, A.J.(1994) Organic and organometallic molecular magnetic-materials – Designer magnets, *Angew. Chem.*, **33**, 385–415.
17. Buchachenko, A.L. (1990) Organic and molecular ferromagnetics: advances and problems, *Russ. Chem. Rev.*, **59**, 307–319.
18. Kahn, O. (1987) Magnetism of the heteropolymetallic systems, *Struct. Bonding*, **68**, 89–167.
19. Zheludev, A., Grand, A., Ressouche, E. *et al.* (1994) Experimental determination of the spin density in the tetracyanoethenide free radical, $[TCNE]^{\cdot -}$, by single-crystal polarized neutron diffraction, a view of a π^* orbital, *J. Am. Chem. Soc.*, **116**, 7243–7249.
20. Zheludev, A., Papoular, R.J., Ressouche, E. *et al.* (1995) A non-uniform reference model for maximum-entropy density reconstructions from diffraction data, *Acta Cryst.*, **A51**, 450–455.
21. Zheludev, A., Bonnet, M., Delley, B. *et al.* (1995) An imino nitroxide free radical: experimental and theoretical spin density and electronic structure, *J. Mag. Mag. Mat.*, **145**, 293–305.
22. Zheludev, A., Bonnet, Luneau, D. *et al.* (1995) The spin density in an imino nitroxide free radical: a polarized-neutron study, *Physica B*, **213–214**, 268–271.
23. Bonnet, M., Luneau, D., Ressouche, E. *et al.* (1995) The experimental spin density of two nitrophenyl nitroxides: a nitronyl nitroxide and an imino nitroxide, *Mol. Cryst. Liqu.Cryst.*, **271–272**, 35–53.
24. Papoular, R.J., Zheludev, A., Ressouche, E. *et al.* (1995) The inverse Fourier problem in the case of poor resolution in one given direction: the maximum-entropy solution, *Acta Cryst.*, **A51**, 295–300.
25. Zheludev, A., Bonnet, M., Ressouche, E. *et al.* (1994) Experimental spin density in the first purely organic ferromagnet: the beta para-nitrophenyl nitronyl nitroxide, *J. Mag. Mag. Mat.*, **135**, 147–160.
26. Ressouche, E., Zheludev, A., Boucherle, J.X. *et al.* (1993) Spin densities in nitronyl nitroxide free radicals, *Mol. Cryst. Liqu. Cryst.*, **232**, 13–25.
27. Kahn, O. and Briat, B. (1976) Exchange interaction in polynuclear complexes – Part 1, *J. Chem. Soc., Faraday II*, **B72**, 268–281.
28. Kahn, O. and Briat, B. (1976) Exchange interaction in polynuclear complexes – Part 2, *J. Chem. Soc., Faraday II*, **B72**, 1441–1446.
29. Okumara, M., Mori W. and Yamaguchi, K. (1993) Theoritical studies of magnetic orderings in the β- and γ-phases of p-NPNN and related nitroxides, *Mol. Cryst. Liqu. Cryst.*, **232**, 35–44.
30. Zheludev, A., Baron, V., Bonnet, M. *et al.* (1994) Spin density in a nitronyl nitroxide free radical. Polarized neutron diffraction investigation and *ab initio* calculation, *J. Am. Chem. Soc.*, **116**, 2019–2027.
31. Brown, P.J., Déportes, J., Neumann, K.U. and Ziebeck, K.R.A. (1992) High-temperature magnetisation distribution in nickel, *J. Mag. Mag. Mat.*, **104–107**, 2083–2084.
32. Dobrzynski, L., Papoular, R.J. and Sakata, M. (1995) Internal magnetization density distributions of iron and nickel by the maximum entropy method, *J. Phys. Soc. Japan*, **65**, 255–263.
33. Brown, P.J., Déportes, J. and Ouladdiaf, B. (1992) Magnetic structure of the Laves phase compound $TiFe_2$, *J. Phys.: Cond. Mat.*, **4**, 10015–10024.
34. Shiga, M., Wada, H. and Nakamura, Y. (1983) Magnetism and thermal expansion anomaly of RMn_2 (R = Y, Gd, Tb, Ho and Er), *J. Mag. Mag. Mat.*, **31–34**, 119–120.

35. Brown, P.J., Ouladdiaf, B., Ballou, R. *et al.* (1992) Mn moment instability in the $TbMn_2$ intermetallic compound, *J. Phys.: Cond. Mat.*, **4**, 1103–1113.
36. Lelièvre-Berna, E., Ouladdiaf, B., Galéra, R.M. *et al.* (1993) Mn moment instability and magnetic structures of $Tb_{1-x}Sc_xMn_2$, *J. Mag. Mag. Mat.*, **123**, L249–L254.
37. Lelièvre-Berna, E., Rouchy, J. and Ballou, R. (1994) Field induced first order magnetic transition and associated volume effect in $TbMn_2$, *J. Mag. Mag. Mat.*, **137**, L6–L10.
38. Warren, P., Forsyth, J.B., McIntyre, G.J. *et al.* (1992) A single-crystal neutron diffraction study of the magnetization density in Fe_2Zr, *J. Phys.: Cond. Mat.*, **4**, 5795–5800.
39. Kennedy, S.J., Brown, P.J. and Coles, B.R (1993) A polarized neutron study of the magnetic form factors in $CeFe_2$, *J. Phys.: Cond. Mat.*, **5**, 5169–5178.
40. Boucherle, J.X., Alekseev, P.A., Gillion, B. *et al.* (1995) Induced magnetic form factor of Sm in mixed valence $^{154}Sm^{11}B_6$, *Physica B*, **206–207**, 374–376.
41. Sanchez, J.P., Lebech, B., Lander, G.H. *et al.* (1992) Examination of the magnetic properities of $NpCo_2$, *J. Phys.: Cond. Mat.*, **4**, 9423-9440.
42. Lander, G.H., Brooks, M.S.S. and Johansson, B. (1991) Orbital band magnetism in actinide intermetallics, *Phys. Rev. B*, **43**, 13672–13675.
43. Lebech, B., Wulff, M. and Lander G.H. (1991) Spin and orbital moments in actinide compounds (invited), *J. Appl. Phys.*, **69**, 5891–5896.
44. Langridge, S., Lander, G.H., Bernhoeft, N. *et al.* (1997) Separation of the spin and orbital moments in antiferromagnetic UAs, *Phy. Rev. B*, **55**, 6392–6398.
45. Paixao, J.A., Lander, G.H., Brown, P.J. *et al.* (1992) Magnetization density in URhA1 – evidence for hybridization effects, *J. Phys.: Cond. Mat.*, **4**, 829–846.
46. Gukasov, A., Wisniewski, P. and Henkie, Z. (1996) Neutron diffraction study of magnetic structure of U_3Bi_4 and U_3Sb_4, *J. Phys.: Cond. Mat.*, **8**, 10589–10600.

16

Magnetisation densities and polarised neutron diffraction: optimised flipping ratio measurements

E. LELIÈVRE-BERNA, M. PORTES DE ALBUQUERQUE*,
F. TASSET and P.J. BROWN

Institut Laue Langevin, BP 156, 38042 Grenoble cedex 9, France (E-mail: lelievre@ill.fr)
**Permanent address: Centro Brasileiro de Pesquisas Físicas, Rio de Janeiro, 22290-180, Brazil*

1. Introduction

The polarised neutron diffraction (PND) technique is applicable to single crystals of systems in which the magnetisation can been partly or completely aligned by an applied magnetic field [1, 2]. If the nuclear structure factors are accurately known, the components, parallel to the field, of the magnetic structure factors of the Bragg reflections can be determined, i.e. the Fourier components of the magnetisation. Several different methods can be used to retrieve the magnetisation distribution. The simplest is a simple Fourier inversion of a set of magnetic structure factors. A much better result can be obtained using the Maximum Entropy method in which both the uncertainty of the data and the fact that not all Fourier components are measured can be taken into account [3–6]. An alternative technique is to compare the measurements to a model which gives the spin and orbital densities. One can express the atomic form factor in the dipolar approximation [7–9] at low $\sin\theta/\lambda$, or the magnetisation density in a multipolar expansion [10, 11]. In the case of molecular magnets [12, 13], e.g. 3d transition metal complexes and organic free radicals, it is usual to fit a Hartree–Fock magnetic wave function constructed from standard Slater orbitals at each magnetic site. A more versatile model is obtained by expanding the spin density in a superposition of aspherical densities (series expansion in real spherical harmonics). To summarise, the PND technique when combined with well adapted methods of analysis, can yield precise quantitative results. In this paper, we present two complementary ways in which the accuracy of the experimental data can be optimised.

2. The flipping ratio

In a PND experiment, the data measured are the (flipping) ratios R_{hkl} between the intensities observed for (+) and (−) polarisation states of the incident neutron beam at the peak of Bragg reflections hkl. These intensities are related to both nuclear and magnetic interactions, and the (+) and (−) states correspond to a neutron magnetic moment, respectively, parallel and antiparallel to the applied magnetic field. The nuclear interaction between the neutrons and the nuclei of the atoms leads to a coherent scattering term which contributes to the Bragg intensity, and to an incoherent scattering term which comes from the isotopic and nuclear spin contributions. In

Paul G. Mezey and Beverly E. Robertson (eds.), Electron, Spin and Momentum Densities and Chemical Reactivity, 235–243

normal experimental conditions, the nuclear spins are randomly oriented and the coherent nuclear scattering does not depend on the incident neutron polarisation direction. The incoherent scattering also depends on the Debye–Waller factor and is subtracted from the Bragg intensity by measuring the background intensity away from the Bragg peak. The magnetic interaction between the neutron magnetic moments and the ferromagnetic component of the unpaired electron spins and orbital moments also contributes to the Bragg peak intensity which can be written as

$$I_{+/-}(\boldsymbol{\kappa}) = NN^* + \mathbf{P}_{+/-}\left(N\mathbf{M}^*_{\perp\boldsymbol{\kappa}} + N^*\mathbf{M}_{\perp\boldsymbol{\kappa}}\right) + \mathbf{M}_{\perp\boldsymbol{\kappa}} \cdot \mathbf{M}^*_{\perp\boldsymbol{\kappa}}, \tag{1}$$

where N is the nuclear structure factor. $\mathbf{M}_{\perp\boldsymbol{\kappa}}$ is the projection of the magnetic structure factor onto a plane perpendicular to the scattering vector $\boldsymbol{\kappa} = hkl$, and $\mathbf{P}_{+/-}$ is the polarisation vector whose magnitude gives the incident beam polarisation. The polarisation dependence of the intensity comes from the nuclear-magnetic interference terms, and these are the origin of the high sensitivity of the PND technique. Experimentally, the flipping ratio is determined by measuring the count rates $r_{\mathrm{p}+}$, $r_{\mathrm{p}-}$ at the peak of a Bragg reflection, and the background count rates $r_{\mathrm{b}+}$ and $r_{\mathrm{b}-}$ on either side of it. Then,

$$R(\boldsymbol{\kappa}) = \frac{I_+}{I_-} = \frac{r_{\mathrm{p}+} - r_{\mathrm{b}+}}{r_{\mathrm{p}-} - r_{\mathrm{b}-}}. \tag{2}$$

Before going further, it may be noted that the flipping ratio does not depend either on the Lorentz factor or on absorption in the sample. Certain instrumental parameters such as the polarisation of the neutron beam for the two spin states, the half wavelength contamination of the neutron beam and the dead-time detector can readily be taken into account when analysing the data. On the other hand, the extinction which may occur in the scattering process is not so easy to assess, but must also be included [14]. Sometimes, it is even possible to determine the magnetisation density of twinned crystals [15].

The variance (3) of the flipping ratio as a function of rates is easily obtained from Equation (2):

$$\frac{V(R)}{R^2} = \frac{V(r_{\mathrm{p}+}) + V(r_{\mathrm{b}+})}{(r_{\mathrm{p}+} - r_{\mathrm{b}+})^2} + \frac{V(r_{\mathrm{p}-}) + V(r_{\mathrm{b}-})}{(r_{\mathrm{p}-} - r_{\mathrm{b}-})^2}. \tag{3}$$

We shall present, in the following sections, two complementary measurement strategies which allow the value of this variance to be minimised. The first concerns the technique by which the measuring time is divided between the two polarisation states, and the second minimises the variance by determining an optimal division of the counting times.

3. Cyclic measurements

The rates are commonly defined by $r = N/M$, i.e. the intensity N normalised by monitor M. The monitor count is proportional to the incident flux, as this normalisation corrects for fluctuations of the neutron source. For simplicity, before discussing the general case, let us consider the case for which the background contribution is negligible compared to the Bragg intensity. In such a situation, the expressions for the flipping ratio and for the variance are very simple

$$R \approx R_0 = \frac{N_{p+} M_{p-}}{N_{p-} M_{p+}}, \tag{4}$$

$$\frac{V(R)}{R^2} \approx \frac{V(R_0)}{R_0^2} = \frac{V(N_{p+})}{N_{p+}^2} + \frac{V(N_{p-})}{N_{p-}^2} + \frac{V(M_{p+})}{M_{p+}^2} + \frac{V(M_{p-})}{M_{p-}^2}. \tag{5}$$

Normally, the dead-time of the monitor is negligible and $V(M) \approx M$. If the monitor counts neutrons at a typical rate of 10^3 n/s, its contribution to the standard deviation $\sigma(R_0)$ of a measurement taking 40 s is of the order of $0.01 \times R_0$. In the case of a strong Bragg intensity, i.e. a rate greater than 10^3 n/s, Equation (5) shows that $\sigma(R_0)$ depends mainly on the monitor values. To make the point clear, suppose now that for a constant flux on the sample, the monitor counts only 10 n/s. Then, the monitor contribution to $\sigma(R_0)$ for the same measurement time is an order of magnitude greater, which means that the value of the standard deviation of R_0 is unnecessarily large.

The high sensitivity of the variance of R to the monitor count rate can be avoided by making use of the fact that the fluctuations of the neutron source are rather slow. At the Institut Laue Langevin (ILL), the fastest fluctuations which can be observed have a period of about 15 min and are negligible. The larger, but still relatively minor, fluctuations of the High Flux Reactor are slower and arise firstly when the weir which retains the cooling water for the reactor is opened (period of 12 h), and secondly from the change in distribution of reactivity which occurs during a reactor cycle as the fuel element is consumed (50 days).

In the measurement technique, which has been used on D3 for many years, the ratio of the time spent counting with the cryoflipper in (+) or (−) mode is controlled by a quartz crystal controlled oscillator with a highly stable output frequency f of 1 MHz. There are two scalers to count the detector pulses (+ and − states), a single monitor scaler and a single time scaler used to end the measurement when the total time is reached (precision of 1 ms).

The logic of the CAMAC unit driving the cryoflipper and gating the scalers (FCU) [16] operates cyclicly every $2 \cdot 10^6 + 2n$ pulses of 1 μs. The operations it performs in each cycle are illustrated in Figure 1. At $t = 0$, all scalers are gated OFF and the cryoflipper is switched (+). At $t = n_0$, the monitor, the time and the (+) scalers are switched on; after n pulses (at $t = n_0 + n$) the flipper is switched (−) and all the scalers are gated off for n_0 pulses. Finally at $t = 2n_0 + n$ the monitor, time and (−) scalers are switched on for $2 \cdot 10^6 - n$ pulses. n_0 is chosen to allow time for the currents in the flipper circuits to stabilise after switching (typically 15 ms), and n can be selected by the program to fix the ratio $n/(10^6 - n)$ of t_{p+}/t_{p-}. If the neutron flux variations are slow compared to the cycle time, the precision of the ratio M_{p+}/M_{p-} is determined by that of the oscillator, and one can write the expressions

$$M_{p+} = M_p \frac{t_{p+}}{t_{p+} + t_{p-}} \quad \text{and} \quad M_{p-} = M_p \frac{t_{p-}}{t_{p+} + t_{p-}}, \tag{6}$$

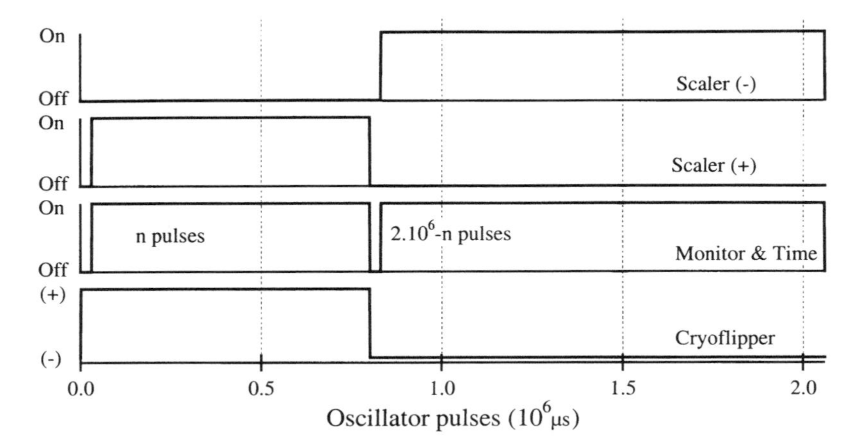

Figure 1. Logic of the flipper control unit (FCU).

where M_{p} is the total monitor value corresponding to the peak measurement for both $(+)$ and $(-)$ states. Substituting (6) in expressions (4) and (5) leads to new formulae for the flipping ratio:

$$R_0 = \frac{N_{\mathrm{p}+}\, t_{\mathrm{p}-}}{N_{\mathrm{p}-}\, t_{\mathrm{p}+}}, \tag{7}$$

$$\frac{V(R_0)}{R_0^2} = \frac{V(N_{\mathrm{p}+})}{N_{\mathrm{p}+}^2} + \frac{V(N_{\mathrm{p}-})}{N_{\mathrm{p}-}^2} + \frac{V(t_{\mathrm{p}+})}{t_{\mathrm{p}+}^2} + \frac{V(t_{\mathrm{p}-})}{t_{\mathrm{p}-}^2}. \tag{8}$$

With a 1 MHz quartz clock, the contribution to the standard deviation $\sigma(R_0)$ is mainly that from the intensities N. For example, with $t_{\mathrm{p}+} = t_{\mathrm{p}-} = 20\,\mathrm{s}$, the time contribution to $\sigma(R_0)$ is equal to $1/\sqrt{2} \cdot 10^{-7} R_0$. This is negligible compared to the values calculated above. This cyclic method reduces the time needed to obtain a given accuracy in flipping ratio measurements particularly in the case of strong reflections.

One may ask whether this method could be generalised to include the background contributions also. The flipping ratio can be written as a function of the counts and the time determined fractions of the monitor count as

$$M_{\mathrm{p}+} = M \frac{t_{\mathrm{p}+}}{t_{\mathrm{p}+} + t_{\mathrm{p}-} + t_{\mathrm{b}+} + t_{\mathrm{b}-}}, \quad M_{\mathrm{p}-} = M \frac{t_{\mathrm{p}-}}{t_{\mathrm{p}+} + t_{\mathrm{p}-} + t_{\mathrm{b}+} + t_{\mathrm{b}-}},$$
$$M_{\mathrm{b}+} = M \frac{t_{\mathrm{b}+}}{t_{\mathrm{p}+} + t_{\mathrm{p}-} + t_{\mathrm{b}+} + t_{\mathrm{b}-}}, \quad M_{\mathrm{b}-} = M \frac{t_{\mathrm{b}-}}{t_{\mathrm{p}+} + t_{\mathrm{p}-} + t_{\mathrm{b}+} + t_{\mathrm{b}-}}. \tag{9}$$

Practically, this requires that crystal be rotated quickly and reproducibly into and out of the reflecting position, so as to allow the peak and background contributions be counted repeatedly for both incident polarisation states. Even with present day technology, making such movements with the required precision is time consuming.

Table 1. Flipping ratios of strong and weak reflections obtained using the standard (std) and cyclic (cyc) methods (total measurement time of 180 s). As expected, an improvement is observed for the strong reflection.

Reflection	200	600
Rate (background)	≈ 7000 n/s (≈ 40 n/s)	≈ 150 n/s (≈ 30 n/s)
t_+/t_-	10%/ 90%	50%/50%
t_b/t_p	20%/ 80%	20%/80%
R_{std}	13.956 ± 0.063 (0.45%)	$1.200 \pm 0.019(3)$ (1.61%)
R_{cyc}	13.986 ± 0.054 (0.38%)	$1.199 \pm 0.019(2)$ (1.60%)

Because of this, cyclic measurements are made first of the background (+) and (−) intensities, and then of the (+) and (−) peak intensities (6). Thus, the (+)/(−) ratios are obtained with a high accuracy (clock precision), but the peak/background ratios still depend on the monitor counts $M_{\text{p}} = M_{\text{p}+} + M_{\text{p}-}$ and $M_{\text{b}} = M_{\text{b}+} + M_{\text{b}-}$. In such conditions, the flipping ratio is defined by Equations (10) and (11) where $t_{\text{p}} = t_{\text{p}+} + t_{\text{p}-}$ and $t_{\text{b}} = t_{\text{b}+} + t_{\text{b}-}$. One may note here that the count rates are no longer independent, so that expression (3) cannot be used to derive the variance of R directly

$$R = \frac{I_+}{I_-} \quad \text{with } I_{+/-} = \frac{N_{\text{p}+,-}\, t_{\text{p}}}{M_{\text{p}}\, t_{\text{p}+,-}} - \frac{N_{\text{b}+,-}\, t_{\text{b}}}{M_{\text{b}}\, t_{\text{b}+,-}}, \tag{10}$$

$$\begin{aligned} \frac{V(R)}{R^2} = {} & \frac{t_{\text{p}}^2}{M_{\text{p}}^3}\left(\frac{N_{\text{p}-}}{I_-\, t_{\text{p}-}} - \frac{N_{\text{p}+}}{I_+\, t_{\text{p}+}}\right)^2 + \frac{t_{\text{p}}^2}{M_{\text{p}}^2}\left(\frac{V(N_{\text{p}+})}{I_+^2\, t_{\text{p}+}^2} + \frac{V(N_{\text{p}-})}{I_-^2\, t_{\text{p}-}^2}\right) \\ & + \frac{t_{\text{b}}^2}{M_{\text{b}}^3}\left(\frac{N_{\text{b}+}}{I_+\, t_{\text{b}+}} - \frac{N_{\text{b}-}}{I_-\, t_{\text{b}-}}\right)^2 + \frac{t_{\text{b}}^2}{M_{\text{b}}^2}\left(\frac{V(N_{\text{b}+})}{I_+^2\, t_{\text{b}+}^2} + \frac{V(N_{\text{b}-})}{I_-^2\, t_{\text{b}-}^2}\right). \end{aligned} \tag{11}$$

With the new VME/UNIX control system on the polarised hot-neutron normal-beam diffractometer D3 at ILL, each measurement cycle for both peak and background intensities lasts 2 s, and the (+)/(−) counting-time fractions are defined with a 1 MHz clock. There are two detector scalers and two monitor scalers ((+) and (−) states). In Table 1, we compare the flipping ratio measured for the strong 200 and the weak 600 Bragg peak reflections of a CoFe sample. As expected, the standard deviation $\sigma(R)$ is improved in the case of the strong reflection (16%).

4. Optimised counting-time distribution

Another way to improve the quality of flipping ratio is to optimise the proportions of time spent counting in the four states: peak (+), peak (−), background (+) and background (−), so as to minimise the variance obtained with a fixed total counting time. For example, in the case of a weak reflection for which the Bragg peak count rates are comparable to those of the background, one has to count the peak and background intensities for the same time. On the other hand, if the peak to background ratio is large, the precision of the peak count and consequently the time spent measuring it, should be greater than that for the background. Finally, if the (+) and (−) intensities are very different, the ratio of the (+)/(−) counting times must be chosen in order to obtain the same total counts for both incident polarisation states [17].

We present here the theory behind the method, which has been used on D3 for some 20 years, to achieve such optimisation. Because (+) and (−) peak count rates and peak/background ratios may differ strongly from one reflection to another and are not known *a priori*, the measurement is divided into a number of steps of duration T. A first flipping ratio measurement is made in predefined conditions ($t_{p+}, t_{p-}, t_{b+}, t_{b-}$). Then, after calculating the counting-time proportions which minimise the variance of the flipping ratio, the time already spent is subtracted and the measurement is made again with times chosen to achieve these optimised proportions. The process of calculation and measurement is repeated in each step.

To determine the expressions for the optimised counting times, we write the expressions (10) and (11) in terms of count-rates and times (count rates are constant quantities for each Bragg reflection). We assume that the incident neutron flux is constant during a flipping ratio measurement, and that no dead-time correction is needed. In these conditions, we have the relations:

$$\frac{M_{p+}}{t_{p+}} = \frac{M_{p-}}{t_{p-}} = \frac{M_{b+}}{t_{b+}} = \frac{M_{b-}}{t_{b-}}, \tag{12}$$

$$V(N_{p+}) = N_{p+}, \quad V(N_{p-}) = N_{p-}, \quad V(N_{b+}) = N_{b+}, \quad V(N_{b-}) = N_{b-}. \tag{13}$$

Using (12) and (13) in (10) and (11), we obtain the simplified expression for the variance:

$$\begin{aligned} V(R) = \frac{1}{(r_{p-} - r_{b-})^4} & \left(\frac{r_{p+}(r_{p-} - r_{b-})^2}{t_{p+}} + \frac{r_{p-}(r_{p+} - r_{b+})^2}{t_{p-}} \right. \\ & \left. + \frac{r_{b+}(r_{p-} - r_{b-})^2}{t_{b+}} + \frac{r_{b-}(r_{p+} - r_{b+})^2}{t_{b-}} \right) \\ & \text{with } r_{p+} = \frac{N_{p+}}{t_{p+}}, \quad r_{p-} = \frac{N_{p-}}{t_{p-}}, \quad r_{b+} = \frac{N_{b+}}{t_{b+}}, \quad r_{b-} = \frac{N_{b-}}{t_{b-}}. \end{aligned} \tag{14}$$

Minimising this expression subject to the constraint $t_{p+} + t_{p-} + t_{b+} + t_{b-} = T$ using the Lagrange multipliers, one obtains the optimal counting-time proportions:

$$\begin{aligned} t_{p+} &\propto (r_{p+} - r_{b+})\sqrt{r_{p+}}, \qquad & t_{p-} &\propto (r_{p+} - r_{b+})\sqrt{r_{p-}}, \\ t_{b+} &\propto (r_{p-} - r_{b-})\sqrt{r_{b+}}, \qquad & t_{b-} &\propto (r_{p-} - r_{b-})\sqrt{r_{b-}}. \end{aligned} \tag{15}$$

On D3, we have tested this method, starting with the conditions $t_b/t_p = 20\%/80\%$ or 50%/50% for $t_+/t_- = 10\%/90\%$, $50\%/50\%$ and 90%/10%. These tests were performed for the 200 and the 600 reflections of a CoFe single crystal for different magnitudes of the applied magnetic field, R varying from 1 (no magnetic field) to about

20. The main results are summarised in Figures 2 and 3. The dashed curves associated with the left axis represent the time variation of the relative standard deviation of the flipping ratio measured with steps of 60 s. The time dependence of the relative improvement using the proposed method is represented by the full curve (right axis). In Figure 2, the predefined conditions are not too far from being optimal, and an improvement of about 7% is observed. In Figure 3, the (+)/(−) starting counting-time proportions are bad and the improvement is more than 17%. This method works

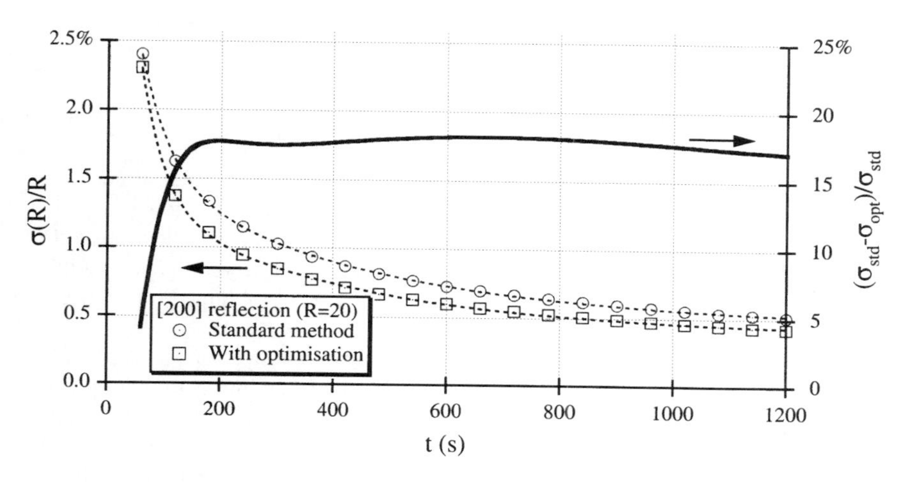

Figure 2. Time variation of $\sigma(R)/R$ (dashed lines) and of the relative improvement (full line) for the 200 reflection. No significant improvement is obtained by going beyond three steps.

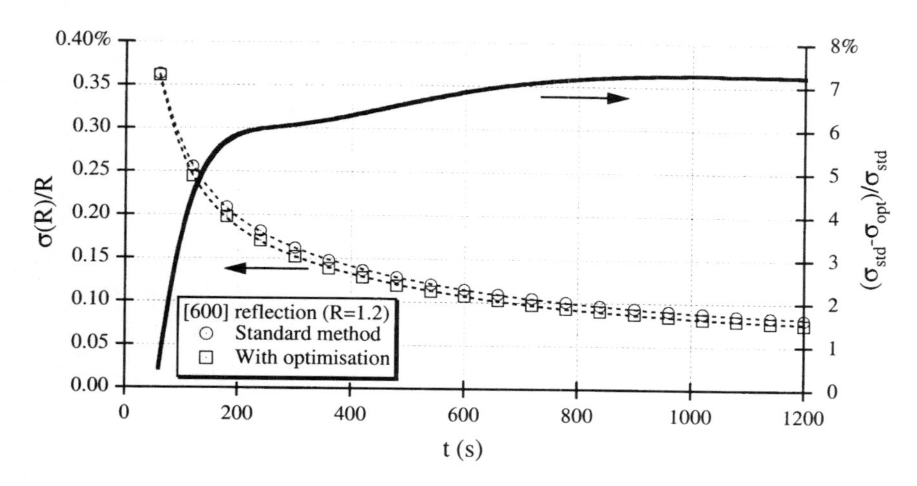

Figure 3. Time variation of $\sigma(R)/R$ (dashed lines) and of the relative improvement (full line) for the 600 reflection. Starting with bad counting-time proportions, the improvement is more than 17%.

Table 2. Flipping ratios of strong and weak reflections measured starting from different time proportions (three steps of 60 s). When using the optimised method, the standard deviation $\sigma(R)$ does not depend on the initial conditions.

t_b/t_p (%/%)	t_+/t_- (%/%)	$\sigma_{std}(R_{200})$ (%)	$\sigma_{opt}(R_{200})$ (%)	$\sigma_{std}(R_{600})$ (%)	$\sigma_{opt}(R_{600})$ (%)
20/80	10/90	0.38	0.35	2.52	1.59
...	50/50	0.42	0.35	1.63	1.59
...	90/10	0.91	0.37	2.67	1.58
50/50	10/90	0.47	0.35	2.93	1.61
...	50/50	0.51	0.36	1.89	1.61
...	90/10	1.11	0.37	3.19	1.60

well, and the main result is that, in all cases examined, no significant improvement is obtained by going beyond three steps.

In Table 2, we present the standard deviation of flipping ratios measured in three steps starting from different timing conditions. As was done previously, we have considered two types of Bragg peaks: the strong polarising 200 reflection and the weak almost non-polarising 600 reflection of a CoFe single crystal. The results are very satisfactory. They show that almost the same accuracy is obtained for each Bragg reflection, independently on the initial conditions.

The only problem with this method is observed for weak reflections where $(+)/(-)$ count-rates are similar (i.e. $R\approx 1$). The $(+)/(-)$ optimised counting-time proportions must be 50%/50%, but with low count-rates, we have observed that the lack of precision may lead to proportions which are not optimum (e.g. 47%/53%). The same behaviour has been observed for peak to background proportions. In fact, when measuring a flipping ratio in many steps, we observe oscillations of the time proportions which slow the decrease of the standard deviation. Of course, these time variations have no sense, and one should calculate the variances of the 'optimised' counting-times (Equations (16)) to avoid such spurious fluctuations:

$$\begin{aligned} V\left(t_{p+/-}\right) &= r_{p+/-}\left(\frac{r_{p-/+}}{t_{p-/+}} + \frac{r_{b-/+}}{t_{b-/+}}\right) + \frac{(r_{p-/+} - r_{b-/+})^2}{4t_{p+/-}}, \\ V\left(t_{b+/-}\right) &= r_{b+/-}\left(\frac{r_{p-/+}}{t_{p-/+}} + \frac{r_{b-/+}}{t_{b-/+}}\right) + \frac{(r_{p-/+} - r_{b-/+})^2}{4t_{b+/-}}. \end{aligned} \tag{16}$$

We propose to use (17) as criteria to determine whether the statistics allow the calculated proportions to be distinguished from the (50%/50%) case. Using the criteria,

$$\begin{aligned} &\text{if } \left|t_{p+} - t_{p-}\right| \le (\sigma t_{p+} + \sigma t_{p-}) \quad \text{then } t_{p+} = t_{p-}, \\ &\text{if } \left|t_{b+} - t_{b-}\right| \le (\sigma t_{b+} + \sigma t_{b-}) \quad \text{then } t_{b+} = t_{b-}, \\ &\text{if } \left|t_{p+} - t_{b+}\right| \le (\sigma t_{p+} + \sigma t_{b+}) \quad \text{then } t_{p+} = t_{b+}, \\ &\text{if } \left|t_{p-} - t_{b-}\right| \le (\sigma t_{p-} + \sigma t_{b-}) \quad \text{then } t_{p-} = t_{b-}, \end{aligned} \tag{17}$$

no more oscillations were observed.

5. Conclusion

We have presented two complementary ways to improve flipping ratios measurements. The first one applies mainly to strong peaks. The $(+)/(-)$ counting-time ratios are fixed with high accuracy and the monitor can then be corrected for slow variations of the neutron flux, without degrading the final variance of the flipping ratio. The second one is more general, and permits flipping ratios to be measured with the best accuracy available in a determined measurement time without any pre-knowledge of the $(+)/(-)$ or peak/background count rates. The counting-time proportions are adjusted optimally during the measurement using an iterative process. An algorithm based on the methods presented here is used to optimise flipping ratio measurements made by the polarised hot- neutron normal-beam diffractometer D3 at ILL.

References

1. Nathans, R., Shull, C.G., Shirane, G. *et al.* (1959) The use of polarised neutrons in determining the magnetic scattering by Iron and Nickel, *J. Phys. Chem. Solids*, **10**, 138–146.
2. Squires, G.L. (1978) *Introduction to the Theory of Thermal Neutron Scattering*, University Press, Cambridge.
3. Papoular, R.J. and Gillon, B. (1990) Maximum entropy reconstruction of spin density maps in crystals from polarised neutron diffraction data, *Europhys. Lett.*, **13**(5), 429–434.
4. Papoular, R.J. (1992) A generalised n-dimensional inverse Fourier transform incorporating experimental error bars, *Acta Cryst.*, **A48**, 244–246.
5. Papoular, R.J. and Delapalme, A. (1994) Model-free polarised neutron diffraction study of an acentric crystal: metamagnetic UCoAl, *Phys. Rev. Lett.*, **72**(10), 1486–1489.
6. Papoular, R.J., Ressouche, E., Schweizer, J. *et al.* (1993) MaxEnt enhancement of 2D projections from 3D phased Fourier data: an application to polarised neutron diffraction, In *Maximum Entropy and Bayesian Methods*, Mohammad-Djafari, A. and Demoments, G. (Eds.), Kluwer Academic Publisher, Dordrecht, Vol. 53, pp. 311–318.
7. Marshall, W. and Lovesey, S.W. (1971) *Theory of Thermal Neutron Scattering*, University Press, Oxford.
8. Lander, G.H., Brooks, M.S.S. and Johansson, B. (1991) Orbital band magnetism in actinide intermetallics, *Phys. Rev. B*, **43**(16), 13672–13675.
9. Lebech, B., Wulff, M. and Lander, G.H. (1991) Spin and orbital moments in actinide compounds (invited), *J. Appl. Phys.*, **69**(8), 5891–5896.
10. Moze, O., Caciuffo, R., Gillon, B. *et al.* (1994) Polarised-neutron diffraction study of the magnetisation density in hexagonal Y_2Fe_{17}, *Phys. Rev. B*, **50**(13), 9293–9299.
11. Kennedy, S.J., Brown P.J. and Coles, B.R. (1993) A polarised neutron study of the magnetic form factors in $CeFe_2$, *J. Phys.: Cond. Mat.*, **5**(29), 5169–5178.
12. Gillon, B. and Schweizer, J. (1989) Study of chemical bonding in molecules: the interest of polarised neutron diffraction, *Mol. Phys. Chem. Bio.*, **3**, 111–147.
13. Schweizer, J. (1996) Spin densities in magnetic molecular compounds, *Physica B* (in press).
14. Schweizer, J. and Tasset, F. (1980) Polarised neutron study of the RCo_5 intermetallic compounds: I. The cobalt magnetisation in YCo_5, *J. Phys. F: Met. Phys.*, **10**, 2799–2818.
15. Muñoz, A., Batallan, F., Boucherle, J.X. *et al.* (1995) Magnetisation density in Ce_3Al_{11}, *J. Phys.: Cond. Mat.*, **7**(46), 8821–8831.
16. Forsyth, J.B., Knight, K.M., Penfold, J. and Saunders, P.W. (1974) Flipper Control Unit, Rutherford Laboratory Report RL-74-114.
17. Brown, P.J. and Forsyth, J.B. (1964) The determination of beam polarisation and flipping efficiency in polarised neutron diffractometry, *Brit. J. Appl. Phys.*, **14**, 1529–1532.

17

Concerning the magnetisation density in magnetic neutron scattering experiments

DYLAN JAYATILAKA

Department of Chemistry, The University of Western Australia, Nedlands 6009, Australia

1. Introduction

The magnetic dipole moment density, or magnetisation density, has assumed a fundamental role for the interpretation of magnetic scattering of neutrons from matter [1]. This is unfortunate, as it is well known that the magnetisation density is not determined to within a gradient of a scalar function [1, 2]. Strictly, the magnetisation density is not an observable. To highlight this fact, note that there are several acceptable formula for the magnetisation density; Steinsvoll *et al.* [3] have proposed

$$\mathbf{M}^{\mathrm{S}}(\mathbf{r}) = \frac{1}{4\pi}\nabla \times \int \frac{\mathbf{J}(\mathbf{r}')}{|\mathbf{r}-\mathbf{r}'|}\,\mathrm{d}\mathbf{r}' - \frac{1}{8\pi}\nabla \int \frac{\nabla \cdot \mathbf{r}' \times \mathbf{J}(\mathbf{r}')}{|\mathbf{r}-\mathbf{r}'|}\,\mathrm{d}\mathbf{r}' \tag{1}$$

while, on the other hand, Trammell [4] gives

$$\mathbf{M}^{\mathrm{T}}(\mathbf{r}) = \int_{1}^{\infty} \mathrm{d}\varsigma\; \varsigma\mathbf{r} \times \mathbf{J}(\varsigma\mathbf{r}). \tag{2}$$

In these equations, $\mathbf{J}$ is the physical current density. Clearly, the magnetisation density is a derived quantity.

Why, then, is the magnetisation density used? The answer is that the magnetisation density is important for certain approximations which are usually made in analysing neutron scattering experiments. In the standard polarised neutron diffraction (PND) experiment [5], only one parameter is measured – the so-called 'flipping ratio'. It is impossible to determine a vector quantity like the magnetisation density from a single number, unless some assumptions are made. The assumptions usually made are:

1. The magnetic field seen by the probe neutron is solely due to the magnetic dipole moment density of the unpaired electrons. In other words, the magnetisation density is simply related to the electron spin density by a multiplicative factor, and there is no ambiguity in its definition.
2. The direction of the spin density can be fixed experimentally, usually to be along the direction of an externally applied magnetic field.

These assumptions define the *collinear approximation.* In this approximation, the PND experiment becomes a powerful tool for seeing what the unpaired electrons are 'doing', which is particularly useful in a chemical context. Although more detailed

Paul G. Mezey and Beverly E. Robertson (eds.), Electron, Spin and Momentum Densities and Chemical Reactivity, 245–251

experiments can fix all components of the magnetisation density (for example, the polarisation analysis experiments [6]), they are rarely carried out.

The question now arises: can we tell when the collinear approximation is likely to break down? A practical indicator of trouble would be anisotropy in the **g** tensor, since the **g** tensor defines the angle of the total magnetic moment relative to an external applied field; systems with highly anisotropic **g** values are known [7]. On the other hand, theoretically, the failure of the collinear approximation is related to a breakdown in assumption (1) above: magnetic interactions between the neutron and internal electronic *orbital currents* become significant. These internal currents are due to spin–orbit coupling effects [8]. Since energy associated with spin–orbit coupling is known to grow as Z^4 , where Z is the nuclear charge, it seem likely that failure of the collinear approximation will be the rule rather than the exception for heavy atoms. There are certainly examples of systems where this approximation breaks down (see, for example, Refs. [9, 10]), but they appear to be rare. It may be true that studies on systems which are not fruitfully analysed within the conventional collinear scheme are avoided, or not published.

In this article I propose a formalism which avoids the magnetisation density and the associated collinear approximation in favour of the observable magnetic field density $\mathbf{B(r)}$ and the current density $\mathbf{J(r)}$. I will show that the magnetic scattering of neutrons is completely determined by the magnetic field density, and that the current density can be determined from the magnetic field density. I will also show that the Fourier components of the magnetic field density determine the magnetic scattering amplitudes for the neutron, so that magnetic field density plays exactly the same role as does the charge density in an X-ray scattering experiment. The key to the formalism is to recognise that the magnetic field density is 'sampled' by the magnetic moment of the neutron, M_{n}, via an interaction term

$$V(\mathbf{r}) = M_{\mathrm{n}} \cdot \mathbf{B}(\mathbf{r}). \tag{3}$$

Halpern and Johnson [11] noted the significance of this form of interaction in 1939, but did not pursue it. Finally, I show how it is possible to retain some connection within the collinear magnetisation density framework by defining a 'canting angle' which describes quantitatively the deviation from the collinear approximation for each measured neutron scattering reflection. We will present calculations for this canting angle for the system $CoCl_4^{2-}$ in the crystal Cs_3CoCl_5.

2. Magnetic neutron scattering in terms of the magnetic field density

In the first Born approximation [12], the scattering cross section for a beam of neutrons incident on a magnetic material, assuming form (3) for the interaction, is given by the square of the scattering amplitude, $F(\mathbf{k}_{\mathrm{f}}, \mathbf{k}_{\mathrm{i}})$, where

$$F(\mathbf{k}_{\mathrm{f}}, \mathbf{k}_{\mathrm{i}}) = -\frac{2\pi M_{\mathrm{n}}}{h^2} \int \chi_{\mathrm{f}}^* \exp(-\mathrm{i}\mathbf{k}_{\mathrm{f}} \cdot \mathbf{r}) \mathbf{V}(\mathbf{r}) \chi_{\mathrm{i}} \exp(\mathrm{i}\mathbf{k}_{\mathrm{i}} \cdot \mathbf{r})\, \mathrm{d}\mathbf{r}, \tag{4}$$

$\mathbf{k}_\mathrm{i}$ and χ_i are the initial wave vector and spin state of the incoming neutron, while $\mathbf{k}_\mathrm{f}$ and χ_f are the same quantities for the scattered neutron. M_n is the mass of the neutron, and $\mathbf{V}$ is the spin dependent scattering potential seen by the neutron. Using form (3), this is [5]

$$\mathbf{V}(\mathbf{r}) = -2\gamma\mu_\mathrm{N}\mathbf{S}_\mathrm{n} \cdot \mathbf{B}(\mathbf{r}), \tag{5}$$

where γ is the magnetic moment of the neutron in nuclear magnetons μ_N, $\mathbf{S}_\mathrm{n}$ is the spin operator for the neutron and $\mathbf{B}(\mathbf{r})$ is the magnetic field density of the electrons seen by the neutron. Note that the non-magnetic isotropic scattering of neutrons from the nuclei has been neglected. It may easily be incorporated in this approach using delta functions to represent the scattering potential of the nuclei. For elastic scattering considered here, the sample (and hence the potential $\mathbf{V}$) will not be modified during the scattering process. The scattering matrix element (4) is seen to involve a Fourier transform of $\mathbf{B}(\mathbf{r})$ in the variable $\mathbf{k} = \mathbf{k}_\mathrm{i} - \mathbf{k}_\mathrm{f}$,

$$F(\mathbf{k}; \chi_\mathrm{f}, \chi_\mathrm{i}) = \frac{4\pi\gamma\mu_\mathrm{N}M_\mathrm{n}}{\hbar^2}\langle\chi_\mathrm{f}|\mathbf{S}_\mathrm{n}|\chi_\mathrm{i}\rangle \cdot \overline{\mathbf{B}}(\mathbf{k}), \tag{6}$$

where the bar indicates the aforementioned Fourier transform. In another work in the literature, the magnetic field density Fourier component $\overline{\mathbf{B}}$ is known as the 'perpendicular structure factor', $\mathbf{F}^\perp$. This hides the fact that we are in fact just dealing with the Fourier transform of the magnetic field density, a concept which is easily grasped.

3. Expressions for the magnetic field density

The following development is devoted to obtaining an expression for $\overline{\mathbf{B}}$ in terms of more familiar quantities. To this end, consider the Fourier transform of Maxwell's equations (in the time independent case)

$$-\mathrm{i}\mathbf{k} \times \overline{\mathbf{B}} = \mu_0\overline{\mathbf{J}}, \tag{7}$$

$$-\mathrm{i}\mathbf{k} \cdot \overline{\mathbf{B}} = 0, \tag{8}$$

$\overline{\mathbf{J}}$ is the Fourier transform of the electron current which gives rise to the magnetic field $\overline{\mathbf{B}}$. Using these equations with the following vector identity,

$$\overline{\mathbf{B}} = (\hat{\mathbf{k}} \cdot \overline{\mathbf{B}})\hat{\mathbf{k}} + \hat{\mathbf{k}} \times \overline{\mathbf{B}} \times \hat{\mathbf{k}}, \tag{9}$$

(where $\hat{\mathbf{k}} = \mathbf{k}/k$ is a unit vector), gives

$$\overline{\mathbf{B}} = -\frac{\mathrm{i}\mu_0}{k^2}\mathbf{k} \times \overline{\mathbf{J}}. \tag{10}$$

In the non-relativistic limit, the electron current is comprised of two contributions [13],

$$\mathbf{J} = \mathbf{J}_\mathrm{L} + \mathbf{J}_\mathrm{S}, \tag{11}$$

$\mathbf{J}_\mathrm{L}$ is due to the linear velocity of the electrons (the 'orbital' component) while $\mathbf{J}_\mathrm{S}$ is due to the spin of the electron. The quantum mechanical expressions for these currents are

$$\mathbf{J}_\mathrm{L}(\mathbf{r}) = -\frac{e}{m}\sum_{\alpha,\beta} \mathrm{Re}\left[\pi\rho(\mathbf{x};\mathbf{x}')\right]_{\mathbf{x}=\mathbf{x}'}, \tag{12}$$

$$\mathbf{J}_\mathrm{S}(\mathbf{r}) = \nabla \times \mathbf{M}_\mathrm{S}(\mathbf{r}), \tag{13}$$

$$\mathbf{M}_\mathrm{S}(\mathbf{r}) = -g\mu_\mathrm{B}\sum_{\alpha,\beta}\left[\mathbf{S}_\mathrm{e}\rho(\mathbf{x};\mathbf{x}')\right]_{\mathbf{x}=\mathbf{x}'}, \tag{14}$$

where π is the gauge invariant momentum operator, $\mathbf{M}_\mathrm{S}$ is the spin magnetisation, and $\mathbf{S}_\mathrm{e}$ is the spin operator for the electrons. $\rho(\mathbf{x};\mathbf{x}')$ is the reduced one particle matrix for the state under consideration, where $\mathbf{x} = (\mathbf{r}, \chi)$ are space–spin coordinates. Note that due to the continuity equation for the total electron current,

$$\nabla \cdot \mathbf{J} = 0, \tag{15}$$

the linear current $\mathbf{J}_\mathrm{L}$ may also be written exactly as the curl of a vector function $\mathbf{M}_\mathrm{L}$. However, as discussed in the introduction, it is not clear how to define a unique $\mathbf{M}_\mathrm{L}$ from a given current density $\mathbf{J}_\mathrm{L}$. The problem does not arise for $\mathbf{M}_\mathrm{S}$ since it is related to the spin operator, which is unique. Now using (10) and (11) and the Fourier transform of (13) gives

$$\overline{\mathbf{B}} = -\frac{\mathrm{i}\mu_0}{k^2}\mathbf{k} \times \left(\overline{\mathbf{J}}_\mathrm{L} + \mathrm{i}\mathbf{k} \times \overline{\mathbf{M}}_\mathrm{S}\right). \tag{16}$$

This is the desired result which may be substituted into the scattering amplitude formula (6). The resulting scattering formula is the same as found by other authors [5], except that in this work SI units are used. The contributions to the Fourier component of magnetic field density are seen to be the physically distinct (i) linear current $\overline{\mathbf{J}}_\mathrm{L}$ and (ii) the magnetisation density $\overline{\mathbf{M}}_\mathrm{S}$ associated with the spin density. A concrete picture of the physical system has been established, in contrast to other derivations which are heavily biased toward operator representations [5]. We note in passing that the treatment here could be easily extended to inelastic scattering if transition one particle density matrices $\rho_{fi}(\mathbf{x};\mathbf{x}')$ were used in Equations (12)–(14).

4. Magnetic neutron scattering in terms of the current density

It will now be shown that the current density is uniquely determined from the magnetic field density. From the Fourier transform of the current conservation condition (15) we have

$$-\mathrm{i}\mathbf{k} \cdot \overline{\mathbf{J}} = 0. \tag{17}$$

Combined with Equation (10) and a decomposition analogous to (9), we obtain the desired result

$$\overline{\mathbf{J}} = -\frac{\mathrm{i}}{\mu_0}\mathbf{k} \times \overline{\mathbf{B}}. \tag{18}$$

Alternatively, we could have derived this from Maxwell's Equation (7). Magnetic neutron scattering can therefore be seen as a probe of the current density in the system. In this view, the magnetic scattering of neutrons is a pleasing complement to the X-ray scattering experiment, which probes the charge density.

5. Non-collinear magnetisation and the canting angle

In view of the central role that a magnetisation density plays for magnetic neutron scattering, it is useful to define a parameter for each reflection, called a *canting angle,* which gives a quantitative estimate of the deviation from the collinear approximation. The idea is as follows.

In the collinear approximation, the direction of the magnetic field would be given by $\hat{\mathbf{k}} \times \mathbf{z} \times \hat{\mathbf{k}}$, assuming that the magnetisation lies along the applied field direction, taken to be the unit vector $\mathbf{z}$ by convention (to see this, look at Equation (16) and ignoring the orbitals currents). On the other hand, the actual magnetic field is given by $\overline{\mathbf{B}}$. The canting angle is therefore defined as the angle between these two directions. An expression for the canting angle is

$$\theta(\mathbf{k}) = \arccos\left(\frac{\overline{B}_3 - (\overline{\mathbf{B}} \cdot \hat{\mathbf{k}})\hat{k}_3}{|\overline{\mathbf{B}}|\sqrt{1 - \hat{k}_3^2}}\right). \tag{19}$$

The maximum value obtainable for the canting angle is 90°. This maximum value occurs when $\overline{\mathbf{B}}$ is in the same direction as the scattering vector. (It may be necesary to use $180° - \theta$ when the magnetic field is against the direction of the scattering vector.) Essentially, this is a canting angle into the plane of the scattering.

We have calculated the canting angle for the system $CoCl_4^{2-}$ in the crystal Cs_3 $CoCl_5$ using the *ab initio* methodology in Ref. [16]. The calculations differ from those previously reported in that (i) experimental (neutron) geometrical, thermal, and scattering length parameters were used for the cobalt complex [15] (note: these are the only experimental data used, and they were not 'refined' to give a better fit to experiment), and (ii) a better basis set was used, from [14], supplemented with an extra p polarisation function on the cobalt atom (exponent 0.141308 atomic units) and d polarisation function on the chlorine atoms (exponent 0.65 atomic units). As in the previous work, to account for the shielding effect of the missing two-electron spin–orbit interactions, the one-electron spin–orbit integrals were scaled by the Slater factor of 1/3. The agreement between the calculated thermally averaged magnetic structure factors and the experimental magnetic structure factors gave $\chi^2 = 6.3$, with an overall scale factor of 1/1.080. (The calculated structure factors are too large.) This χ^2 agreement statistic is an improvement over the previous work. Table 1 shows values of the calculated canting angle for various reflections together with, for comparison, the nuclear structure factors F_{N}, and scaled effective scalar magnetic structure factors F_{M}. We have only included those reflections from the experiment which our calculations show have canting angles greater than 0.1°. Clearly, out of the original 98 calculated

Table 1. Canting angles $\theta(hkl)$ (degree) for various reflections (hkl), with nuclear structure factors F_N and scaled scalar magnetic structure factors F_M (Bohr magneton).

h	k	l	F_N	F_M	θ (hkl)	h	k	l	F_N	F_M	θ (hkl)
1	2	3	−54.306	−0.040	9.47	3	11	2	27.648	0.636	0.14
1	4	3	28.720	0.129	5.84	4	12	2	−11.845	−0.289	0.31
1	6	3	35.176	0.038	5.35	5	11	2	11.318	0.680	0.16
1	8	3	−42.943	0.008	14.71	2	10	4	21.637	1.174	0.10
2	3	3	25.121	−0.150	1.78	2	12	4	14.668	0.244	0.47
2	5	3	−63.881	−0.139	1.93	3	11	4	46.191	−0.677	0.25
2	9	3	51.361	−0.002	51.41	5	9	4	23.464	1.622	0.12
3	4	3	10.920	0.188	0.39	6	10	4	29.852	1.056	0.16
3	6	3	19.138	0.071	0.93	1	2	1	11.476	0.072	18.05
3	8	3	−18.186	0.001	79.77	1	6	1	−16.273	−0.011	22.13
5	12	3	−46.183	−0.020	7.72	2	3	1	−12.105	0.060	11.79
1	2	5	−2.666	−0.110	3.16	2	5	1	15.554	0.061	8.67
1	4	5	6.942	−0.095	4.42	3	6	1	−10.175	−0.033	5.86
2	7	5	11.160	0.024	2.62	5	6	1	19.564	−0.019	1.85
2	3	5	−4.295	0.197	0.35						

reflections, very few are significantly canted. Further, the few magnetic structure factors that show large canting angles (say, greater than 10°) are associated with very small magnetic structure factors, so are not well determined by either experiment or calculation. There are only four reflections with a canting angle greater than 1° for which F_M is above 0.1 Bohr magneton. It would appear that for this system, except for a few reflections, the calculations show that the collinear approximation is good. This is in agreement with previous conclusions that the system is, essentially, collinear. We plan to report calculations on systems where non-collinear effects are suspected to be important in the near future.

Acknowledgements

The author acknowledges many useful discussions with Dr. S.K. Wolff. Financial support was received from the Australian Research Council, QEII fellowship scheme.

References

1. Coppens P. and Becker P.J. (1987) In *International Tables for Crystallography*, Vol. A, Hahn T. (Ed.) p. 637, 2nd rev. Edn., Kluwer.
2. Balcar, E. (1975) *J. Phys. C*, **8**, 1581.
3. Steinsvoll, O., Shirane, G., Nathans R. and Blume, M. (1967) *Phys. Rev.*, **161**, 499.
4. Trammell, G.T. (1953) *Phys. Rev.*, **92**, 1387.
5. Marshall W. and Lovesey, S.W. (1971) *Theory of Thermal Neutron Scattering*, Oxford University Press.
6. See, for example, Moon R.M. and Koehler W.C. (1969) *Phys. Rev*, **181**, 883 for an early example.
7. See, for example, Abragam A. and Bleaney, B. (1970) *Electron Paramagnetic Resonance of Transition Metal Ions*, Clarendon Press, Oxford, pp. 650–653.
8. Leoni, F. (1972) *Phys. Rev. B*, **6**, 178.
9. Barnes, L.A., Chandler G.S. and Figgis, B.N. (1989) *Mol. Phys.*, **68**, 711.

10. Figgis, B.N., Williams, G.A., Forsyth J.B. and Mason R. (1981), *J. Chem. Soc., Dalton Trans.*, 1837.
11. Halpern O. and Johnson M.H. (1938) *Phys. Rev.*, **55,** 898.
12. Merzbacher, E. (1970) *Quantum Mechanics*, Ch. 11, 2nd edn., Wiley, New York.
13. See § 115, Landau L.D. and Lifshitz, A.M. (1977) *Quantum Mechanics*, 3rd edn., Pergamon, Oxford.
14. Schafer, A., Huber C. and Ahlrichs R. (1994) *J. Chem. Phys.*, **100**, 5829.
15. Figgis, B.N., Mason, R., Smith A.R.P. and Williams, G.A. (1980) *Acta. Cryst.*, **B36**, 509.
16. Wolff, S.K., Jayatilaka D. and Chandler, G.S. (1995) *J. Chem. Phys.*, **103**, 4562.

18

A wave function for beryllium from X-ray diffraction data

DYLAN JAYATILAKA

Department of Chemistry, The University of Western Australia, Nedlands 6009, Australia

1. Introduction

The fundamental object in the quantum theory of matter is the wave function, which is the most compact way to represent all the information contained in a system. Exact wave functions are usually not available, so if we want to know certain properties of the system the procedure is to set up some model Hamiltonian and get an approximate wave function, from which the desired properties can be extracted. This program can be represented by

$$\text{Model Hamiltonian wave function(s)} \rightarrow \text{Properties} \quad (1)$$

The reverse program is also quite common:

$$\text{Properties} \rightarrow \text{Model Hamiltonian wave function(s)} \quad (2)$$

The reason for pursuing the reverse program is simply to condense the observed properties into some manageable format consistent with quantum theory. In favourable cases, the model Hamiltonian and wave functions can be used to reliably predict related properties which were not observed. For spectroscopic experiments, the properties that are available are the *energies of many different* wave functions. One is not so interested in the wave functions themselves, but in the eigenvalue spectrum of the fitted model Hamiltonian. On the other hand, diffraction experiments offer information about the *density* of a particular property in some coordinate space for *one single* wave function. In this case, the interest is not so much in the model Hamiltonian, but in the fitted wave function itself.

In this article I will be concerned with determining wave functions from charge densities measured by X-ray diffraction experiments. The extraction of wave functions (or density matrices) directly from diffraction related experiments, although not as well developed as extracting model Hamiltonians in spectroscopy, has nevertheless had a long history [3–8]. Indeed, Massa *et al.* have recently coined the term 'Quantum Crystallography' to describe the field where using crystallographic techniques are used to enhance quantum mechanical calculations [2].

The technique used to extract the wave function in this work is conceptually simple: the wave function obtained is a single determinant which reproduces the observed experimental data to the desired accuracy, while minimising the Hartree–Fock (HF) energy. The idea is closely related to some interesting recent work by Zhao *et al.* [1]. These authors have obtained the Kohn–Sham single determinant wave function of density functional theory (DFT) from a theoretical electron density.

Paul G. Mezey and Beverly E. Robertson (eds.), Electron, Spin and Momentum Densities and Chemical Reactivity, 253–263

(The Kohn–Sham determinant is the single determinant which reproduces the electron density and minimises the kinetic energy [1, 9].) They observed that for the Be atom, the Kohn–Sham orbitals were nearly indistinguishable from the HF orbitals, and on this evidence they claim that the problem of finding a physically meaningful wave function from an electron density is 'solved'. Here, we merely note that there are a number of desirable features for our model:

- The problem of having sufficient data to fit does not arise, unlike previous work [3, 7].
- The model gives a unique answer in the limit of an infinite basis set, whereas density matrix fitting methods do not [10, 11].
- Since the HF model already gives good charge densities, and reliably predicts many other diverse properties, it seems reasonable to expect that the charge densities produced from this model will be better than those from conventional least squares fitting. In other words, 'quantum knowledge' is built into the model.
- Comparison between the model and *ab initio* calculations are greatly facilitated since exactly the same basis sets and methodology are used in both.
- The form that the equations take involves a straightforward modification of the self-consistent HF or density functional methods currently used in the oretical chemistry.

There are, however, two new issues which arise when using real data. First, because of experimental errors, our wave function should not exactly produce the experimental charge density. This has some important consequences. Second, because data for a periodic system will be used, the orbitals obtained should be orthogonal throughout the crystal if a single determinant wave function is to be constructed. The resolution of the orthogonality problem is not critical for the purposes of charge density modelling, but will allow useful results from formal density functional theory to be used. These problems will be discussed and resolved. The method will be demonstrated to work by extracting the wave function for beryllium crystal, for which accurate experimental charge density data is available.

2. Theory

2.1. *Review of the Zhao–Parr technique*

Consider a single determinant wave function whose orbitals ϕ_i are obtained from a model hamiltonian h,

$$h\phi_i = \left(-\tfrac{1}{2}\nabla^2 + v\right)\phi_i = \varepsilon_i \phi_i. \tag{3}$$

The density is given by

$$\rho(\mathbf{r}) = 2 \sum_i^{N_e/2} |\phi_i(\mathbf{r})|^2. \tag{4}$$

If ρ is constrained to be the same as the exact ground state density ρ_0, then the orbitals will satisfy the equation

$$h_{\text{eff}}\phi_i = (h + \lambda v_c)\phi_i = \varepsilon_i \phi_i, \tag{5}$$

where λ is the Lagrange multiplier attached to the constraint. The form of v_c depends on the specific choice of the constraint (or penalty) function C. Zhao *et al.* [1] propose

$$C = \frac{1}{2} \iint \frac{[\rho(\mathbf{r}) - \rho_0(\mathbf{r})][\rho(\mathbf{r}') - \rho_0(\mathbf{r}')]}{|\mathbf{r} - \mathbf{r}'|} \, \mathbf{dr}\, \mathbf{dr}', \tag{6}$$

$$v_c(\mathbf{r})\phi_i(\mathbf{r}) = \frac{\partial C}{\partial \phi_i^*} = \left(\int \frac{\rho(\mathbf{r}') - \rho_0(\mathbf{r}')}{|\mathbf{r} - \mathbf{r}'|} \, \mathbf{dr}' \right) \phi_i(\mathbf{r}). \tag{7}$$

Provided the potential v is local in $\mathbf{r}$, in the limit that $\lambda \to \infty$ we will have $\rho \to \rho_0$ *independent of the choice of* v. In this limit then, Equation (5) gives the Kohn–Sham orbitals and eigenvalues. The determinant formed from these orbitals is a wave function obtained from the density ρ_0.

In this work I choose a different constraint function. Instead of working with the charge density in real space, I prefer to work directly with the experimentally measured *structure factors*, $F_{\mathbf{h}}$. These structure factors are directly related to the charge density by a Fourier transform, as will be shown in the next section. To constrain the calculated cell charge density to be the same as experiment, a Lagrange multiplier technique is used to minimise the χ^2 statistic,

$$\chi^2 = \frac{1}{M} \sum_{\mathbf{h}} \frac{1}{\sigma_{\mathbf{h}}^2} \left(F_{\mathbf{h}} - F_{\mathbf{h}}^{\text{c}} \right)^2, \tag{8}$$

where $F_{\mathbf{h}}^{\text{c}}$ are the calculated structure factors, $\sigma_{\mathbf{h}}$ is the error associated with each measured structure factor $F_{\mathbf{h}}$, and M is the number of observations. Our choice is motivated by the fact that the χ^2 statistic is often used in crystallography as a measure of error. Unlike Parr and coworkers, I do *not* constrain χ^2 to be zero: it does not make sense to exactly reproduce an experimental density which contains errors. Instead, χ^2 is constrained to be equal to a certain value ξ_p.

One way to choose the value of ξ_p is as follows. Assume that the distribution of squared residuals is normal, as is often done in crystallography. Then tables are available [17] which give the probability p that a particular experiment will give a χ^2 less than ξ_p. The value of ξ_p can be chosen according to the desired confidence level, p. Of course, other ways to choose ξ_p are possible. Indeed, other choices for the agreement of statistic are possible.

As a consequence of the non-zero value demanded for the ξ^2 statistic, the solution to (5) (if possible at all) will occur at a *finite* value of λ. Because of this, the choice of v is no longer arbitrary. Clearly, v must be chosen using the best possible model, so that one does not have to constrain the orbitals very much to obtain the observed charge density. The HF model is chosen for this work, $v = v^{\text{HF}}$, because studies have already indicated that very good results are obtained at this level [18], and, as already noted, the HF orbitals appear to be very similar to the Kohn–Sham orbitals for the case examined here, beryllium.

However, one feature of the HF potential is that it is not a local potential. In the case of perfect data (i.e. zero experimental error), the fitted orbitals obtained are no longer Kohn–Sham orbitals, as they would have been if a local potential (for example, the local exchange approximation [27]) had been used. Since the fitted orbitals can be described as 'orbitals which minimise the HF energy and are constrained produce the real density', they are obviously quite closely related to the Kohn–Sham orbitals, which are 'orbitals which minimise the kinetic energy and produce the real density'. In fact, Levy [16] has already considered these kind of orbitals within the context of 'hybrid' density functional theories.

2.2. The charge density and the X-ray diffraction experiment

Real data is often available only for periodic systems, so only the density in the crystal unit cell need to be considered. Now the X-ray experiment gives structure factors $F_{\mathbf{h}}$ (along with errors $\sigma_{\mathbf{h}}$) which are related to the unit cell charge density via a Fourier transform,

$$F_{\mathbf{h}} = \int \rho^{\text{cell}}(\mathbf{r}) \exp(2\pi i \mathbf{r} \cdot \mathbf{Bh})\, d\mathbf{r}, \tag{9}$$

where $\mathbf{B}$ is the reciprocal lattice matrix dependent only on the crystal morphology, and $\mathbf{h}$ is an integer vector (the Miller indices) labelling the reflection. So constraining the calculated and experimental charge densities to be the same is equivalent to constraining the calculated and experimental structure factors to be the same. Restricting our attention to systems which are centrosymmetric (so the structure factors are real) and composed of one symmetry unique molecule in each unit cell, it follows that the cell charge density can be decomposed into a sum of N_{m} molecular charge densities ρ^j, each related by unit cell symmetry operations $\{\mathbf{S}_j, \mathbf{r}_j\}$ to a reference charge density for the molecule ρ_0,

$$\begin{aligned} \rho^{\text{cell}}(\mathbf{r}) &= \sum_{j=1}^{N_{\text{m}}} \rho^j(\mathbf{r}), \\ \rho^j(\mathbf{r}) &= \rho^0(\mathbf{S}_j^{-1}(\mathbf{r} - \mathbf{r}_j)), \end{aligned} \tag{10}$$

It is usually a good approximation to take ρ^0 to be the isolated molecule charge density, but within the above restrictions, no approximation has yet been made. For practical calculations ρ^0 is usually obtained in a basis set. If we write

$$\phi_i = \sum_{\mu} c_{\mu i} g_{\mu}, \tag{11}$$

where $C_{\mu i}$ are the orbital expansion coefficients, then, using (4), the reference molecule charge density basis set expansion is

$$\rho^0(\mathbf{r}) = \sum_{\mu,\nu} D_{\mu i} g_{\mu}(\mathbf{r}) g_{\nu}(\mathbf{r}), \tag{12}$$

where

$$D_{\mu\nu} = 2 \sum_{i}^{N_e/2} c_{\mu i} c_{\nu i} \tag{13}$$

is the (closed shell) density matrix. Using this in (10) and substituting in (9) yields the desired result for the calculated structure factors,

$$F_{\mathbf{h}}^{c} = \mathrm{Tr}(\mathbf{DI_h}). \tag{14}$$

$\mathbf{I_h}$ are the thermally smeared Fourier transforms of the basis function pairs summed over all the equivalent unit cell sites,

$$\begin{aligned} I_{\mu\nu,\mathrm{h}} = & \sum_{j}^{N_m} \exp(2\pi \mathrm{i}\mathbf{r}_j) t_{\mu\nu}((\mathbf{B})^{-1}\mathbf{S}_j^{\mathrm{T}}\mathbf{Bh}) \\ & \times \int g_\mu(\mathbf{r}) g_\nu(\mathbf{r}) \exp(2\pi \mathrm{i}\mathbf{r} \cdot \mathbf{B}[(\mathbf{B})^{-1}\mathbf{S}_j^{\mathrm{T}}\mathbf{Bh}])\, \mathrm{d}\mathbf{r}. \end{aligned} \tag{15}$$

Fast methods for evaluating these integrals for the case of gaussian basis functions are known [12]. Also, Hall has described how to get the symmetry operators $\{(\mathbf{B})^{-1}\mathbf{S}_j^{\mathrm{T}}\mathbf{B}, \mathbf{r}_j\}$ for any crystal space group [13]. The parameters $t_{\mu\nu}$ account for thermal smearing of the charge density. In this work I use the form recommended by Stewart [14],

$$t_{\mu\nu}(\mathbf{h}) = \exp(-2\pi^2 g(\mathbf{Bh}) \cdot (U^\mu + U^\nu)\mathbf{Bh}), \tag{16}$$

which is expressed in terms of the thermal vibration parameters U^μ (also obtained from the X-ray experiment) for the atom on which basis function g_μ is centred. The factor g is $\frac{1}{2}$ if the motions of atoms μ and ν are 'correlated', or $\frac{1}{4}$ if 'uncorrelated'. In this work, atoms were deemed correlated if they were less than 2.5 Bohr radii apart. The formula for the temperature factors is model dependent, but the use of a different thermal smearing model [15] makes little difference. Additional $\mathbf{h}$ dependent factors which account for extinction may also be incorporated in (15), but I have not done that in this work, because extinction was shown to be very small for the case of beryllium crystal. However, an overall ($\mathbf{h}$ independent) scale factor is used, since the absolute scale is not always well defined in the X-ray experiment.

2.3. A wave function for the entire crystal based on localised orbitals

It is possible to ensure that the orbitals we extract for one molecule in the crystal are orthogonal to all other orbitals on all other molecules in the crystal. If this is the case, a determinant wave function can be constructed for the entire crystal. To ensure the required orthogonality, a projection operator is used:

$$\kappa = \sum_{k,\mathrm{neighbours}} |\phi_i^k\rangle\langle\phi_i^k|. \tag{17}$$

It is assumed that the orbitals on the reference system ϕ_i are fairly localised. All the other orbitals in the crystal ϕ_i^k are related to the reference orbitals by translations and crystal symmetry operations as in (10). The assumption of locality means that only a finite number of neighbours near to the reference molecule need to be included in the above summation. Adding the above projector to any equation for the reference system orbitals and choosing the Lagrange multiplier κ large enough will ensure orthogonality. The projection operator ensures that the parts of the orbitals which are not orthogonal are energetically unfavourable, and are thus removed.

2.4. *Working equations in a finite basis set*

The matrix form for (5) expressed in a finite basis set is easily shown to be

$$\tilde{\mathbf{f}}\mathbf{c} \equiv (\mathbf{f} - \lambda \mathbf{v}_\mathrm{c} + \kappa \mathbf{p})\mathbf{c} = \mathbf{S}\mathbf{c}\epsilon, \tag{18}$$

where the usual definitions hold,

$$\begin{aligned} S_{\mu\nu} &= \langle g_\mu | g_\nu \rangle \\ f_{\mu\nu} &= \left\langle g_\mu \right| - \tfrac{1}{2}\nabla^2 + v \left| g_\nu \right\rangle, \end{aligned} \tag{19}$$

S is the overlap matrix, while **f** is taken to be the Fock matrix in this work. The matrix of the χ^2 constraint term, $\mathbf{v}_\mathrm{c}$, is given by

$$\mathbf{v}_\mathrm{c} = \frac{2}{M} \sum_{\mathbf{h}}^{n_h} \frac{1}{\sigma_{\mathbf{h}}^2} (F_{\mathbf{h}} - F_{\mathbf{h}}^\mathrm{c}) \mathbf{I}_{\mathbf{h}}. \tag{20}$$

It is essentially the derivative of χ^2 with respect to the orbital coefficients **c** (see Equation (7)). The matrix of the projection term **p** which ensures orthogonality to neighbouring molecules is

$$\mathbf{p} = \sum_{k,\text{neighbours}} (\mathbf{S}\mathbf{c}^k)(\mathbf{S}\mathbf{c}^k)^\mathrm{T}. \tag{21}$$

These equations are solved in the usual self-consistent way, the Lagrange multipliers λ and κ being chosen large enough to give, respectively, the desired agreement with experiment, or the desired orthogonality to near neighbours. As for normal HF equations, there will be $N_\mathrm{e}/2$ 'occupied' orbitals $\mathbf{c}_\mathrm{o}$ and a number of 'virtual' orbitals $\mathbf{c}_\mathrm{u}$. It is usual to write

$$\mathbf{c} = (\mathbf{c}_\mathrm{o}\mathbf{c}_\mathrm{u}). \tag{22}$$

2.5. *Convergence issues*

In practice, convergence problems are observed, because as λ becomes larger, **f** becomes small compared with $\mathbf{v}_\mathrm{c}$, and the solution of (18) becomes like a least squares fit, which is a singular problem if there are less data than parameters. Using real data with normal basis sets, $\mathbf{v}_c$ is unlikely to go to zero as λ gets larger, so the equations become increasing ill-conditioned. The convergence acceleration technique of Pulay [19] improves the situation. Alternatively, Equation (18) can be recast as follows. The occupied–occupied and virtual–virtual block of $\tilde{\mathbf{f}}$ are arbitrary and can be scaled by λ. Now divide the scaled $\tilde{\mathbf{f}}$ in (18) by λ and substitute $\epsilon/\lambda \to \epsilon$ and $\kappa/\lambda \to \kappa$ (this follows because they are Lagrange multipliers). The result is

$$\tilde{\mathbf{f}}^{\lambda}\mathbf{c} = \mathbf{S}\mathbf{c}\epsilon, \tag{23}$$

where $\tilde{\mathbf{f}}^{\lambda}$ has had its occupied–unoccupied blocks scaled by $1/\lambda$,

$$\mathbf{c}^{\mathrm{T}}\tilde{\mathbf{f}}^{\lambda}\mathbf{c} = \mathbf{c}_{\mathrm{o}}^{\mathrm{T}}\tilde{\mathbf{f}}\mathbf{c}_{\mathrm{o}} + \mathbf{c}_{\mathrm{v}}^{\mathrm{T}}\tilde{\mathbf{f}}\mathbf{c}_{\mathrm{v}} + \frac{1}{\lambda}(\mathbf{c}_{\mathrm{v}}{}^{\mathrm{T}}\tilde{\mathbf{f}}\mathbf{c}_{\mathrm{o}} + \mathbf{c}_{\mathrm{o}}^{\mathrm{T}}\tilde{\mathbf{f}}\mathbf{c}_{\mathrm{v}}). \tag{24}$$

Later, the occupied–occupied block will be diagonalised using different effective potentials to get eigenvalues.

3. Results

The theory described in the previous section is now applied to beryllium metal. Accurate low temperature data was taken from the paper of Larsen and Hansen [20]. (But note that in (20) I used the structure factors multiplied by 1000, as given in their paper.) For the orthogonalisation, all nearest neighbours we included within the first shell. There were 12 atoms. A 'triple zeta' basis set from Ref. [21] was used. There are 182 basis functions and 361 independent parameters in the wave function, whereas there are 58 experimental measurements. Figure 1 shows a plot of the χ^2 agreement statistic as a function of the parameter λ, for $\kappa = 0.2$. Larger values of κ caused

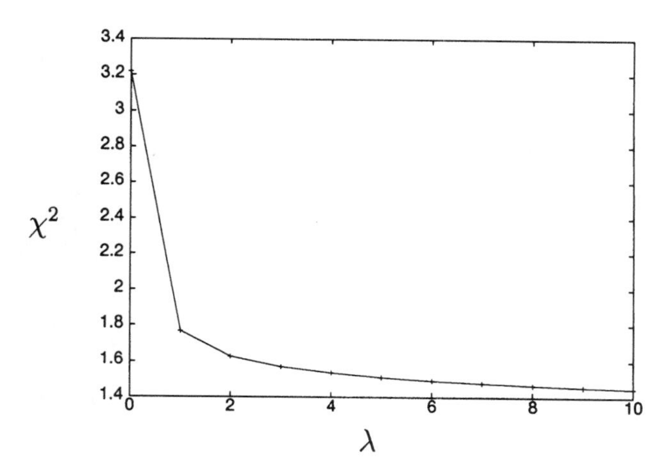

Figure 1. χ^2 agreement statistic versus λ for $\kappa = 0.2$.

numerical instability. For $\lambda = 10$ the overlap with the near neighbours was 0.004 and 0.002 for κ equal to 0.1 and 0.2, respectively. The plot for $\kappa = 0.1$ is indistinguishable on this scale. The value of χ^2 at $\lambda = 10$ was 1.44, with the overall scale factor being 0.997. A straight atomic density gave a χ^2 of 2.37: the atomic model is already very good. It seems clear that a χ^2 much lower than 1.44 is not practically obtainable, either because the energy penalty is too high, or the basis set is inadequate. To test the latter, calculations were performed with an additional d function (exponent 0.32 atomic units) on the beryllium atom, but to make the calculations practical, orthogonalising the orbitals to the near-neighbours was not performed. (This approach corresponds to using up to $l = 4$ in a normal least squares multipole moment approach.) The value of χ^2 obtained was 1.40, indicating the basis set is not the problem. It is perhaps a good time to note that considerable computational effort is expended with regard to the near neighbours in the projection term in (23). If orthogonalisation is neglected, as above, the method is no more time consuming than a normal HF calculation, but the orbitals obtained are no longer suitable for constructing a single determinant for the entire crystal. A scatter plot in Figure 2 as a function of scattering angle shows that the deviations of the fitted results are random. There are no obvious systematic errors. Figure 3 is a plot of the thermally smeared deformation densities, calculated from our structure factors. Interestingly, there is hardly any buildup of charge in the tetrahedral and octahedral holes, although there are depletions of charge similar to those observed by Larsen and Hansen [20], in plots that they give which are derived from Fourier summation techniques.

We now consider more interesting properties that can be extracted in our approach which cannot be extracted in a standard X-ray charge analysis. For a system at equilibrium, the virial theorem gives the total energy as

$$E = -T. \tag{25}$$

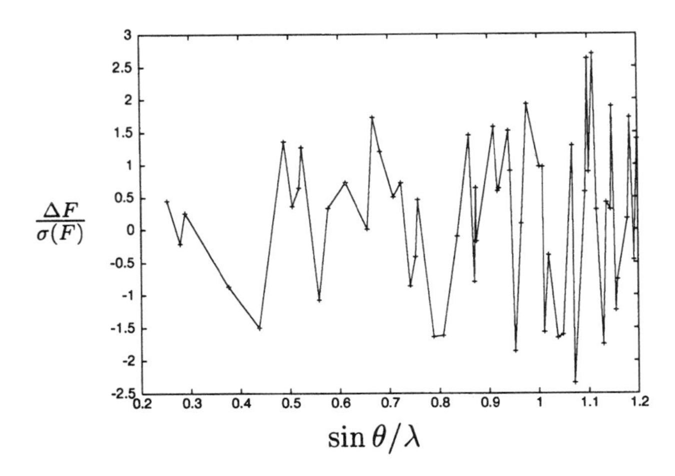

Figure 2. Errors in fitted structure factors.

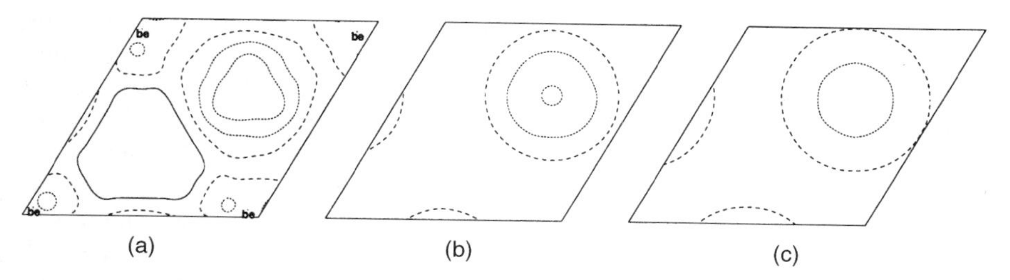

Figure 3. Thermally averaged deformation maps. Section (a) at $z = 0.75$ in the basal plane, (b) at $z = 0.625$ (containing the tetrahedral position) and (c) at $z = 0.5$ (containing the octahedral position). The origin is at the top left corner and the a and b axes increase across and down the page respectively. Contours are at intervals of 0.015 a.u. and dotted lines are negative, solid lines are positive.

The 'non-interacting' kinetic energy T_s from our determinant is 14.7861 a.u. for $\lambda = 10$, $\kappa = 0.2$. From this and the exact result for the Be atom ground state energy of -14.6674 a.u. [22], a binding energy of 312 kJ/mol is obtained. At $\lambda = 5$ and $\kappa = 0.2$, the binding energy is 390 kJ/mol. The observed value is 318 kJ/mol [23]. However, it should be noted that the error $T_c = T - T_s$ in the Kohn–Sham theory is known to be 194 kJ/mol for the Be atom [1].

An approximate ionization energy or work function can also be extracted. In density function theory, the ionization energy is given by the highest eigenvalue [24–26], which is governed by the long range behaviour of the density. If our fitted orbitals are a good approximation to the Kohn–Sham orbitals, they can be used to define the highest eigenvalue for an *approximate* one-particle effective potential [see comments below Equation (24)]. With three popular choices, the HF potential, the local density approximation (comprised of Dirac's exchange [27] plus local correlation energy functional of Ref. [28]), and the 'BLYP' approximation (comprised of the Becke exchange functional [29] plus correlation functional from [30]), the results are, respectively, 6.0, 3.7 and 4.9 eV. These are to be compared with the experimental value of 4.98 eV [23]. Even though there is a considerable spread in the results, as expected for such a crude calculation, they are all better than the free atom HF value of 8.4 eV. From these results it would seem that our fitted wave function is not unreasonable.

4. Conclusions

A new and accurate quantum mechanical model for charge densities obtained from X-ray experiments has been proposed. This model yields an approximate experimental single determinant wave function. The orbitals for this wave function are best described as HF orbitals constrained to give the experimental density to a prescribed accuracy, and they are closely related to the Kohn–Sham orbitals of density functional theory. The model has been demonstrated with calculations on the beryllium crystal.

There is no reason why the technique cannot be applied to extract wave functions for larger systems. Calculations have recently been completed in our group on the oxalic acid dihydrate system; in this case, hydrogen bonding effects were examined in detail.

Finally, a few general remarks are in order regarding systematic errors. The view adopted in this work is that the X-ray date 'enhances' the reliability of quantum mechanical calculations. The alternate view, that the quantum mechanical model introduces unwanted and unquantifiable systematic errors, could also be put forward. To counter this, the general merits of the HF method are mentioned; but there is no impediment, in principle, to using a theory more accurate than HF model. Secondly, according to the study of Frishberg [4], 'X-ray fitted wave functions give better results than HF for most expectation values ... with little sacrifice in the energy'. The last part of this statement would support the view that the HF wave function is a reasonable starting point for X-ray diffraction models. Thirdly, one should keep in mind that any model contains assumptions which can lead to systematic errors.

In the case of beryllium specifically, there has been considerable controversy recently regarding the existence of 'non-nuclear attractors' in this system [31, 32]. That is, regions in space (not at the nucleus) where there is a buildup of charge. Features like this are not observed in our maps. However, with an atom-based fitting technique used in this work, it could be argued that systematic errors will work against such feature appearing. Studies are planned which will allow for cluster-based fitted wave functions, which will provide a more realistic model for the crystal. That is why no statements were made in this work concerning the existence of the non-nuclear attractors in beryllium crystal. Nevertheless, it is hoped that techniques like this will stimulate more accurate charge density measurements.

Acknowledgements

The author acknowledges many stimulating discussions with attendees of the Sagamore XII conference, including with Prof. Lou Massa, Prof. Mel Levy, Prof. Robert Parr, Prof. Wolf Weyerich, Prof. Vedene Smith, Dr. Mark Spackman, Prof. Graham Chandler, Prof. Brian Figgis, Dr. Stephen K. Wolff, and Dr. Harmut Schmider. Thanks to the Australian Research Council QEII fellowship scheme for funds.

References

1. Zhao, Q., Morrison, R.C. and Parr, R.G. (1994) *Phys. Rev. A*, **50**, 2138; Zhao, Q. and Parr, R.G. (1992) *Phys. Rev. A*, **46**, 2337; Zhao, Q. and Parr, R.G. (1993) *J. Chem. Phys.*, **98**, 543.
2. Massa, L., Huang, L. and Karle, J. (1995) *Int. J. Quantum Chem. Symp.*, **29**, 371.
3. Massa, L. Goldberg, M. Frishberg, C. Boehme, R.F. and La Placa, S.J. (1985) *Phys. Rev. Lett.*, **55**, 622.
4. Frishberg, C. (1986) *Int. J. Quant. Chem.*, **30**, 1.
5. Schmider, H., Smith, V.H., Weyrich, W. (1990) *Trans. Am. Cryst. Assoc.*, **26**, 92.
6. Schmider, H. Smith, V.H. Weyrich, W. (1992) *J. Chem. Phys.*, **96**, 8986.
7. Howard, S.T., Hursthouse, M.B., Lehmann, C.W., Mallinson, P.R. and Frampton, C. S. (1992) *J. Chem. Phys.*, **97**, 5616; Howard, S. T., Huke, J. P., Mallinson, P. R. and Frampton, C. S. (1994) *Phys. Rev. B*, **49**, 7124.
8. Schwarz, W.H.E. and Müller, B. (1990) *Chem. Phys. Lett.*, **166**, 621.
9. Kohn, W. and Sham, L.J. (1965) *Phys. Rev.*, **140**, A1133.
10. Levy M. and Goldstein, J.A. (1987) *Phys. Rev. B*, **35**, 7887.
11. Gilbert, T.L. (1975) *Phys. Rev. B*, **12**, 2111.
12. Jayatilaka, D. (1995) *Chem. Phys. Lett.*, **230**, 228.
13. Hall, S.R. (1981) *Acta Cryst. A*, **37**, 517.
14. Stewart, R.F. (1969) *J. Chem. Phys.*, **51**, 4569.

15. Coppens, P., Willoughby T.V. and Csonka, L.N. (1971) *Acta Cryst. A*, **27**, 248.
16. Levy, M. In *Recent Developments and Applications of Modern Density Functional Theory*, Theoretical and Computational Chemistry, Vol. 4, Seminario, J.M. (Ed.), p. 17.
17. *International Tables for Crystallography*, Vol. A, Hahn, T. (Ed.), 2nd rev. edn., Kluwer, 1987, pp. 618–619
18. See, for example, Chandler, G.S., Figgis, B.N., Reynolds P.A. and Wolff, S.K. (1994) *Chem. Phys. Lett.*, **225**, 421.
19. Pulay, P. (1982) *J. Comp. Chem.*, **3**, 556.
20. Larsen, F.K. and Hansen, N.K. (1984) *Acta Cryst. B*, **40**, 169.
21. Dunning, T.H. (1971) *J. Chem. Phys.*, **55**, 716.
22. Szaz, L. and Bryne, J. (1967) *Phys. Rev.*, **158**, 34.
23. *CRC Handbook of Chemistry and Physics*, 73rd edn., CRC Press, Boca Raton, 1993.
24. Parr R.G. and Yang, W. (1989) *Density Functional Theory of Atoms and Molecules*, Oxford University Press.
25. Morrell, M.M., Parr, R.G. and Levy, M. (1975) *J. Chem. Phys.*, **62**, 549.
26. Levy, M., Perdew J.P. and Sahni, V. (1984) *Phys. Rev. A*, **30**, 2745.
27. Dirac, P.A.M. (1930) *Proc. Cambridge Phil. Soc.*, **26**, 376.
28. Vosko, S.H. Wilk L. and Nusair, M. (1980) *Can. J. Phys.*, **58**, 1200.
29. Becke, A.D (1988) *Phys. Rev. A*, **38**, 3098.
30. Lee, C., Yang, W. and Parr, R.G. (1988) *Phys. Rev. B*, **37**, 785.
31. Iversen, B.B., Larsen, F.K., Souhassou M. and Takata, M. (1995) *Acta Cryst. B*, **51**, 580.
32. de Vries, R.Y., Briels W.J. and Feil, D. (1996) *Phys. Rev. Lett.*, **77**, 1719.

19

Spin density in interacting nitronyl nitroxide radicals

Y. PONTILLON[1], A. CANESCHI [2], D. GATTESCHI [2], E. RESSOUCHE [1], F. ROMERO [3], J. SCHWEIZER [1], R. SESSOLI[2] and R. ZIESSEL[3]

[1] *DRFMC-SPSMS-MDN, 17 rue des Martyrs, 38054 Grenoble Cedex 9, France*

[2] *Dipartimento di Chimica, Universitá degli studi di Firenze, Via Maragliano 77, 50144 Firenze, Italy*

[3] *Laboratoire de Chimie, Ecole de Chimie, Université Louis Pasteur, 1 rue Blaise Pascal, 67008 Strasbourg, France*

Since the reported synthesis [1] of stable 2-phenyl-4,4,5,5-tetramethylimidazoline-1-oxyl-3-oxide, NitPh, considerable effort has been spent in the physical characterization of nitronyl nitroxide compounds. Their general formula is presented in Figure 1A. These compounds carry a delocalized, $S = 1/2$, unpaired electron. Among them, some derivatives were found to be paramagnetic at low temperature (NitPh, [1]), others were found to exhibit an antiferromagnetic or a ferromagnetic behavior [2–7].

This magnetic behavior is very sensitive to the chemical structure of the spin carrier and to the crystal packing. For instance, for the *para*-nitro substituted derivative, Nit(p-NO_2)Ph, which crystallizes in four different phases, the β phase only orders ferromagnetically ($T_c = 0.6$ K) [8, 9]. Moreover, attaching the nitro group in the *meta*-, rather than the *para*-position of the phenyl, leads to an antiferromagnetic compound [2].

Furthermore, the NitR radicals have been shown to behave as valuable bridging ligands for obtaining low-dimensional, strongly coupled magnetic systems. Examples of high nuclearity spin clusters, magnetic chains and magnetic planes have been reported in the course of a rather exhaustive investigation of complexes with 3d and 4f transition-metal hexafluoroacetylacetonates, $M(hfac)_n$ [10–14].

We report herein a single crystal polarized neutron investigation of the spin density of two purely organic nitronyl nitroxide free radicals which present ferromagnetic interactions: the 2-(6-ethynyl-2-pyridyl)-4,4,5,5-tetramethylimidazoline-1-oxyl-3-oxide, NitPy(C≡C–H), and the 2-(4-methylthiophenyl)-4,4,5,5-tetramethylimidazoline-1-oxyl-3-oxide, Nit(SMe)Ph. We have compared these results with the spin density of the NitPh [15] where the molecules are isolated one from the other, with practically no intermolecular magnetic interaction. The aryl group R of the NitPh, NitPy(C≡C–H) and Nit(SMe)Ph compounds are depicted in Figure 1B–D, respectively.

NitPh. The $P2_1/c$ form of the NitPh has been investigated by polarized neutron diffraction (Zheludev *et al.* [15]). We present herein only the main results of this study. (i) Most of the spin density is equally shared between the four atoms of the two NO groups (Figure 2). (ii) The bridging sp^2 carbon atom carries a significant negative spin density (the ratio of its spin population to the spin populations of the oxygen or nitrogen atoms of the two NO groups is approximately $-1/3$). (iii) Delocalization of

Paul G. Mezey and Beverly E. Robertson (eds.), Electron, Spin and Momentum Densities and Chemical Reactivity, 265–274

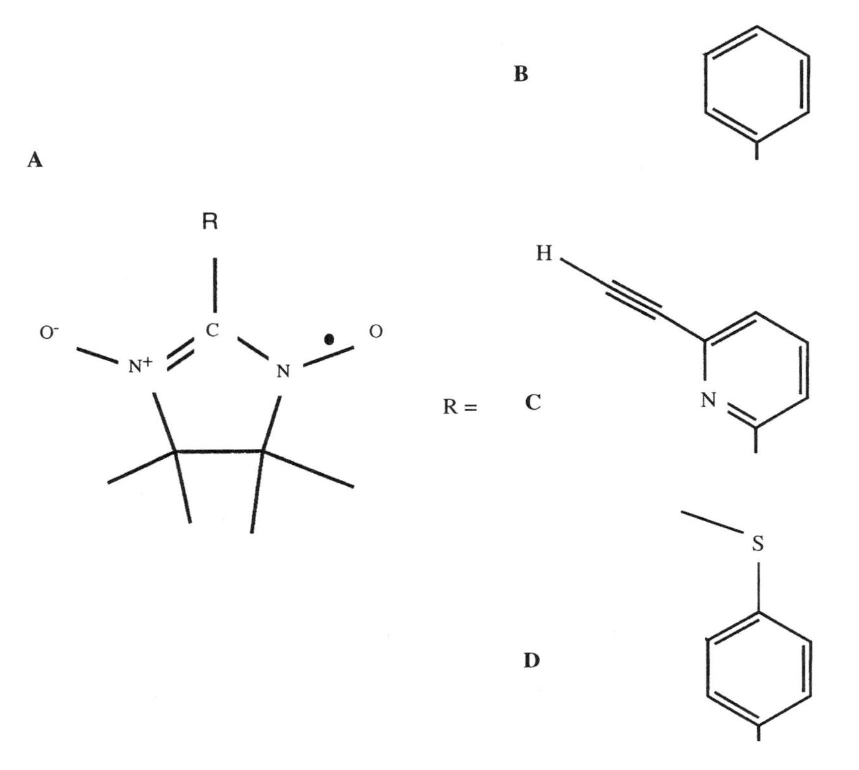

Figure 1. General formula of nitronyl nitroxide free radicals (A). The phenyl (B), 2-(6-ethynyl-2-pyridyl) (C) and 2-(4-methylthiophenyl) (D) groups are possible substituents for R.

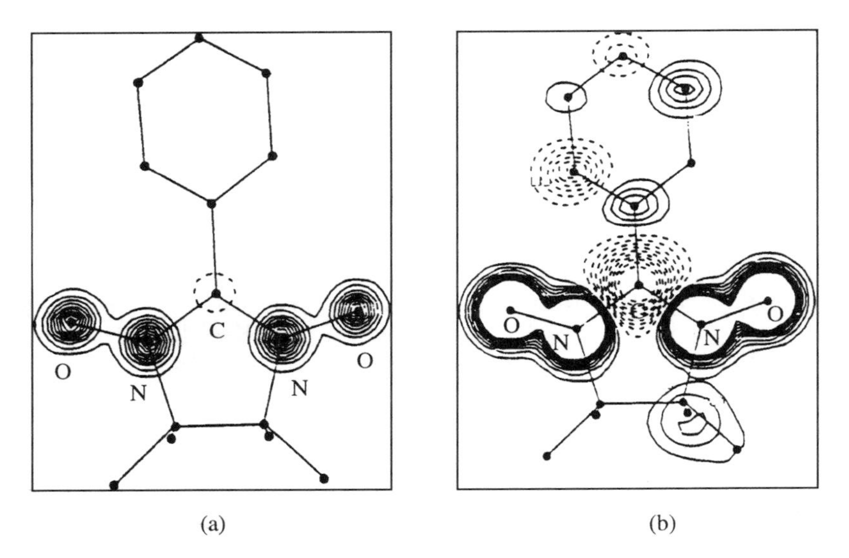

Figure 2. Projection onto the nitroxide mean plane of the spin density (NitPh). Negative contours are dashed: (a) high-level contours (step $0.1\mu_B/\text{Å}^2$); (b) low-level contours (step $0.01\ \mu_B/\text{Å}^2$).

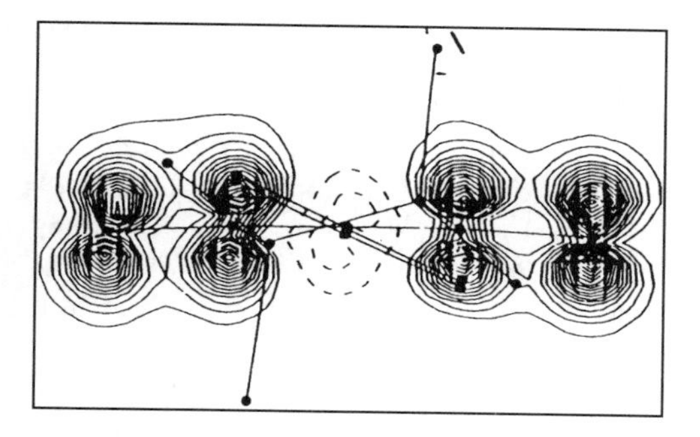

Figure 3. Projection onto a plane perpendicular to the nitroxide mean plane of the contours (step $0.04\,\mu_B/\text{Å}^2$) of the spin density (NitPh). Negative contours are dashed.

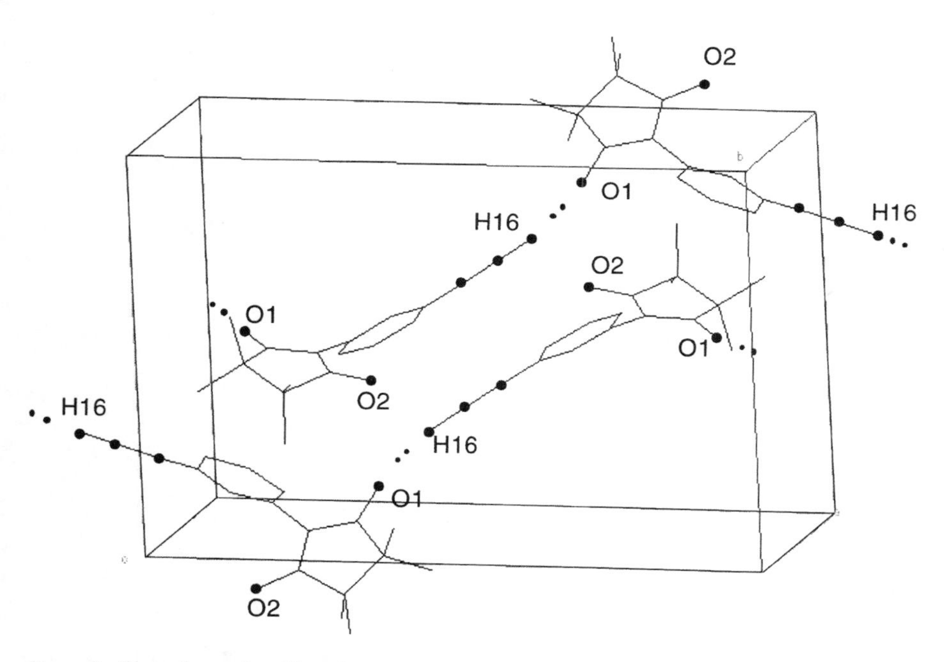

Figure 4. View of crystal packing of NitPy(C≡C–H).

the unpaired electron onto the phenyl ring is weak. The corresponding spin populations are at the limit of the experimental accuracy. (iv) The magnetic orbitals of the O–N–C–N–O atoms are p_z orbitals, mainly perpendicular to the O–N–C–N–O plane (Figure 3).

NitPy(C≡C–H). In the solid state, the NitPy(C≡C–H) molecules (space group $P2_1/n$ crystallize in chains via a hydrogen bond –C≡C–H···O–N– (Figure 4), with positive intrachain magnetic coupling [16]. A single crystal ($5.0 \times 4.5 \times 1.7\ \text{mm}^3$) was investigated by polarized neutron diffraction (DN2 diffractometer, SILOE reactor) and 216 independent flipping ratios were collected ($H = 8\,T$, $T = 4.65\,K$). We have analyzed the data in two steps. First, we have performed a 3D Maximum Entropy (MaxEnt) reconstruction [17]. Figures 5(a) and 5(b) show the projected spin density, respectively, onto the O–N–C–N–O plane and the pyridine cycle. As in the NitPh, the main part of the spin density is located on the two NO groups. But in the present case, a depletion of the spin density on the O1 atom (oxygen atom involved in the hydrogen bond) in favor of O2 (oxygen atom not involved in the hydrogen bond) is observed (Figure 5(a)). Moreover, the spin density on O1 and O2 is not exactly centered on these atoms, but is slightly shifted away from the center of the NO bond. This illustrates the antibonding character of the singly occupied magnetic molecular π^* orbital. Besides this, a noticeable positive spin density is found on the acetylenic hydrogen H16 (Figure 5(b)).

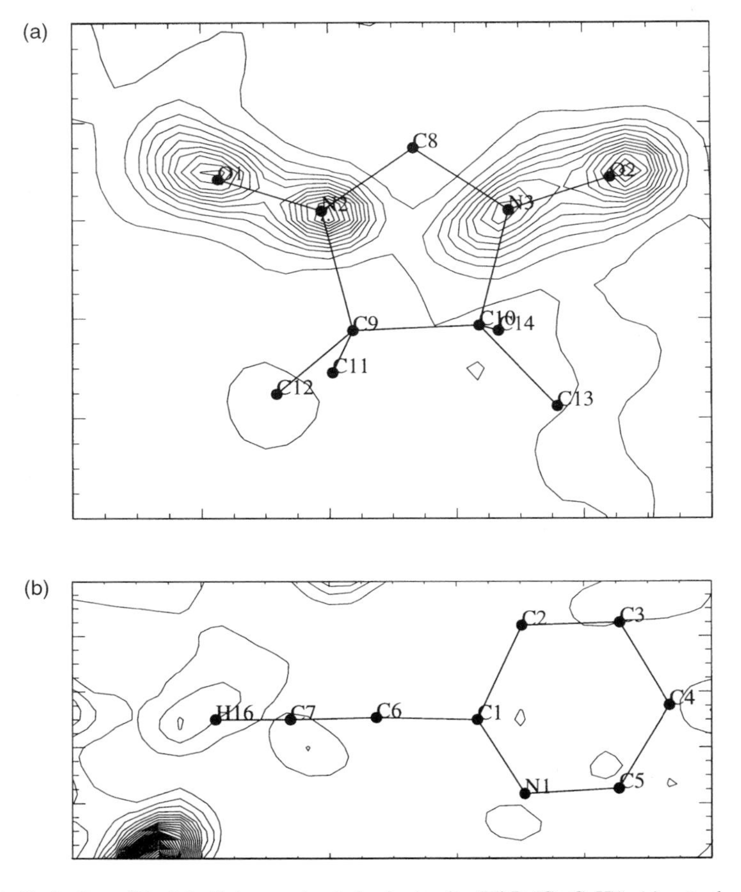

Figure 5. Projection of the MaxEnt reconstructed spin density (NitPy(C≡C–H)): (a) onto the nitroxide mean plane (step 0.02 $\mu_B/Å^2$); (b) onto the pyridine ring (step 0.006 $\mu_B/Å^2$).

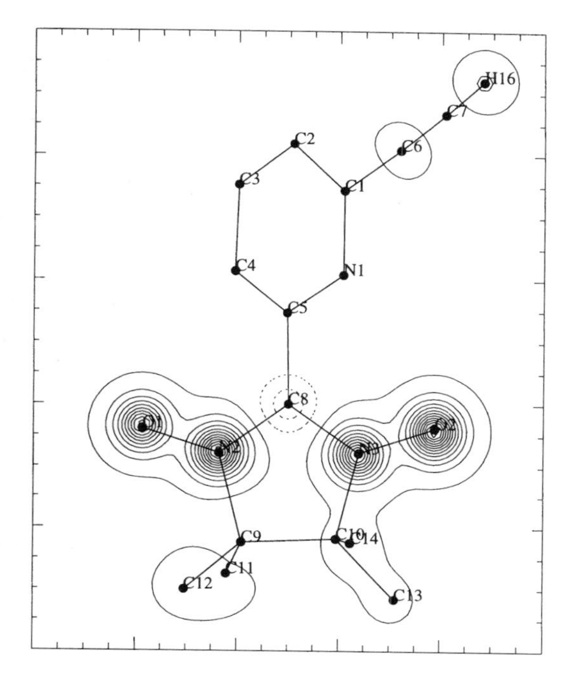

Figure 6. Projection onto the nitroxide mean plane of the spin density as analyzed by wave-function modeling (NitPy(C≡C–H)). Negative contours are dashed, contour step 0.04 $\mu_B/\text{Å}^2$.

To obtain quantitative data in terms of individual atomic spin populations and to extract information concerning the polarization of the aryl fragment, we have then refined the atomic orbital coefficients of the unpaired electron molecular orbital [18]. Figure 6 shows the reconstructed spin density projected onto the Nit cycle of the molecule. As expected, the strongest spin populations are carried by the O–N–C–N–O fragment and the bridging sp^2 carbon carries a negative spin density. The ratio of its spin population to the average of the spin populations of the O1, N2, N3 and O2 atoms is, as in the NitPh, approximately $-1/3$. In the present case, the main difference lies in the oxygen atoms. They were equivalent in NitPh but not here: the transfer from O1 (0.203(10) μ_B) to O2 (0.278(9) μ_B) is confirmed. Besides this, a significant and positive contribution is found on the hydrogen atom H16 (0.045(9) μ_B), much higher than on the other atoms (except the O–N–C–N–O fragment).

The spin density on the acetylenic hydrogen H16 is a sign of the active role played by the hydrogen bond in the intrachain coupling. Moreover, the hydrogen bond corresponds to C7–H16$\cdots$O1 and the spin population is much less on O1 than on O2: O1 participates to the magnetic interaction, and O2 does not. We have then strong evidence that the hydrogen bond is involved in the path of the ferromagnetic interactions.

Nit(SMe)Ph. The unit cell of Nit(SMe)Ph (space group $P2_1/a$) comprises four molecules of free radicals which are arranged in pairs. These pairs form a layered structure (Figure 7). This radical is a ferromagnet with $T_c = 0.6$ K [19]. The experiment was performed on the DN2 polarized neutron diffractometer (SILOE reactor, Grenoble), with a vertical field provided by a cryomagnet (8 T at $T = 5.3$ K). In this case, 350 independent reflections have been measured with a $9.25 \times 2.5 \times 1.4\,\text{mm}^3$ crystal. As done previously, we have analyzed the data in two ways: the three-dimensional MaxEnt method [17] and the magnetic wave function refinement method [18]. Figure 8 represents the spin density map reconstructed by the MaxEnt method, and projected onto the Nit ring. As in the NitPh, one clearly sees the major part of spin density localized on the O–N–C–N–O fragment and this spin density is equally shared between the O1, N1, N2 and O2 atoms. A negative contribution is also found on the central carbon atom C8.

Then, to obtain quantitative data in terms of individual atomic spin populations, we have refined the magnetic wave function. Figure 9 shows the spin density projected onto the Nit cycle of the molecule. As expected from the MaxEnt analysis, most of the spin density is carried by the ONCNO fragment and the bridging sp^2 carbon atom carries a negative spin density which corresponds, as in NitPh, at $-1/3$ of those carried by oxygen and nitrogen atoms of the NO groups. The main difference between NitPh and Nit(SMe)Ph concerns the shape of the molecular magnetic orbital of the two oxygen

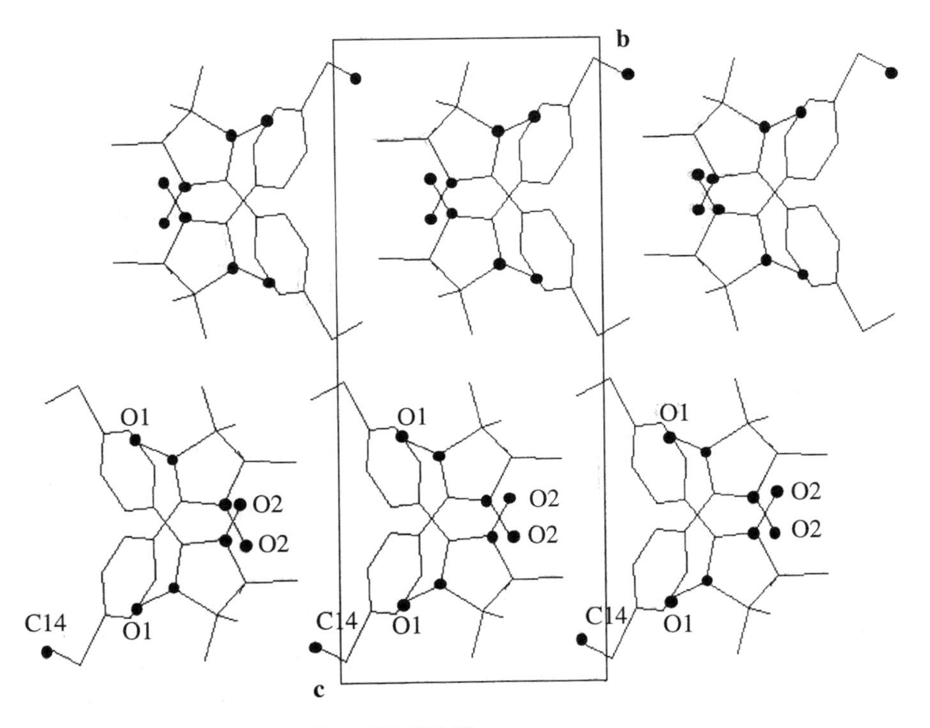

Figure 7. View of the crystal packing of Nit(SMe)Ph.

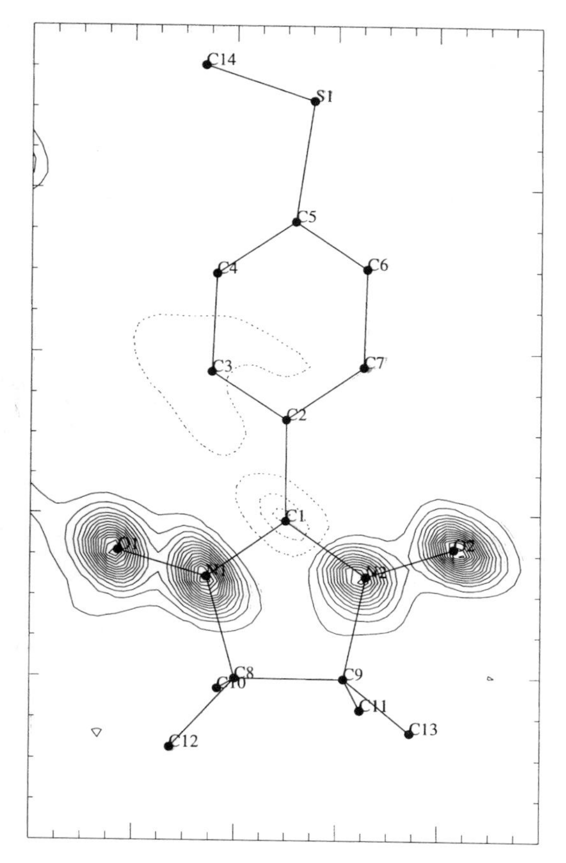

Figure 8. Projection of the MaxEnt reconstructed spin density onto the nitroxide mean plane (Nit(SMe)Ph). Negative contours are dashed, contour step 0.02 $\mu_B/\text{Å}^2$.

atoms O1 and O2 of the NO groups (Figure 10). In Nit(SMe)Ph a clear rotation, from z-axis, is evidenced. Besides this, a hybridization of this magnetic orbital is observed (one lobe is stronger than the other). Moreover, some significant spin density is found on the C4 (0.016(6) μ_B), C5 ($-0.018(7)$ μ_B) and C6 (0.027(6) μ_B) carbon atoms of the phenyl ring and on the terminal carbon C14 (0.028(6) μ_B) of the methylthio group.

The carbon atoms which are involved in the short intermolecular contacts between molecules (less than 4 Å: O2/C14 and O1/C4, C5, C6, Figure 7) carry a significant spin density. The magnetic molecular orbitals of the corresponding oxygen atoms (O1 and O2) are twisted and hybridized. Thus we have evidence that the intermolecular exchange involves these contacts.

In these two examples where the magnetic coupling between adjacent molecules exist and are positive, we have evidenced specific features which reflect the role of the spin density in the magnetic interaction propagation. These specific features

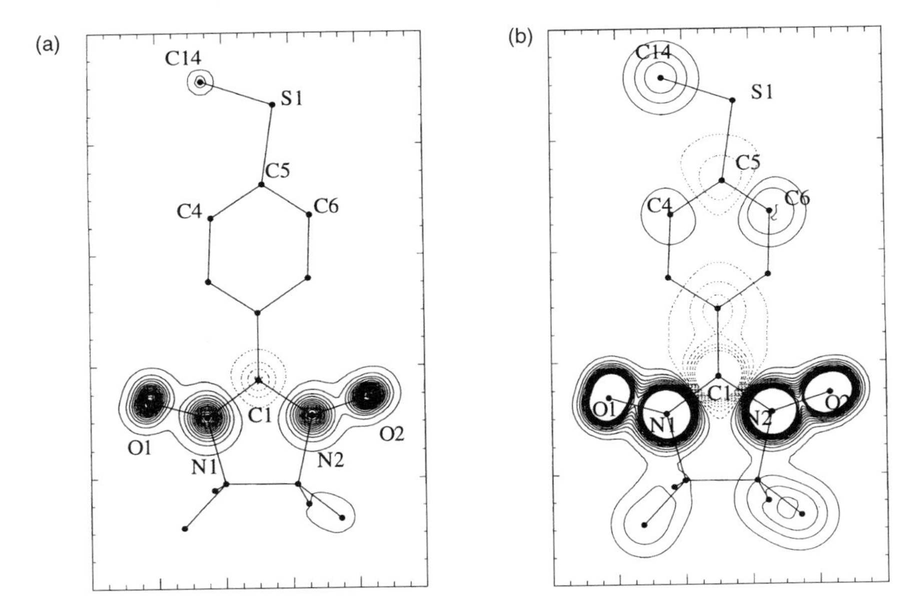

Figure 9. Projection onto the nitroxide mean plane of the spin density as analyzed by wave-function modeling (Nit(SMe)Ph). Negative contours are dashed: (a) high-level contours (step 0.04 $\mu_B/\text{Å}^2$); (b) low- level contours (step 0.006 $\mu_B/\text{Å}^2$).

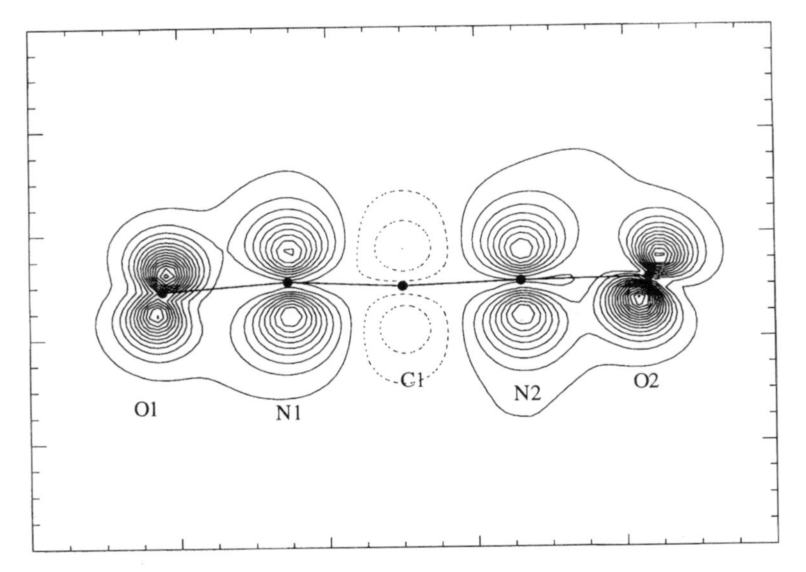

Figure 10. Projection onto a plane perpendicular to the nitroxide mean plane of the contours (step 0.03 $\mu_B/\text{Å}^2$) of the spin density as analyzed by wave- function modeling (Nit(SMe)Ph). Negative contours are dashed.

Table 1. Spin populations of the atomic sites in μ_B (scaled to 1 μ_B/molecule) of the O–N–C–N–O fragment for the NitPh, NitPy(C≡C–H) and Nit(SMe)Ph.

	NitPh [15]	NitPy(C≡C–H)	Nit(SMe)Ph
O	0.267	0.278	0.239
N	0.272	0.225	0.238
C	−0.109	−0.071	−0.098
N	0.272	0.242	0.265
O	0.267	0.203	0.233
Sum of the spin populations on O–N–C–N–O	0.969	0.877	0.877

concern, to a large extent, the spin density which is located outside of the O–N–C–N–O fragment. We have summarized in Table 1, for the two radicals NitPy(C≡C–H) and Nit(SMe)Ph, as well as for the reference radical NitPh, the spin density which remains on the O–N–C–N–O group. The sum of the spin populations on these atoms amounts to 0.969 μ_B for NitPh, indicating that almost all the spin density is localized on that fragment. Contrarily, this sum is equal to 0.877 μ_B only for each of the two other radicals. We have here clear evidence that the magnetic interactions imply a noticeable delocalization of the spin density out of the O–N–C–N–O fragment. This delocalization happens to be equal for the two ferromagnetic radicals investigated here.

References

1. Osiecky, J.H. and Ullman, E.F. (1968) *J. Am. Chem. Soc.*, **90**, 1078.
2. Awaga, K. Inabe, T. Okayama, T. and Maruyama, Y. (1993) *Mol. Cryst. Liq. Cryst.*, **232**, 79.
3. Awaga, K. Inabe, T. and Maruyama, Y. (1992) *Chem. Phys. Lett.*, **190**, 349.
4. Awaga, K. Inabe, T. Maruyama, Y. Nakamura, T. and Matsumoto, M. (1992) *Chem Phys. Lett.*, **195**, 21.
5. Sugano, T. Tamura, M. Kinoshita, M. Sakai, Y. and Ohashi, Y. (1992) *Chem. Phys. Lett.*, **200**, 235.
6. Turek, P. Nozawa, K., Shiomi, D. Awaga, K. Inabe, T. Maruyama, Y. and Kinoshita, M. (1991) *Chem. Phys. Lett.*, **180**, 327.
7. Hosokoshi, Y. Tamura, M. and Kinoshita, M. (1993) *Mol. Cryst. Liq. Cryst.*, **232**, 45.
8. Awaga, K. and Maruyama, Y. (1989) *Chem. Phys. Lett.*, **158**, 556.
9. Awaga, K. Inabe, T. Nagashima, U. and Maruyama, Y. (1989) *J. Chem. Soc., Chem Commun.*, 1617.
10. Caneschi, A. Gatteschi, D. and Rey, P. (1991) *Prog. Inorg. Chem*, **39**, 331.
11. Caneschi, A. Gatteschi, D. Laugier, J. Rey, P. Sessoli, R. and Zanchini, C. (1988) *J. Am. Chem. Soc.*, **110**, 2795.
12. Caneschi, A. Gatteschi D. Rey, P. and Sessoli, R., (1988) *Inorg. Chem.*, **27**, 1756.
13. Caneschi, A. Gatteschi, D. and Sessoli, R. (1993) *Inorg. Chem.*, **32**, 4612.
14. Benelli, C. Caneschi, A. Gattcshi, D. and Sessoli, R. (1993) *Inorg. Chem.*, **32**, 4797.
15. Zheludev, A. Barone, V. Bonnet, M. Delley, B. Grand, A. Ressouche, E. Rey, P. Subra, R. and Schweizer, J. (1994) *J. Am. Chem. Soc.*, **116**, 2019.
16. Romero, F. De Cian, A. Fisher, J. Turek, P. and Ziessel, R. (1996) *New. J. Chem.*, **20**, 919.
17. Papoular, R.J. Zheludev, A. Ressouche, E. and Schweizer, J. (1995) *Acta. Crystallogr.*, **A51**, 295.

18. Gillon, B. Schweizer, J. (1989) Study of chemical bonding in molecules: the interest of polarized neutron diffraction, In *Molecules in Physics, Chemistry and Biology*, Maruani, Jean(Ed.), Kluwer, Dordrecht, The Netherlands, Vol. II, p. 111.
19. Ganeschi, A. Ferraro, F. Gatteschi, D. le Lirzin, A. and Rentschler, E. (1994) *Inorg. Chemica Acta*, **217**, 7.

20

Electrostatic potential of a new angiotensin II receptor antagonist from X-ray diffraction and *ab initio* calculations

RAFFAELLA SOAVE[1], RICCARDO DESTRO[1], LAURA BELVISI[2] and CARLO SCOLASTICO[2]

[1]*Dipartimento di Chimica Fisica ed Elettrochimica e Centro CNR, Università degli Studi, Via Golgi, 19 Milano, Italy*

[2]*Dipartimento di Chimica Organica e Industriale e Centro CNR, Università degli Studi, Via Venezian, 21 Milano, Italy*

1. Introduction

The hormone angiotensin II (AII), an octapeptide of sequence Asp-Arg-Val-Tyr-Ile-His-Pro-Phe produced by the renin–angiotensin system (RAS), participates in a number of physiological functions associated with the regulation of blood pressure. Drugs that inhibit the RAS have been shown to be effective in treating human hypertension [1]. A possible approach to interfering with the RAS is to inhibit the binding of AII to its receptor, and recent pharmacological research is currently directed towards non-peptide AII receptor antagonists that, unlike peptide AII receptor inhibitors, can exhibit oral activity, long-plasma half-life and no partial agonism.

In a recent paper by Salimbeni *et al.* [2], a novel series of such AII antagonists has been presented: on the basis of a comparative analysis of theoretical distributions of the electrostatic potential ($\Phi(\mathbf{r})$) of active and inactive compounds and overlay studies, employing a computational model of an AII active conformation, it was found that the compound named LR-B/081 [3, 4] ($C_{30}H_{30}N_6O_3S$), i.e. 2-[(6-butyl-2-methyl-4-oxo-5-{4-[2-(1H-tetrazol-5-yl)phenyl] benzyl}-3H-pyrimidin-3-yl)methyl]-3-thiophenecarboxylate (Scheme 1), was one of the most potent in the series, and was selected as a candidate for further studies.

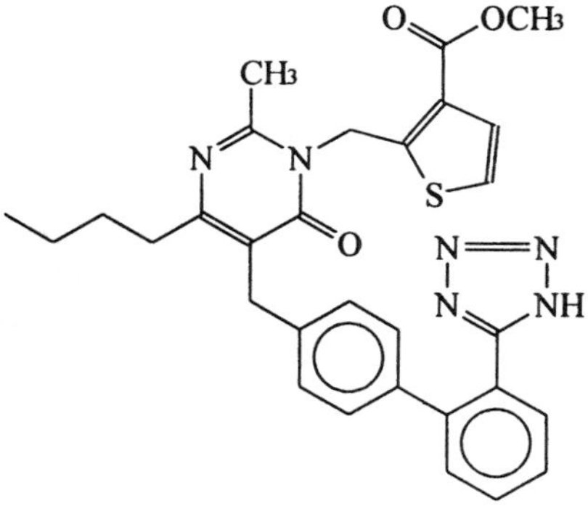

Scheme 1.

Paul G. Mezey and Beverly E. Robertson (eds.), Electron, Spin and Momentum Densities and Chemical Reactivity, 275–283

For a better understanding of the electrostatic requirements for LR-B/081 activity, its total experimental electron distribution $\rho(\mathbf{r})$ and electrostatic potential $\Phi(\mathbf{r})$ have been obtained from an extensive set of X-ray diffraction data collected at 18 K. Indeed, the electrostatic potential that the nuclei and electrons of a molecule create in the surrounding space is well established as a useful analytical tool for the study of molecular reactivity. It is this potential that is first "seen" or "felt" by another approaching chemical species. Thus, key features of a molecule that are necessary for a successful interaction with a receptor can be identified through an analysis of its electrostatic potential [5]. This property has been used extensively for interpreting and predicting the reactive behaviour of a variety of chemical systems, and in the study of biological recognition processes, such as drug–receptor and enzyme–substrate interactions [6–9].

$\Phi(\mathbf{r})$ was also computed from *ab initio* wave functions in the framework of the HF/SCF method using 3-21G and 6-31G* basis sets: due to the large size of LR-B/081, the calculation has as yet been performed on isolated molecular fragments, adopting a geometry based on molecular dimensions from X-ray diffraction studies.

The preliminary comparison between these experimental and theoretical results shows satisfactory agreement, both qualitatively and quantitatively, and clearly demonstrates that the two methods support one another well.

2. Methods

Experimental

Crystal data and details of data collection, data reduction and final refinement are reported in Table 1. The procedure for data collection and processing, which included a correction for scan-truncation effects, were similar to those recently described for syn-1,6:8,13-biscarbonyl[14]annulene [10] and citrinin [11]. Figure 1 shows the numbering scheme adopted in the present analysis.

A preliminary least-squares refinement with the conventional, spherical-atom model indicated no disorder in the low-temperature structure, unlike what had been observed in a previous room-temperature study [4], which showed disorder in the butylic chain at C1. The intensities were then analysed with various multipole models [12], using the VALRAY [13] set of programs, modified to allow the treatment of a structure as large as LR-B/081; the original maximum number of atoms and variables have been increased from 50 to 70 and from 349 to 1200, respectively. The final multipole model adopted to analyse the X-ray diffraction data is described here.

According to the aspherical-atom formalism proposed by Stewart [12], the one-electron density function is represented by an expansion in terms of rigid pseudoatoms, each formed by a core-invariant part and a deformable valence part. Spherical surface harmonics (multipoles) are employed to describe the directional properties of the deformable part. Our model consisted of two monopole (three for the sulfur atom), three dipole, five quadrupole, and seven octopole functions for each non-H atom. The generalised scattering factors (GSF) for the monopoles of these species were computed from the Hartree–Fock atomic functions tabulated by Clementi [14].

Table 1. LR-B/081: crystal data and summary of the X-ray diffraction experiment.

Name	LR-B/081	Refl. for cell determination	240 ($23^\circ \leq 2\theta \leq 32^\circ$)
Dimensions/mm	$0.45 \times 0.45 \times 0.125$	No. of 2θ values	15
Temperature/K	18	Standard	820, 829, $41\bar{3}$
Crystal system	Orthorhombic	$2\theta^{\mathrm{Mo}}_{\mathrm{max}}$/degrees	75
Space group	Pbca	Decay	none
a/Å	29.831(4)	Time of exposure/h	~ 600
b/Å	15.505(2)	No. intensities	51,485
c/Å	11.985(1)	No. independent	14,699
$V/\text{Å}^3$	5543(1)	refl. with $I < 0$	886
Z	8	No. of parameters	1161
$F(000)$	2336	$R(F)$	0.0308
$D_x/\mathrm{Mg\,m^{-3}}$	1.329	$wR(F)$	0.0219
$\mu/\mathrm{mm^{-1}}$	0.145	$R(F^2)$	0.0254
Diffractometer	Syntex $P\bar{1}$	$wR(F^2)$	0.0421
Radiation	MoKα	Goodness-of-fit	1.2088
Scan technique	ω–2θ	$\lvert\varepsilon_n - \varepsilon_{n-1}\rvert/\varepsilon_n$ with $\varepsilon = \sum w(F_o^2 - F_c^2)^2$	5.98×10^{-7}
Scan range (2θ)/deg	$1.2 + S_{\alpha_1-\alpha_2}$		
Scan speed (2θ)/deg min^{-1}	3	Extinction $\times 10^{-4}/\mathrm{rad}^{-2}$	0.530(44)

Electron population parameters of inner monopoles were constrained to be equal for all 40 non-H atoms. Single exponentials $r^n \exp(-\alpha r)$ were adopted as radial functions for the higher multipoles, with $n = 2,\ 2,\ 3$ respectively for dipole, quadrupole, and octopole of the species C, N and O, and $n = 4,\ 4,\ 4$ for the same multipoles of the S atom. A radial scaling parameter κ, to shape the outer shell monopoles, and the exponential parameter α of all non-H atomic species were also refined. H atoms were initially given scattering factors taken from the H_2 molecule [15] and polarised in the direction of the atom to which they are bonded.

In the final stages of the refinement the positional parameters of the H atoms were kept fixed, and these atoms too were described with multipoles, up to the dipole level. For both poles of the H pseudoatoms the radial functions were again single exponentials, with $n = 0,\ 1$ for monopole and dipole respectively, and the α value was 2.48 bohr^{-1}.

An isotropic extinction parameter, of type I and Lorentzian distribution (in the formalism of Becker and Coppens [16]), was also refined. The motions of the non-H atoms were described by anisotropic parameters, while those of the H atoms by isotropic B's. All these displacement parameters were included among the refinable quantities of the model, for a total of 1161 variables in a single least-squares matrix.

Within this model, the electron density $\rho(\mathbf{r})$ is analytically represented by a finite multipole expansion [12] about the equilibrium nuclear configuration $\mathbf{Q}_\mathrm{e}$

$$\rho\left(\mathbf{r}, \mathbf{Q}_\mathrm{e}\right) = \sum_p \sum_l \sum_m C_{plm} B_{plm}\left(\mathbf{r} - \mathbf{R}_\mathbf{p}\right), \tag{1}$$

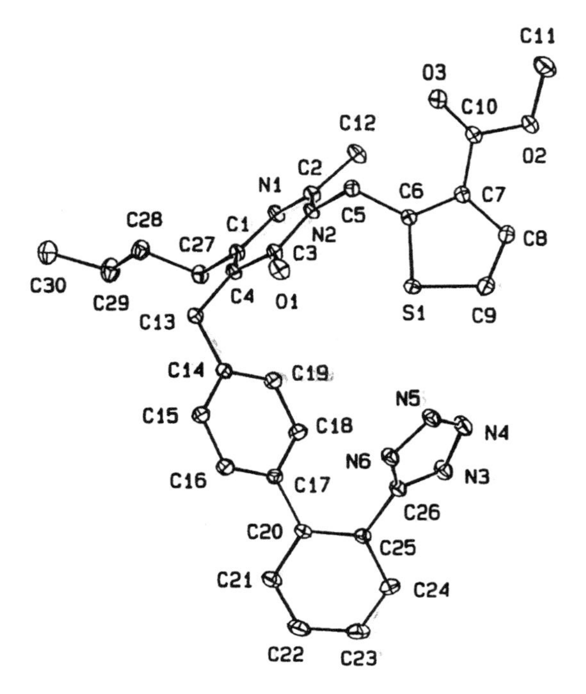

Figure 1. ORTEP plot of LR-B/081 at 18 K. Ellipsoids at 95% probability level. H atoms are omitted for clarity.

where C_{plm} are electron population coefficients and B_{plm} are basis functions. Population coefficients for each basis function are obtained through a least-squares fit of the calculated, model structure factors (F_c) to the observed ones (F_o).

The electrostatic potential $\Phi(\mathbf{r})$ is then computed in direct space as

$$\Phi(\mathbf{r}) = \int \rho(\mathbf{r}')|\mathbf{r} - \mathbf{r}'|^{-1}\,\mathrm{d}^3\mathbf{r}'. \tag{2}$$

A detailed description of the method has been presented by Stewart and Craven [17]. The procedure has been applied, in our case, over discrete molecular fragments removed from the cell and in isolation. Estimated standard deviations (esd's) of $\Phi(\mathbf{r})$ were also calculated [18].

Ab initio *calculations*

Equation (2) was also used to calculate $\Phi(\mathbf{r})$ in the quantum chemical approach. On the basis of previous results [19], calculated electrostatic potentials were computed from *ab initio* wave functions obtained in the framework of the HF/SCF method using a split-valence basis set (3-21G) and a split-valence basis set plus polarisation functions on atoms other than hydrogen (6-31G*). The GAUSSIAN 90 software package [20] was used. Since *ab initio* calculations of the molecular wave function for the whole

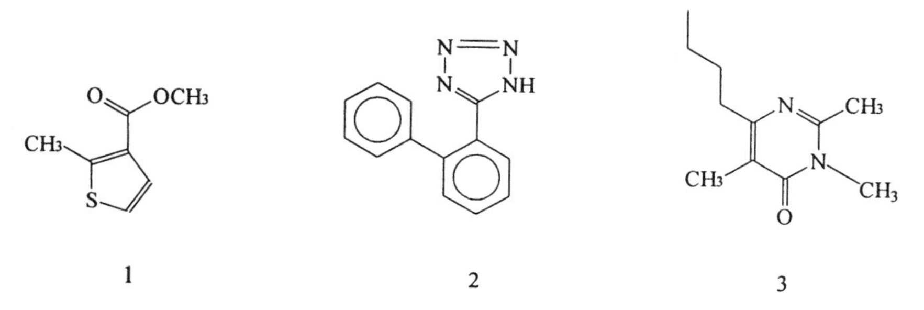

Figure 2. The three molecular fragments constituting LR-B/081.

molecule of LR-B/081 required computer time and storage memory greater than the available computational resources, three molecular fragments constituting the AII antagonist (Figure 2) were here separately analysed, adopting the experimental molecular geometry.

Two-dimensional isopotential maps in meaningful planes were adopted as electrostatic potential representation techniques. For the three fragments, the $\Phi(\mathbf{r})$ values were calculated with 0.2 Å spacing at points belonging to regular two-dimensional grids in the planes containing the thiophene ring, the tetrazole ring and the pyrimidinone ring. The isopotential lines over the two-dimensional maps were constructed by an interpolation technique with the SURFER program [21] on a personal computer.

3. Results and discussion

Results are presented in terms of maps of experimental and theoretical $\Phi(\mathbf{r})$ for the three molecular fragments. Only regions with $\Phi(\mathbf{r}) < 50$ kcal/mol are mapped. A table of $\Phi(\mathbf{r})$ minima (Table 2) is also given. Experimental results refer to the multipole model adopted to analyse the X-ray diffraction data, the theoretical ones to the 6-31G* basis set. Calculations have been performed on the fragments reported in Figure 2, so that Figure 3(a) and (b) are relative to fragment 1, Figure 4(a) and (b) to fragment 2 and Figure 5(b) to fragment 3. In Figure 5(a), on the other hand, the electrostatic potential includes the contributions of all atoms in the molecule.

We see overall qualitative agreement between experiment and theory. In Figure 3(a) and (b) the position of the zero contours of $\Phi(\mathbf{r})$ and the shape and height or depth of the constant potential lines agree quantitatively, the maximum difference between $\Phi(\mathbf{r})^{exp}_{min}$ and $\Phi(\mathbf{r})^{th}_{min}$ being less than $1.3\sigma[\Phi(\mathbf{r})^{exp}_{min}]$, for the minimum in the proximity of the oxygen atom O3.

In the maps of Figure 4(a) and (b) the three minima near the three nitrogen atoms (N3, N4 and N5) agree very well, while the zero contour level appears somehow different. Moreover, in the experimental map (and not in the theoretical one) a negative region of $\Phi(\mathbf{r})$ is found between the two phenyl rings, probably due to the π electrons of the aromatic systems. A 6-31G** calculation has been performed on this fragment to check if the lack of this negative region in the *ab initio* map was due to the

Table 2. Experimental (final multipole model) and theoretical (6-31G*) values of $\Phi_{min}(\mathbf{r})$ and of $|\mathbf{r}(\Phi_{min}) - \mathbf{r}_{closest\ atom}|$ in the most meaningful planes of LR-B/081.

Plane	$\Phi_{min}(\mathbf{r})$, kcal/mol		$\|\mathbf{r}(\Phi_{min}) - \mathbf{r}_{closest\ atom}\|$, Å	
	Exp.[a]	Theor.	Exp.[b]	Theor.[b]
Thiophene	−37 (12)	−51.71	1.28 (O3)	1.23 (O3)
(Figures 3(a) and (b))	−20 (13)	−28.31	1.26 (O2)	1.20 (O2)
	−12 (11)	−3.86	1.78 (S1)	1.72 (S1)
Tetrazole	−84 (14)	−65.13	1.20 (N3)	1.32 (N3)
(Figures 4(a) and (b))	−80 (12)	−63.43	1.27 (N4)	1.33 (N4)
	−76 (13)	−50.87	1.25 (N5)	1.33 (N5)
	−24 (17)	—	1.64 (C20)	—
Pyrimidinone	−62 (22)	−63.54	1.21 (N1)	1.29 (N1)
(Figures 5(a) and (b))	−42 (26)	−68.07	1.21 (O1)	1.20 (O1)

[a]Esd's in parentheses.
[b]Closest atom in parentheses.

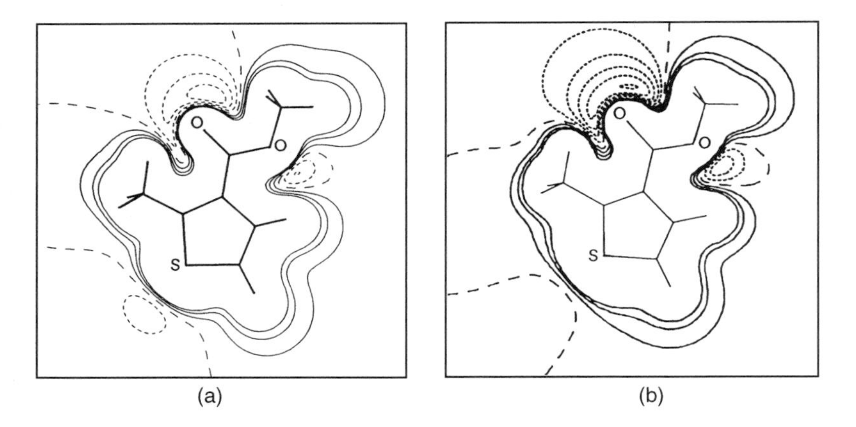

Figure 3. $\Phi(\mathbf{r})$ contour plots in the plane of the thiophene ring. Sections of 12 Å × 12 Å. Contour interval: 10 kcal/mol. (a) Experimental (minimum contour −30 kcal/mol), (b) theoretical (minimum contour −50 kcal/mol). Negative and zero contours: short and long dashed lines, respectively, positive contours: solid lines.

theoretical treatment of the H atoms, but it provided a result identical to that reported in Figure 4(b).

Figure 5(a) and (b) show excellent agreement as to the depth of the minimum and the position of the zero contour in the region near the N1 atom. The maps are not so similar in the bottom part, the difference being due to the presence, in the experimental map, of all the rest of the molecule.

Figure 6(a) and (b) report the experimental $\Phi(\mathbf{r})$ maps in the planes of thiophene and tetrazole, respectively, including the contributions of all atoms in the molecule: they are to be compared with those calculated on the corresponding fragments (Figures 3(a) and 4(a)) and are presented to show the consequences implied by breaking the molecule into fragments and separately analysing them. It may be

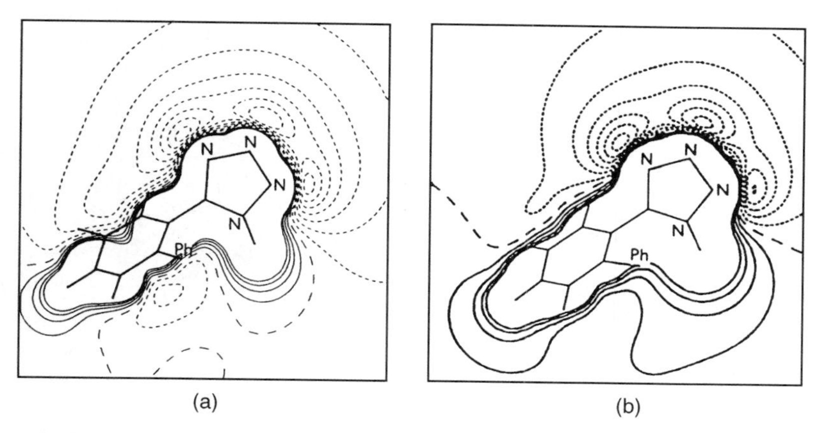

Figure 4. $\Phi(\mathbf{r})$ contour plots in the plane of the tetrazole ring. Dimensions and contour interval as in Figure 3. (a) Experimental (minimum contour -80 kcal/mol), (b) theoretical (minimum contour -60 kcal/mol).

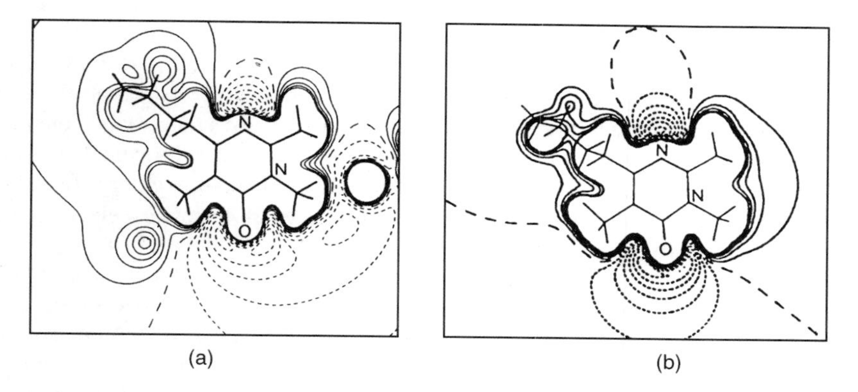

Figure 5. $\Phi(\mathbf{r})$ contour plots in the plane of the pyrimidinone ring. Sections of 16 Å × 14 Å. Contour interval as in Figure 3, minimum contour -60 kcal/mol. (a) Experimental (including contributions of all atoms in the molecule), (b) theoretical.

seen that differences are appreciable only in the regions where the fragment lies in close proximity to the remaining part of the molecule, i.e. below the sulfur atom in Figure 6(a) ($\Phi(\mathbf{r})^{\mathrm{exp}}_{\mathrm{min}}$ of -68 (26) kcal/mol) and above the N4 and N5 atoms in Figure 6(b). In order to complete this comparison, an *ab initio* calculation of the electrostatic potential of the whole molecule of LR-B/081 is planned for the future.

4. Conclusions

Our results indicate an overall qualitative and quantitative agreement between theoretical and experimental $\Phi(\mathbf{r})$, discrepancies between $\Phi(\mathbf{r})^{\mathrm{exp}}_{\mathrm{min}}$ and $\Phi(\mathbf{r})^{\mathrm{th}}_{\mathrm{min}}$ never exceeding $2\sigma[\Phi(\mathbf{r})^{\mathrm{exp}}_{\mathrm{min}}]$. As previously reported [2, 19], it appears that the $\Phi(\mathbf{r})$

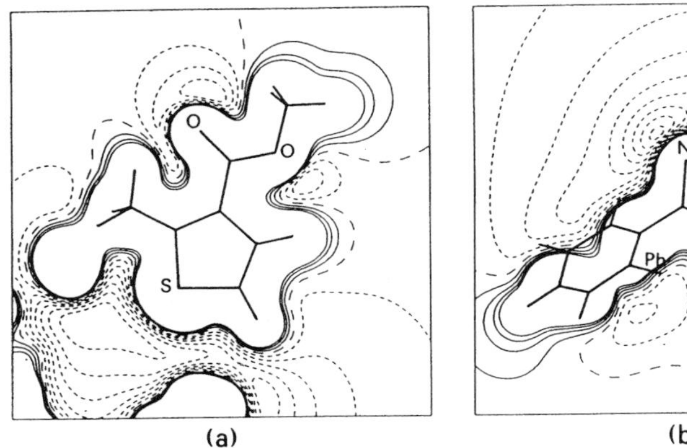

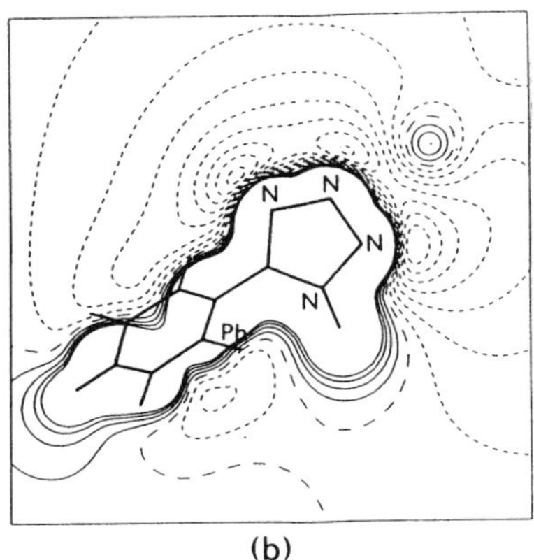

Figure 6. Experimental $\Phi(\mathbf{r})$ contour plots including the contributions of all atoms in the molecule. Dimensions and contour interval as in Figure 3. (a) Plane of the thiophene ring, minimum contour −60 kcal/mol. To be compared with Figure 3(a). (b) Plane of the tetrazole ring, minimum contour −80 kcal/mol. To be compared with Figure 4(a).

features which play a key role for binding to the AII receptor are the strongly electrophile attractive regions bulging out from the central heterocycle −N= type nitrogen atom and from the lactamic oxygen atom, and a region of positive long-range potential around the 6-butyl chain, the latter being believed to fit into a lipophilic pocket of the receptor that accommodates the Ile^5 side chain of AII. These features emerge from the maps reported here, confirming that LR-B/081 possesses all necessary electrostatic requirements for activity.

The determination of the bioactive conformation of AII should significantly advance the rational design of potent AII antagonists; its extreme flexibility has heretofore hindered attempts at determining its structure by either spectroscopic or crystallographic methods [22]. Garcia *et al.* [23] have recently proposed a receptor-bound conformation of AII developed from crystallographic data of the complex between AII and a high-affinity monoclonal antibody: it has been seen that the X-ray structure of the complex reveals an AII structure consistent with the overlay hypothesis [2] employed to develop this series of AII antagonists.

On the basis of these findings, and of recent *in vivo* and *in vitro* biological studies [3], the compound LR-B/081 was selected as a candidate for development, and is now undergoing clinical investigation for the treatment of hypertension.

References

1. Reid, J.L. and Rubin, P.C. (1987) Peptides and central neural regulation of the circulation, *Physiol. Rev.*, **67**, 725–745.
2. Salimbeni, A., Canevotti, R., Paleari, F., Poma, D., Caliari, S., Fici, F., Cirillo, R., Renzetti, A.R., Subissi, A., Belvisi, L., Bravi, G., Scolastico, C. and Giachetti, A. (1995) N-3-substituted pyrimidinones as potent, orally active, AT_1 selective angiotensin II receptor antagonists, *J. Med. Chem.*, **38**, 4806–4820.

3. Cirillo, R., Renzetti, A.R., Cucchi, P., Guelfi, M., Salimbeni, A., Caliari, S., Castellucci, A., Evangelista, S., Subissi, A. and Giachetti, A. (1995) Pharmacology of LR-B/081, a new highly potent, selective and orally active nonpeptide angiotensin II AT1 receptor antagonist, *Brit. J. Pharmac.*, **114**, 1117–1124.
4. Destro, R. and Soave, R. (1995) A new non-peptide angiotensin II receptor antagonist, *Acta Crystallogr.*, **C51**, 1383–1385.
5. Politzer, P. and Murray, J.S. (1990) Chemical applications of molecular electrostatic potentials, *Transactions ACA*, **26**, 23–39.
6. Politzer, P. and Truhlar, D.G. (1981) *Chemical Applications of Atomic and Molecular Electrostatic Potential*, Plenum Press, New York.
7. Zhu, N. and Klein, C.L. (1994) Electrostatic properties of (−)-norcocaine by X-ray diffraction, *J. Phys. Chem.*, **98**, 10699–10705.
8. Lecomte, C. (1995) Experimental electron densities of molecular crystals and calculation of electrostatic properties from high resolution X-ray diffraction, *Adv. in Molec. Struct. Res.*, **1**, 261–302.
9. Náray-Szabó, G. and Ferenczy, G.G. (1995) Molecular electrostatics, *Chem. Rev.*, **95**, 829–847.
10. Destro, R. and Merati, F. (1995) Bond lengths, and beyond, *Acta Crystallogr. B*, **51**, 559–570.
11. Roversi, P., Barzaghi, M., Merati, F. and Destro, R. (1996) Charge density in crystalline citrinin from X-ray diffraction at 19 K, *Can. J. Chem.*, **74**, 1145–1161.
12. Stewart, R.F. (1976) Electron population analysis with rigid pseudoatoms, *Acta Crystallogr.*, *A*, **32**, 565–574.
13. Stewart, R.F. and Spackman, M.A., (1983). VALRAY users manual. Department of Chemistry, Carnegie *Mellon University, Pittsburgh.*
14. Clementi, E. (1965) *Tables of atomic functions*, Supplement to *IBM J. Res. Dev.*, **9**, 2.
15. Stewart, R.F., Bentley, J.J. and Goodman, B. (1975) Generalized X-ray scattering factors in diatomic molecules, *J. Chem. Phys.*, **63**, 3786–3793.
16. Becker, P.J. and Coppens, P. (1974) Extinction within the limit of validity of the Darwin transfer equations. I. General formalisms for primary and secondary extinction and their application to spherical crystals, *Acta Crystallogr.*, *A*, **30**, 129–147.
17. Stewart, R.F. and Craven, B.M. (1993) Molecular electrostatic potentials from crystal diffraction: The Neurotransmitter γ-aminobutyric acid, *Biophys. J.*, **65**, 998–1005.
18. Stewart, R.F. (1991) *The Application of Charge Density Research to Chemistry and Drug Design*, Jeffrey, G.A. and Piniella, J.F.(eds.), Plenum Press, New York, pp. 63–101.
19. Belvisi, L., Bonati, L., Bravi, G., Pitea, D., Scolastico, C. and Vulpetti, A. (1993) On the role of the molecular electrostatic potential in modelling the activity of non-peptide angiotensin II receptor antagonists, *J. Mol. Struct. (Teochem.)*, **281**, 237–252.
20. Frisch, M.J., Head-Gordon, M., Trucks, G.W., Foresman, J.B., Schlegel, H.B., Raghavachari, K., Robb, M.A., Binkley, J.S., Gonzalez, C., Defrees, D.J., Fox, D.J., Whiteside, R.A., Seeger, R., Melius, C.F., Baker, J., Martin, R.L., Kahn, L.R., Stewart, J.J.P., Topiol, S. and Pople, J.A. (1990) GAUSSIAN 90, Inc., Pittsburgh, PA.
21. Surfer V4, Golden Software, Inc., Golden CO.
22. Woody, R.W. (1985) *Conformation in Biology and Drug Design*, Academic Press, Orlando, pp. 15–114.
23. Garcia, K.C., Ronco, P.M., Verroust, P.J., Brünger, A.T. and Mazel, L.M. (1992) Three-dimensional structure of an angiotensin II-Fab complex at 3 Å: hormone recognition by an anti-idiotypic antibody, *Science*, **257**, 502–507.

21

Hat matrix and leverages in charge density refinements: example of atomic net charges determination in a natural zeolite, the scolecite

SANDRINE KUNTZINGER, NOUR EDDINE GHERMANI, CLAUDE LECOMTE and YVES DUSAUSOY

Laboratoire de Cristallographie et Modélisation des Matériaux Minéraux et Biologiques, LCM³B, UPRES ACNRS 7036, Université Henri Poincaré, Nancy 1, Faculté des Sciences, Boulevard des Aiguillettes, BP 239, 54506 Vandoeuvre-lès-Nancy Cédex, France

1. Introduction

The zeolites are aluminosilicate materials with a framework formed by oxygen-connected tetrahedral SiO_4 and AlO_4 building blocks. The framework of such compounds surrounds channels or large cavities in the solid state and is negatively charged due to the difference between the formal charges of Si^{4+} and Al^{3+}. Therefore, cations like Na^+, Ca^{2+}, Li^+, etc. are present in the channels or in the cavities of the zeolite to ensure the electroneutrality of the system, and are linked to some framework oxygen atoms. The well-known interest of zeolites lies on their large adsorption capability which is related to the electrostatic properties of these porous compounds. In order to understand the electrostatic interactions which drive the properties of these materials, we have focused on the basic observable which is the electron density distribution. The connection of the SiO_4 and AlO_4 tetrahedra in the framework of the zeolite makes the electron density in the Si–O–T (T = Si, Al) bridges of a particular and fundamental interest in this investigation.

For the crystalline materials, high resolution X-ray diffraction experiment is a powerful tool to derive accurate electron density even for large systems like zeolites. In this study, we are interested in the experimental electron density distribution in the scolecite $CaAl_2Si_3O_{10} \cdot 3H_2O$ in order to make comparison with its sodium analogue natrolite $Na_2Al_2Si_3O_{10} \cdot 2H_2O$ for which the electron density has been reported recently [1, 2].

Scolecite gave the opportunity to relate the electron density features of Si–O–Si and Si–O–Al bonds to the atomic environment and to the bonding geometry. After the multipolar density refinement against Ag Kα high resolution X-ray diffraction data, a kappa refinement was carried out to derive the atomic net charges in this compound. Several least-squares fit have been tested. The hat matrix method which is presented in this paper, has been particularly efficient in the estimation of reliable atomic net charges in scolecite.

Paul G. Mezey and Beverly E. Robertson (eds.), Electron, Spin and Momentum Densities and Chemical Reactivity, 285–299

2. Data collection and processing

A good quality natural sample was used in the X-ray diffraction experiment. The X-ray diffraction data were collected at room temperature on an Enraf-Nonius CAD4 diffractometer with graphite-monochromatized Ag Kα radiation (0.5609 Å). The orientation matrix and the unit cell parameters were determined by least- squares fit to the optimized setting angles of 25 reflections in the range of $20^\circ \leq 2\theta \leq 55^\circ$. The Fd setting was chosen for scolecite instead of Cc in order to make comparison with the crystallographically related compound, the natrolite ($Na_2Al_2Si_3O_{10} \cdot 2H_2O$) which crystallizes in Fdd2 space group. A total of 29,794 intensities were measured consistently with the Fd space group and recorded as ω–2θ scan Bragg profiles up to a maximum reciprocal resolution of $S_{\max} = (\sin\theta/\lambda)_{\max} = 1.28$ Å^{-1}.

DREADD programs package [3, 4] was used to achieve the data reduction and the error analysis of the measurements. The absorption correction was performed with ABSORB program (De Titta [5]). The spherical harmonics empirical method of Blessing [6] was also applied to correct for the inhomogeneity of the X-ray beam. Since the space group of the scolecite is acentric and in order to take into account the anomalous dispersion effect, sorting and averaging of the data was performed in point group m giving 12,959 reflections with $I > 3\sigma(I)$ used in the first stages of refinements. Table 1 gives the main informations about the data collection and processing.

3. Use of the hat matrix in the least-squares fit procedure

In crystallography, the least-squares fit is based on the minimization of the sum $\Delta = \sum w(|F_o| - |F_c|)^2$, where $w = 1/\sigma^2(F_o)$, $\sigma^2(F_o)$ being the variance of the observed structure factor F_o and F_c is the calculated structure factor. Following Hamilton [7], if m is the number of observations, the changes of the n model parameters are determined by the resolution of the m linear equations

$$\mathbf{X}\Delta\mathbf{P} = \Delta\mathbf{F}, \tag{1}$$

where $\Delta\mathbf{P} = (\mathbf{P}^1 - \mathbf{P}^0)$ is the n-dimensional vector of the changes of the n parameters $\{\Delta p_j = p_j^1 - p_j^0\}_{j=1,n}$, $\{p_j^0\}_{j=1,n}$ are the initial parameters; $\Delta\mathbf{F} = (\mathbf{F}_o - \mathbf{F}_c^0)$ is the m-dimensional vector of the m differences $(|F_{o,\mathbf{h}_i}| - |F_{c,\mathbf{h}_i}(p_j^0)|)_{i=1,m}$, between the observed and calculated structure factors moduli at the Bragg vector $\mathbf{h}_i\,(i = 1, m)$, $\mathbf{X}$ is the $m \times n$ matrix of the derivatives:

$$[\mathbf{X}_{ij}] = \mathbf{w}^{1/2}\left[\frac{\partial F_{c,\mathbf{h}_i}}{\partial p_j}\right], \quad \mathbf{w} = \frac{1}{\sigma^2(F_o)}. \tag{2}$$

The least-squares normal equations are obtained by the transformation:

$$^{t}\mathbf{X}\mathbf{X}\Delta\mathbf{P} = {}^{t}\mathbf{X}\Delta\mathbf{F}, \tag{3}$$

where $^{t}\mathbf{X}$ is the transposed matrix then the estimation $\Delta\mathbf{P}$ (by the least-squares method) of the change of parameters $\Delta\mathbf{P}$ is given by

$$\Delta\mathbf{P} = \left[{}^{t}\mathbf{X}\mathbf{X}\right]^{-1}{}^{t}\mathbf{X}\Delta\mathbf{F}, \tag{4}$$

Table 1. Experimental details of the data collection.

Chemical formula	$CaAl_2Si_3O_{10} \cdot 3H_2O$
Chemical formula weight (g)	3138.7
Cell setting	'pseudo-orthorhombic' (monoclinic)
Space group	Fd (Cc)
a (Å)	18.489(2)
b (Å)	18.959(2)
c (Å)	6.519(1)
β (°)	90.611(13)
V (Å^3)	2284.8(5)
Z	8
Dx (Mg m^{-3})	2.28
Radiation type (graphite monochromator)	Ag Kα
Wavelength (Å)	0.5609
No. of reflections for cell parameters	25
θ range (°)	10–28
μ (mm^{-1})	0.542
Temperature	room
Crystal form	parallelopiped
Crystal size (mm)	0.32 × 0.16 × 0.20
Crystal color	colorless
Data collection	
Diffractometer	Enraf-Nonius CAD-4
Data collection method	ω–2θ scans
Absorption correction	Gaussian quadrature (De Titta, 1984)
T_{min}	0.864
T_{max}	0.924
No. of measured reflections	29,794
No. of observed reflections	6610
Criterion for observed reflections	$I > 3\sigma(I)$
R_1, R_2, R_w, S	0.0205, 0.0293, 0.0232, 1.045
θ_{max} (°)	45.9
Range of h, k, l	$-47 \rightarrow h \rightarrow 47$
	$-48 \rightarrow k \rightarrow 48$
	$-16 \rightarrow l \rightarrow 16$
No. of standard reflections	6
Frequency of standard reflections (min)	120
Intensity decay (%)	None

The internal-agreement factors are defined as:

$$R_1 = \frac{\sum_{\mathbf{H}} |I(\mathbf{H}) - \langle I(\mathbf{H})\rangle|}{\sum_{\mathbf{H}} |I(\mathbf{H})|}, \quad R_2 = \left(\frac{\sum_{\mathbf{H}} (I(\mathbf{H}) - \langle I(\mathbf{H})\rangle)^2}{\sum_{\mathbf{H}} (I(\mathbf{H}))^2}\right)^{1/2},$$

$$R_w = \left(\frac{\sum_{\mathbf{H}} w(I(\mathbf{H}) - \langle(I(\mathbf{H}))\rangle)^2}{\sum_{\mathbf{H}} w(I(\mathbf{H}))^2}\right)^{1/2}, \quad \text{and} \quad S = (M/M - N)\left(\frac{\sum_{\mathbf{H}} w\,(I(\mathbf{H} - \langle I(\mathbf{H})\rangle)^2}{\sum_{\mathbf{H}} w}\right),$$

the intensity $I(\mathbf{H}) = K^{-2}|F_o(\mathbf{H})|^2$ ($|F_o(\mathbf{H})|$ is the modulus of the observed structure factor and K^{-1} is the structure factor scale factor), M and N are respectively the number of the equivalent reflections and the number of the unique reflections, $w = 1/\sigma^2\,(I(\mathbf{H}))$ is the statistical weight related to the standard deviation of the intensity.

then

$$\mathbf{X}\Delta\mathbf{P} = \mathbf{X}\left[{}^{\mathrm{t}}\mathbf{X}\mathbf{X}\right]^{-1}{}^{\mathrm{t}}\mathbf{X}\Delta\mathbf{F} = \mathbf{H}\Delta\mathbf{F}, \tag{5}$$

where the hat matrix $\mathbf{H}$ is an $m \times m$ matrix defined by

$$\mathbf{H} = \mathbf{X}\left[{}^{\mathrm{t}}\mathbf{X}\mathbf{X}\right]^{-1}{}^{\mathrm{t}}\mathbf{X}. \tag{6}$$

Since $\mathbf{X}\mathbf{P}^0 = \mathbf{F}_{\mathrm{c}}^0$, it is obvious that $\mathbf{H}\mathbf{F}_{\mathrm{c}}^0 = \mathbf{F}_{\mathrm{c}}^0$, consequently

$$\mathbf{X}\mathbf{P}^1 = \mathbf{F}_{\mathrm{c}}^1 = \mathbf{H}\mathbf{F}_{\mathrm{o}}, \tag{7}$$

where $\mathbf{F}_{\mathrm{c}}^1$ is the estimated vectorial value of $\mathbf{F}_{\mathrm{o}}$. Geometrically the $\mathbf{H}$ matrix projects the observation vector $\mathbf{F}_{\mathrm{o}}$ on the n-dimensional space of solutions [8, 9] (derivatives of parameters) where lie the estimated vectors F_{c}, the $\mathbf{H}$ matrix is therefore called the projection matrix. Furthermore the elements H_{ij} of the $\mathbf{H}$ matrix denote the relation between the estimated value $F_{\mathrm{c},\mathbf{h}_i}^1$ and the observation $F_{\mathrm{o},\mathbf{h}_i}$. It means that the estimation of $F_{\mathrm{c},\mathbf{h}_j}^1$ through the least-squares determination of the n parameters can be explained, in the statistical sense, by more than one observation (the diagonal element H_{ii} is lower than 1). H_{ii} represents the weight of the observation i on the prediction of $F_{\mathrm{c},\mathbf{h}_i}$, the matrix $\mathbf{H}$ is also called the prediction matrix and H_{ii} the leverage.

In crystallography, the number of observations m is larger than the number of parameters n to be determined, therefore H_{ii} is in average very low (about n/m [8]). Velleman and Welsh [9] suggest that the observation with $H_{ii} > 3n/m$ (when $n > 6$ and $(m - n) > 12$) can be considered as influential in the determination by the least-squares fit procedure. In practice during the refinements, the calculations of such H_{ii} will reveal the observations which have an influence on the estimation of some parameters of the model. However, this analysis must be carried out only for parameters of the same type to be efficient. Now, if the H_{ii} value is significantly high, the inspection of the residues $e_i = |F_{\mathrm{o},\mathbf{h}_i}| - |F_{\mathrm{c},\mathbf{h}_i}|$, or weighted residues $e_i/\sigma(F_{\mathrm{o},\mathbf{h}_i})$ is an indicator of the eventual aberration of this observation with respect to the model, when no experimental problems occur for this observation.

4. Pseudo-atom model and refinement strategies

The least-squares Molly program based on the Hansen–Coppens model [10] was used to determine atomic coordinates, thermal parameters and multipolar density coefficients in scolecite. In the Hansen–Coppens model, the electron density of unit cell is considered as the superposition of the pseudo-atomic densities. The pseudo-atom electron density is given by

$$\rho(\mathbf{r}) = \rho_{\mathrm{core}}(r) + P_{\mathrm{val}}\kappa^3\rho_{\mathrm{val}}(\kappa r) + \sum_l \kappa'^3 R_{nl}(\kappa' r) \sum_m P_{lm}\, y_{lm\pm}(\theta, \phi), \tag{8}$$

where ρ_{core} and ρ_{val} are respectively Hartree–Fock spherical core and valence densities, ρ_{val} is normalized to one electron; then the refined valence population parameter P_{val} gives the net atomic charge q with respect to the number of electrons N_{val} in the free atom valence orbitals, $q = N_{\text{val}} - \rho_{\text{val}}$. The $y_{lm\pm}$'s are spherical harmonic angular functions of order l in real form, and $R_{nl}(r)$ are Slater-type radial functions

$$R_{nl}(r) = N_l r^{nl} \exp(-\zeta r), \tag{9}$$

N_l is the normalization factor, n_l and ζ are parameters depending on the atomic type. P_{lm} are the multipolar population parameters and κ and κ' are the contraction–expansion coefficients [11] for, respectively, spherical and multipolar valence densities. We have chosen orthogonal reference axes which respect the tetrahedral (23) T^l point group for Si and Al atoms of the scolecite in order to reduce the number of multipolar parameters; only the cubic harmonic multipoles (one octupole $l = 3$ and two hexadecapoles $l = 4$) have been refined for these two atoms. The pseudo-atom expansion was extended to the octupoles ($l = 3$) for O including oxygen of water, and to the dipoles ($l = 1$) for H. The best radial functions of Si and Al atoms were obtained by inspection of the residual maps [12], ($n_l = 4, 4, 4, 4$ ($l = 1$–4)); ζ's were taken from Clementi and Raimondi [13]: $\zeta_{\text{Si}} = 3.05\,\text{bohr}^{-1}$, $\zeta_{\text{Al}} = 2.72\,\text{bohr}^{-1}$. For O atoms, $\zeta_{\text{O}} = 4.5\,\text{bohr}^{-1}$ and the multipole exponents were respectively $n_l = 2,\ 3,\ 4$ up to the octupole level.

With data averaged in point group m, the first refinements were carried out to estimate the atomic coordinates and anisotropic thermal motion parameters U^{ij}'s. We have started with the atomic coordinates and equivalent isotropic thermal parameters of Joswig *et al.* [14] determined by neutron diffraction at room temperature. The high order X-ray data ($0.9 \leq s \leq 1.28\,\text{Å}^{-1}$) were used in this case in order not to alter these parameters by the valence electron density contributing to low order structure factors. Hydrogen atoms of the water molecules were refined isotropically with all data and the distance O–H were kept fixed at 0.95 Å until the end of the multipolar refinement. The inspection of the residual Fourier maps has revealed anharmonic thermal motion features around the Ca^{2+} cation. Therefore, the coefficients up to order 6 of the Gram–Charlier expansion [15] were refined for the calcium cation in the scolecite.

In Molly program, the asymmetric unit is constrained to be neutral during the refinement of P_{val}. In the case of scolecite, the P_{val}'s of the water molecules have been refined separately to the framework in order to respect the H_2O electroneutrality. The starting P_{val} parameters for the ten oxygen atoms were 6.2 instead of 6 in order to take into account the fixed net charge of the calcium cation (2+). The multipolar refinements were carried out over all reflections up to a resolution of $(\sin\theta/\lambda)_{\max} = 1.28\,\text{Å}^{-1}$. We started with a natrolite-like constraint [2] which consists in imposing to the almost symmetry related atoms of scolecite to have the same electron density. This chemical constraint was kept unchanged until the last refinements of the multipolar parameters; then we removed the chemical constraint, first P_{val} and κ parameters, then all multipole populations. Isotropic extinction was corrected during all refinements cycles and the final extinction parameter value is $g = 0.41(2) \times 10^{-4}$ corresponding

Table 2. Least-squares statistical factors R, R_w and GoF of the refinement strategies.

Refinement number	$s = (\sin\theta)/\lambda$ (Å^{-1})	R (%)	R_w (%)	GoF	n	m	Type of refinement
1	$0.0 \leq s \leq 1.28$	3.77	3.68	1.32	1	12959	Spherical (m averaged data)
2	$0.0 \leq s \leq 1.28$	3.37	3.05	1.11	507	12959	Multipolar (m averaged data)
3	$0.0 \leq s \leq 1.28$	3.25	3.39	1.19	1	6610	Spherical ($2/m$ averaged data)
4	$0.0 \leq s \leq 1.28$	2.65	2.37	0.85	507	6610	Multipolar ($2/m$ averaged data)
5	$0.0 \leq s \leq 1.28$	2.83	2.67	0.94	41	6610	Kappa ($2/m$ averaged data)
6	$0.0 \leq s \leq 1.28$	2.82	2.61	0.92	41	6603	Kappa without the 7 reflections – see Table 4 – ($2/m$ averaged data)

$$R(F) = \frac{\sum (K^{-1}|F_o| - |F_c|)}{\sum K^{-1}|F_o|}, \quad R_w(F) = \left[\frac{\sum_w (K^{-1}|F_o| - |F_c|)^2}{\sum_w k^{-2}|F_o|^2}\right]^{1/2},$$

$$\mathrm{GoF} = \left[\frac{\sum_w (K^{-1}|F_o| - |F_c|)^2}{m - n}\right]^{1/2},$$

where $|F_o|$ and $|F_c|$ are respectively the modulus of the observed and the calculated structure factor, w the statistical weight, K^{-1} the scale factor, n the number of refined parameters and m the number of data.

to a maximum intensity loss of 17% for the (3 5 1) reflection. After convergence of the unconstrained refinements, the atomic positions and thermal motion parameters were relaxed and refined with all multipoler parameters in the last cycles. Then the data were corrected for anomalous dispersion [12], merged and averaged in the Laue group $2/m$ (6610 reflections) and new cycles of refinements of all parameters were carried out again.

At each stage of the refinement of a new set of parameters, the hat matrix diagonal elements were calculated in order to detect the influential observations following the criterium of Velleman and Welsh [8, 9]. The inspection of the residues of such reflections revealed those which are aberrant but progressively, these aberrations disappeared when the pseudo-atoms model was used (introduction of multipoler coefficients). This fact confirms that the determination of the phases in acentric structures is improved by sophisticated models like the multipole density model.

Final residual indices of the refinement strategies are given in Table 2. On the residual density maps shown in Figure 1, the maxima and minima do not exceed 0.2 e Å^{-3}.

5. Crystal structure of scolecite

Figure 2 gives a view of the scolecite structure along c-axis. Table 3 gives a selected set of bond lengths and angles in the scolecite structure. The framework of scolecite

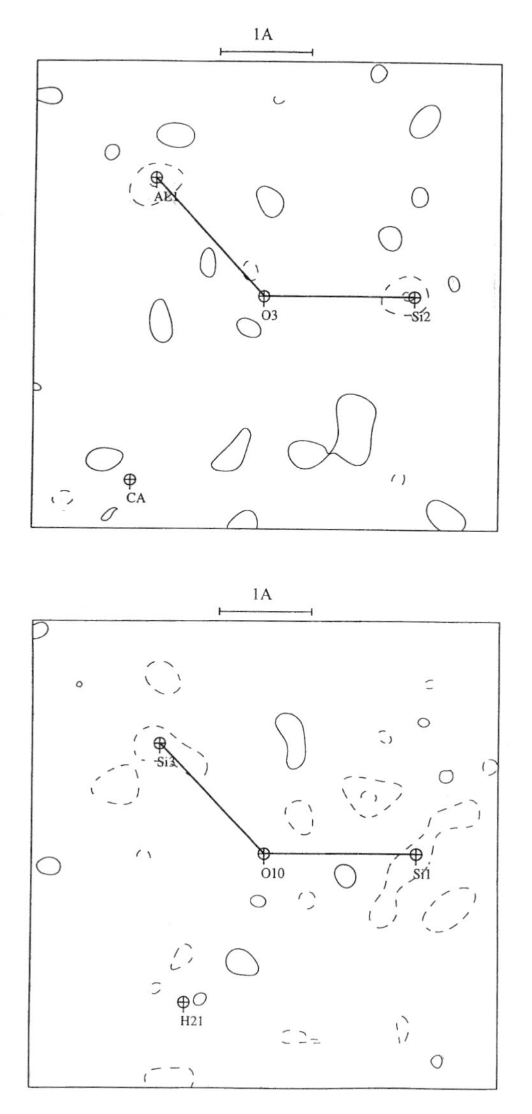

Figure 1. Residual density in Si_2–O_3–Al_1 and Si_1–O_{10}–Si_3 planes of scolecite after the multipole refinement 4 in Table 2. Contour interval ± 0.1 e Å^{-3}; negative contours are dashed, zero contour omitted.

is isotypical with that of natrolite [2] with a slightly distorted framework: the two Na^+ cations of natrolite are replaced by one Ca^{2+} and one water molecule (Ow_3). As in the natrolite, the structure of scolecite presents parallel channels along the *c*-axis containing the Ca^{2+} cations and water molecules. Ow_1 and Ow_2 are equivalent to the symmetry related water molecules in the natrolite structure. The aluminosilicate

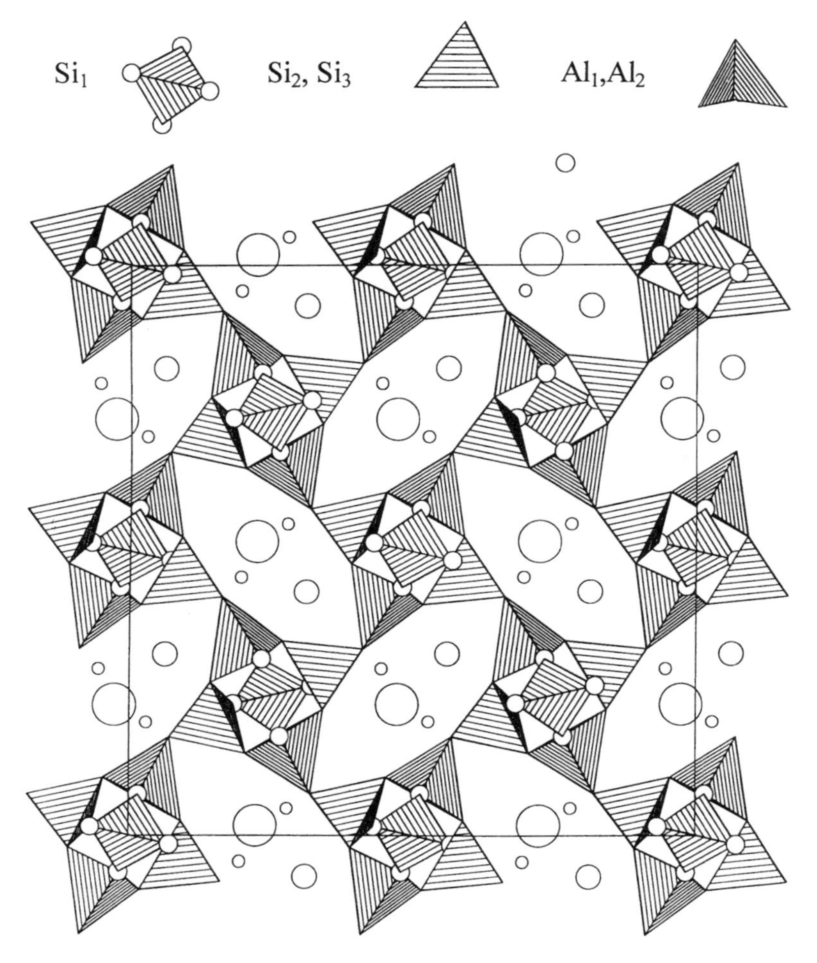

Figure 2. View of the structure of scolecite along *c*-axis using STRUPLO'90 [21]. The small circles correspond to Ow_1 and Ow_2 water molecules, the medium circle to Ow_3 water molecule and the big circles to Ca^{2+} cation.

framework results from oxygen-connected four-membered rings of tetrahedra. A quarter of the ring is composed by a central silicon tetrahedron surrounded by, respectively, two opposing tetrahedra of SiO_4 and AlO_4. Except for the central silicon tetrahedron, all other silicon or aluminum tetrahedra are surrounded by, respectively, three aluminum and one silicon tetrahedra or three silicon and one aluminum tetrahedra. The Ca^{2+} cations are seven-coordinated by four oxygen atoms of the framework (O_3, O_4, O_5, O_7) and the three oxygen atoms of the water molecules are in a distorted pentagonal bipyramid as reported by Fälth and Hansen [17]. The water molecules are closer to the cation [Ow_1–Ca^{2+} = 2.3117(12) Å, Ow_2–Ca^{2+} = 2.3484(11) Å,

Table 3. Selected bond lengths (Å) and angles (°) in scolecite. The esd's are given in parentheses.

Si_1–O_2	1.6008(8)	Si_2–O_6	1.6028(8)	Si_3–O_8	1.5972(8)
Si_1–O_1	1.6101(8)	Si_2–O_3	1.6121(7)	Si_3–O_4	1.6077(7)
Si_1–O_9	1.6196(8)	Si_2–O_9	1.6295(8)	Si_3–O_5	1.6289(8)
Si_1–O_{10}	1.6418(7)	Si_2–O_7	1.6313(8)	Si_3–O_{10}	1.6409(8)
	Al_1–O_3	1.7349(7)	Al_2–O_2	1.7399(8)	
	Al_1–O_1	1.7409(8)	Al_2–O_6	1.7411(8)	
	Al_1–O_5	1.7516(8)	Al_2–O_4	1.7419(7)	
	Al_1–O_7	1.7667(8)	Al_2–O_8	1.7462(8)	
Si_1–O_1–Al_1	134.75(5)	Si_1–O_2–Al_2	143.82(5)	Si_3–O_5–Al_1	135.06(5)
Si_2–O_7–Al_1	127.48(5)	Si_3–O_8–Al_2	133.24(5)	Si_3–O_4–Al_2	138.27(5)
Si_1–O_9–Si_2	150.79(6)	Si_2–O_6–Al_2	134.15(5)		
Si_3–O_{10}–Si_1	134.05(5)	Si_2–O_3–Al_1	133.19(4)		
	Ca–O_3	2.4900(7)	Ca–Ow_1	2.3117(12)	
	Ca–O_5	2.5066(8)	Ca–Ow_2	2.3484(11)	
	Ca–O_7	2.5288(8)	Ca–Ow_3	2.3500(10)	
	Ca–O_4	2.6057(8)			
		$H_{32} \cdots O_2$	1.7518(8)		
		$H_{11} \cdots O_1$	1.7900(8)		
		$H_{21} \cdots O_{10}$	1.8652(8)		
		$H_{22} \cdots O_8$	2.1428(9)		
		$H_{12} \cdots O_6$	2.1654(8)		
		$H_{31} \cdots O_9$	2.3076(9)		

Ow_3–Ca^{2+} = 2.3500(10) Å] than the framework oxygen atoms [the closest is O_3 with O_3–Ca^{2+} = 2.4900(7) Å]. In natrolite, the sodium cation are linked to four oxygen of the framework and two oxygens of the water molecules [the closest Ow being at Ow–Na^+ = 2.371(2) Å and the closest oxygen of the framework is at 2.367(2) Å]. In the two compounds, the water molecules are bound to the aluminosilicate framework by hydrogen bonds, the strongest involving O_2 and $H_{32}(Ow_3)$ [O_2–H_{32} = 1.7518(8) Å] in scolecite.

6. Static deformation densities in Si–O and Al–O bonds

STATDENS program [18] was used to calculate the static deformation density in scolecite. As reported in the previous study of natrolite [2], the main feature of the deformation densities is the concentration of the electron density around the oxygen atoms and the more or less depletion of this density near the silicon or the aluminum atoms. Figure 3 shows the static deformation densities in the plane of Si_1–O_9–Si_2 and Si_1–O_{10}–Si_3 where Si_1 is the near-origin central silicon atom. In Si_1–O_9–Si_2 bridge [150.79(6)°], the density peak-height values are, respectively, 0.70 e Å^{-3} in the Si_1–O_9 bond [Si_1–O_9 = 1.6196(8) Å] and 0.60 e Å^{-3} in the Si_2–O_9 bond [Si_2–O_9 = 1.6295(8) Å], both peaks are localized 0.45 Å from the oxygen O_9. The electron density peaks in Si_1–O_{10}– Si_3 bonds [134.05(5)0°] are enhanced with a peak height reaching 0.8–0.9 e Å^{-3} at 0.45 Å from O_{10} [Si_1–O_{10} = 1.6418(7) Å, Si_3–O_{10} = 1.6409(8) Å]. If we compare the two bridges, the deformation density has a tendency

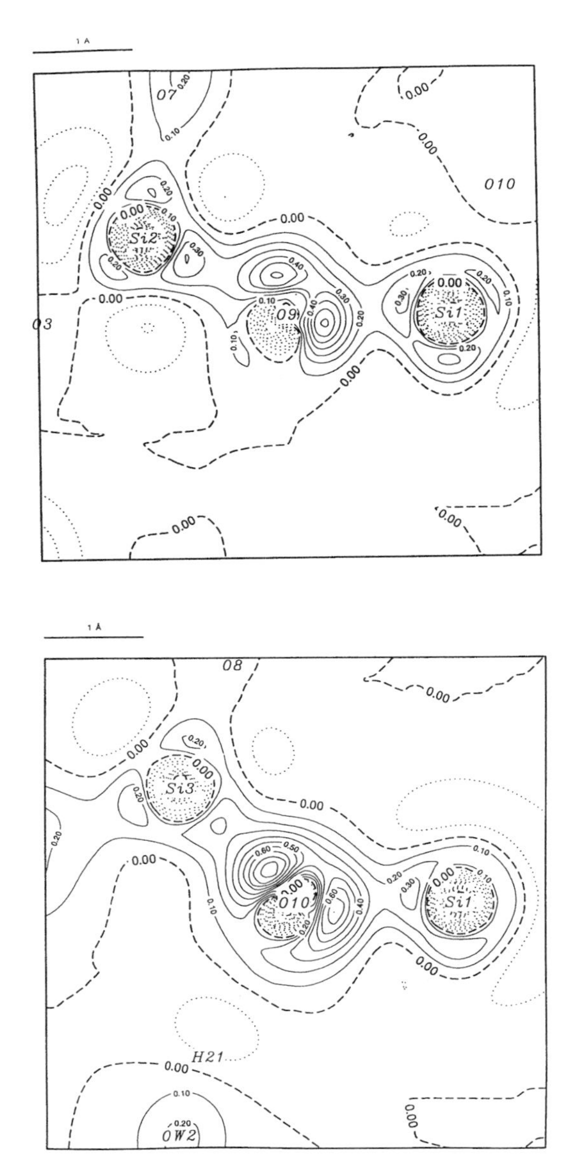

Figure 3. Static deformation density in Si–O–Si bridge planes: Si_1–O_9–Si_2 and Si_1–O_{10}–Si_3. Contours as in Figure 1.

to be shifted toward the interior of the Si–O–Si angle as shown for Si_1–O_9–Si_2. The same remark has been done for natrolite [2] (Si–O–Si = 144.31(6)°). This feature is also in close agreement with the results in coesite [19]. The interaction of these two Si–O–Si bridges in scolecite with the water molecules involves more strongly the O_{10} oxygen atom [O_{10}–$H_{21}(Ow_2)$ = 1.8652(8) Å, H_{21} is at 0.5 Å above the plane of the Si_1–O_{10}–Si_3] than O_9 [O_9–$H_{31}(Ow_3)$ = 2.3076(9) Å]. This feature is in agreement with the observed slight polarization of the electron density lone pair of O_{10} towards the hydrogen atom. The static deformation density in Si_1–O_1–Al_1 and Si_2–O_3–Al_1 bonds is shown in Figure 4. The electron density is also more concentrated around the oxygen atoms with a visible distortion or polarization of their lone pairs either in presence of hydrogen bonds [O_1–$H_{11}(Ow_1)$ = 1.7900(8) Å, compared to O–H = 1.8871(8) Å in natrolite] or when a strong interaction occurs with a cation [O_3–Ca^{2+} = 2.4900(7) Å compared to O–Na^+ = 2.367(2) Å for the closest oxygen to Na^+ in natrolite [2]]. This last feature has also been observed in forsterite by R.J. Van Der Wal and Vos [20] between the oxygens of the SiO_4 tetrahedra and the Mg^{2+} cation. In the Si–O–Al bridges of scolecite, the electron density peak heights are in the range of 0.8 e $Å^{-3}$ for Si_1–O_1 [Si_1–O_1 = 1.6101(8) Å] at 0.45 Å from O_1 and 1.0 e $Å^{-3}$ for Si_2–O_3 [Si_2–O_3 = 1.6121(7) Å] at 0.40 Å from O3. In the Al–O bonds, the peak height varies from 0.4 e $Å^{-3}$ in Al_1–O_1 bond [Al_1–O_1 = 1.7409(8) Å] to 0.7 e $Å^{-3}$ in Al_1–O_3 bond [Al_1–O_3 = 1.7349(7) Å]. The distances between the electron density peaks and the oxygen atoms in the Al–O bonds are respectively 0.50 Å in Al_1–O_1 bond and 0.40 Å in Al_1–O_3 link. On average for Si–O–Al bridges, the electron bond peaks in Al–O bond are lower (0.4 to 0.8 e $Å^{-3}$) than those of Si–O link (0.8–1.0 e $Å^{-3}$) but the distances between the electron peaks and the oxygen atoms are almost the same in the two types of bond.

7. Kappa refinements and atomic net charges

After the multipolar fit (refinement 4 in Table 2), we have carried out a kappa refinement to determine the atomic net charges in scolecite. In order to have reliable values of these parameters, we have analyzed carefully the data using the hat matrix method described in the previous section. At the end of natrolite-like constraint multipole refinement (not reported in Table 2), we have calculated the diagonal elements of the prediction matrix i.e. the leverages H_{ii}. According to the criteria of Velleman and Welsh [9], 201 observations have their leverages greater than $3n/m$, where $n = 7$ (number of refined P_{val} of the framework atoms minus 1 for the electroneutrality constraint) and $m = 6610$ reflections. As expected from the valence contribution, the maximum value of $(\sin\theta)/\lambda$ resolution for these 201 reflections does not exceed 0.4 $Å^{-1}$. We have reported in Table 4, the 7 influential reflections detected by the hat matrix method with significant weighted residues during the fit of the P_{val} parameters. A new statistical analysis was performed in the end of the unconstrained refinement (number 4 in Table 2) over the P_{val} parameters and led to 212 influential reflections. Table 4 shows that these 7 reflections do not have high weighted residues in refinement 4 (Table 2) and that the reflection (0 2 0) has an abnormal leverage of 0.85 and a very low residue. We have, therefore, carried out two kappa refinements (5 and 6 in Table 2), with and without the 7 reflections given above in Table 4.

The values obtained at the convergence for κ, P_{val} parameters and the atomic net charges are reported in Table 5. In these two refinements, the κ parameters are

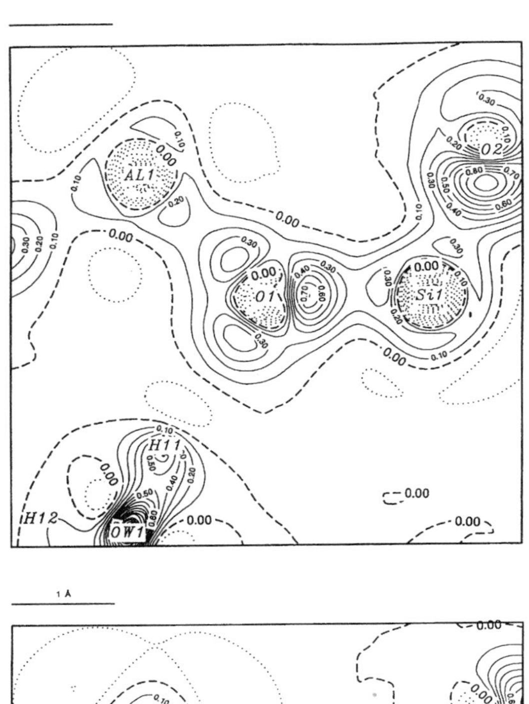

1 Å

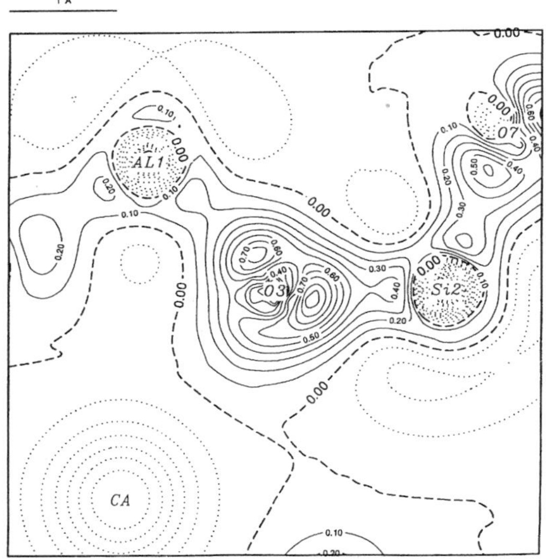

Figure 4. Static deformation density in Si–O–Al bridge planes: $Si_1–O_1–Al_1$ and $Si_2–O_3–Al_1$. H_{11} is at 0.98 Å and Ca^{2+} at 0.44 Å. Contours as in Figure 1.

greater than 1 for the silicon and aluminum atoms revealing the contraction of their valence electron densities. The contraction of the aluminum atoms is, however, more pronounced (in average κ = 1.4–1.5) than for the silicon atoms (in average κ = 1.2). The oxygen atoms have, in turn, κ parameter values of about 0.90–0.96 showing a

Table 4. Diagonal elements (leverages H_{ii}) of the hat matrix and weighted residues ($\|F_o| - |F_c\|/\sigma(F_o)$) of the pertinent data in the determination of the atomic net charges in scolecite.

Reflections (hkl)	Constraint refinement (not reported in Table 2)		Non-constraint refinement (refinement 4 in Table 2)	
	Leverages H_{ii}	Weighted residues	Leverages H_{ii}	Weighted residues
(−153)	0.040	8.20	0.084	4.39
(020)	0.004	14.97	0.854*	0.87
(191)	0.030	3.99	0.063	2.36
(242)	0.061	4.44	0.179	2.16
(331)	0.068	4.75	0.223	0.06
(731)	0.135	3.05	0.163	1.10
(911)	0.013	3.34	0.029	1.01

* Abnormal reflection.

Table 5. κ, P_{val} and the atomic net charges q from the kappa refinements 5 and 6 (Table 2). The esd's are given in parentheses.

Refinement atoms	5			6			Differences of net charges
	κ	P_{val}	q	κ	P_{val}	q	5–6
Si_1	1.16(3)	2.52(13)	1.48(13)	1.18(3)	2.33(13)	1.67(13)	−0.19(13)
Si_2	1.31(4)	1.70(13)	2.30(13)	1.25(4)	1.91(14)	2.09(14)	0.21(14)*
Si_3	1.23(3)	2.25(14)	1.75(14)	1.25(3)	2.21(14)	1.80(14)	−0.05(14)
Al_1	1.53(11)	0.70(11)	2.30(11)	1.55(7)	1.05(12)	1.95(12)	0.35(12)*
Al_2	1.36(5)	1.49(13)	1.51(13)	1.50(8)	1.06(12)	1.94(12)	−0.43(13)*
O_1	0.0946(6)	6.85(8)	−0.85(8)	0.934(6)	7.15(9)	−1.15(9)	−0.30(9)*
O_2	0.910(6)	7.69(9)	−1.69(9)	0.924(6)	7.39(9)	−1.39(9)	0.30(9)*
O_3	0.947(6)	7.28(8)	−1.28(8)	0.939(6)	7.31(8)	−1.31(8)	0.03(8)
O_4	0.922(6)	7.22(8)	−1.22(8)	0.925(6)	7.21(8)	−1.21(8)	−0.01(8)
O_5	0.937(6)	7.22(9)	−1.22(9)	0.932(6)	7.31(9)	−1.31(9)	0.09(9)
O_6	0.949(6)	7.08(8)	−1.08(8)	0.947(6)	6.95(8)	−0.95(8)	−0.13(8)
O_7	0.965(6)	6.70(8)	−0.70(8)	0.945(6)	7.00(9)	−1.00(9)	0.30(9)*
O_8	0.951(6)	7.08(7)	−1.08(7)	0.952(6)	7.00(7)	−1.00(7)	−0.08(7)
O_9	0.960(6)	6.89(7)	−0.89(7)	0.961(6)	6.89(7)	−0.89(7)	0.00(7)
O_{10}	0.923(6)	7.34(8)	−1.34(8)	0.933(6)	7.24(8)	−1.24(8)	−0.10(8)
Ca	1.00	0.00	2.00	1.00	0.00	2.00	0.00
Ow_1	0.988(8)	6.33(8)	−0.33(8)	0.996(8)	6.31(8)	−0.31(8)	−0.02(8)
H_{11}	1.16	0.98(5)	0.02(5)	1.16	0.96(5)	0.04(5)	−0.02(5)
H_{12}	1.16	0.70(5)	0.30(7)	1.16	0.74(7)	0.26(7)	0.04(7)
Ow_2	0.970(8)	6.49(8)	−0.49(8)	0.959(7)	6.63(8)	−0.63(8)	0.14(8)*
H_{21}	1.16	0.75(5)	0.25(5)	1.16	0.60(5)	0.40(5)	−0.15(5)*
H_{22}	1.16	0.76(5)	0.24(7)	1.16	0.78(7)	0.22(7)	0.02(7)
Ow_3	0.951(7)	0.96(8)	−0.96(8)	0.952(7)	6.96(8)	−0.96(8)	0.00(8)
H_{31}	1.16	0.51(6)	0.49(6)	1.16	0.51(6)	0.49(6)	0.00(6)
H_{32}	1.16	0.53(5)	0.47(7)	1.16	0.53(7)	0.45(7)	0.00(7)

*The higher net charge differences between the refinements 5 and 6.

slight expansion of their valence electronic clouds. The last column of Table 5 gives the differences between the atomic net charge values obtained respectively, in the last cycles of the two refinements. With respect to the P_{val}'s esd's, the more significant changes reaching about 0.4 electrons (i.e. more than 3σ's) deal with Si_2, Al_1, Al_2,

O_1, O_2, O_7 atoms and those of the water molecule Ow_2. The main feature is that the net charges become almost equal for the pairs (Si_2, Si_3), (Al_1, Al_2) and (O_1, O_2) in refinement 6 of Table 2. Thus, the 7 reflections, detected by the hat matrix, are those among the 6610 data which differentiate between the symmetrically non-equivalent atoms in scolecite. The number of 7 pertinent observations is obviously very low to give a significant meaning to the atomic charges of refinement 5. In the words of statisticians, these 7 reflections are suspicious in the determination of the atomic net charges. Therefore, it is more reasonable to consider kappa refinement 6 (Table 2) as finally the best estimation of the atomic net charges in scolecite.

8. Conclusion

We have derived electron density distribution in scolecite from high-resolution X-ray diffraction data. Electron densities in Si–O and Al–O bonds have been related to the atomic environment and geometries of the bonds. Careful strategies in the acentric space group of scolecite using the hat matrix analysis in the least-squares refinements have avoided hazardous values of the atomic net charges. The charges obtained after the leverage analysis of the data are consistent with the SiO_4 and AlO_4 building blocks environment. Developments of the hat matrix method could be helpful in difficult least-squares refinements. Further applications of this analysis in the crystallographic field are underway.

Acknowledgements

The financial support of the CNRS, and of Université Henri Poincaré, Nancy 1 is gratefully acknowledged. The graphics have been drawn under a grant of CNI/MAT (CIRIL, Nancy).

References

1. Stuckenschmidt, E., Joswig, W. and Baur, W.H. (1994) Natrolite, Part II: determination of deformation electron densities by X–X Method, *Phys. Chem. Minerals*, **21**, 309– 316.
2. Ghermani, N.E., Lecomte, C. and Dusausoy, Y. (1996) Electrostatic properties in zeolite-type materials from high-resolution X-ray diffraction: the case of natrolite, *Phys. Rev. B*, **53**, 5231–5239.
3. Blessing, R.H. (1987) Data reduction and error analysis for accurate single crystal diffraction intensities, *Cryst. Rev.*, **1**, 3–58.
4. Blessing, R.H. (1989) Data reduction and error analysis for single crystal diffractometer data, *J. Appl. Cryst.*, **22**, 396–397.
5. De Titta, G. (1985) An absorption correction program for crystals enclosed in capillaries with trapped mother liquor, *J. Appl. Cryst.*, **18**, 75–79.
6. Blessing, R.H. (1995) An empirical correction for absorption anisotropy, *Acta Cryst. A*, **51**, 33–38.
7. Hamilton, W.C. (1964) *Statistics in Physical Science*, Ronald Press, New York.
8. Antoniadis, A., Berruyer, J. and Carmona, R. (1992). In Regression non linéaire et Application, Economica, Paris, France.
9. Velleman, P.F., and Welsh, R.E. (1981) Efficient computing of regression diagnostics, *The American Statistician*, **35**, 234–242.
10. Hansen, N.K. and Coppens, P. (1978) Testing aspherical atom refinements on small-molecule data sets, *Acta Cryst. A*, **34**, 909–921.

11. Coppens, P., Guru, T.N., Leung, P., Stevens, E.D., Becker, P. and Yang, Y. (1979) Atomic net charges and molecular dipole moments from spherical-atom X-ray refinement and the relation between atomic charge and shape, *Acta Cryst. A*, **35**, 63–72.
12. Souhassou, M., Espinosa, E., Lecomte, C. and Blessing, R. H. (1995) Experimental electron density in Crystalline H_3PO_4. *Acta Cryst. B*, **51**, 661–668.
13. Clementi, E. and Raimondi, D.L. (1963). Atomic screening constants for SCF functions. *J. Chem. Phys.*, **38**, 2686–2689.
14. Joswig, W., Bartl, H. and Fuess, H. (1984) Structure refinement of scolecite by neutron diffraction, *Z. Kristallogr.*, **171**, 219–223.
15. Kuhs, W.F. (1983) Statistical description of multimodal atomic probability densities, *Acta Cryst. A*, **39**, 148–158.
16. Cromer, D.T. (1974) *International Tables for X-ray Crystallography*, Ibers, J.A. and Hamilton, W.E. (Eds.), Kynoch Press, Birmingham, England, pp. 148–151.
17. Fälth, L. and Hansen, S. (1979) Structure of scolecite from Poona, India, *Acta Cryst. B*, **35**, 1877–1880.
18. Ghermani, N.E., Bouhmaida, N. and Lecomte, C. (1992) ELECTROS, STATDENS: computer programs to calculate electrostatic properties from high resolution X-ray diffraction, Internal report URA CNRS 809, Université Henri Poincaré, Nancy 1, France.
19. Downs, J.W. (1994) The electron density in coesite, *Trans. Am. Geophy Union*, 187.
20. van Der Val, R.J. and Vos, A. (1987) Conflicting results for the deformation properties of forsterite, *Acta Cryst. B*, **43**, 132–143.
21. Fisher, R. X., le Lirzin, A., Kassner, D., Rüdinger, B. (1991) STRUPLO'90: eine neue Version des Fortran Plotprograinins zur Darstellung von Krisfallstructuren, *Z. Kristallogr.* Suppl. issue No. 3, 75.

22

The β decay in anapole crystal

LIU XIAODONG

China Institute of Atomic Energy, P.O. Box 275(10), Beijing 102413, China

1. Introduction

During the β decay process, there exists anapole moment along the spin axis of the parent nuclei [1]. The anapole moment presents a new kind of dipole moment which is invariant under time reversal and odd under parity. A pseudoscalar $p(\bar{V} \times H \cdot \sigma)$ exists between the anapole moment and the spin of the emitted electrons, where p is the interaction strength. This interaction breaks parity conservation.

In recent years, some anapole structures have been discussed [2, 3], but most of them are impossible to distinguish from other existing electro-weak processes. In this paper, it is shown that the anapole moment in crystal [3] can be easily distinguished from other mechanisms since its magnitude is adjustable.

2. Crystal anapole moment and its coupling to electro-weak process

Crystal anapole moment is composed of the atomic magnetic moments which array in anapole structure [3]. It has the same intrinsic structure as Majorana neutrino [2]. If we plant a β decay atom into this anapole lattice, the crystal anapole moment will couple to the nuclear anapole moment of the decaying nuclei. So the emitted electron will be given an additional pseudoscalar interaction by the presence of the crystal anapole moment. Then the emission probability will be increased. This is a similar process to that assumed by Zel'dovich [1]. The variation of the decay rate may be measured to tell whether the crystal anapole moment has an effect on the β decay or not.

For example, let us consider a typical crystal anapole moment of Mn_3NiN [4]. Its anapole moment can be adjusted by temperature. The β source ^3H may be permeated into this lattice without destroying the crystal structure. When the temperature is higher than 266 K, the atomic magnetic moments of Mn do not array in anapole structure. Then the crystal anapole moment is zero. The β emission probability of ^3H is normal. Contributions from other electro-weak processes may be measured at this temperature. When the temperature is lower than 184 K, the atomic magnetic moments of Mn array in the anapole structure and the crystal present anapole moment to the ^3H nuclei. Then the electron's emission rate of ^3H will be increased.

The magnitude of anapole moment can be calculated by Equation (6.6) of Ref. [3]. For a nuclear anapole moment of ^3H, $\mu_e \approx \mu_B$, the anapole radius r could be adapted as Compton wavelength $\lambda = h/m_e c \approx 2.4 \times 10^{-10}$ cm [1], the interaction strength p is the square of the dimensionless constant of weak interaction $f^2 \approx 10^{-12}$ [1]. So the magnitude of nuclear anapole moment is about $1.0 \times 10^{-26}\,\mathrm{J \cdot T^{-1} \cdot cm^{-1}}$. On the

Paul G. Mezey and Beverly E. Robertson (eds.), Electron, Spin and Momentum Densities and Chemical Reactivity, 301–302

other hand, the interaction strength for crystal anapole moment is $\alpha^2 = (1/137)^2$. The anapole radius in this lattice is $\approx 1.5 \times 10^{-8}$ cm, $\mu_{\mathrm{Mn}} = 1.17\,\mu_{\mathrm{B}}$ [4]. Then the magnitude of the crystal anapole moment is about $1.3 \times 10^{-20}\,\mathrm{J \cdot T^{-1} \cdot cm^{-1}}$. So the magnitude of crystal anapole moment is about 6 orders bigger than that of nuclear anapole moment.

The coupling strength for the nuclear anapole moment coupled to the weak process is still unknown. None of the previous authors have calculated the strength in theory because the concept of anapole moment is not consistent with the present schemes of electro-weak theory [5]. Even if this coupling may be too small to be detected in normal ways, the coupling effect would be amplified by about 6 orders of magnitude by the crystal anapole moment. Thus a coupling strength as small as 10^{-6} orders of magnitude would be detectable via the 'amplifier' of the crystal anapole moment. That is to say an upper limit 10^{-6} for the coupling strength will be given by this experiment.

As the anapole interaction is the candidate which directly breaks parity conservation in electromagnetic interaction [1], it is very desirable to test whether the anapole moment could couple to the β decay or not. This experiment can be performed by solid state detectors as well as by a magnetic spectrometer. There are also other choices for the crystal samples [3] and β sources. Since the anapole moment has the same intrinsic structure as for Majorana neutrinos, its coupling is valid to both β^- decay and β^+ decay.

3. Summary

A new experimental method has been introduced to measure the effect of the crystal anapole moment on β decay. The basic hypothesis is very similar to that assumed by Zel'dovich. The special idea is to introduce the description of solid-state physics (crystallography) into the process of weak interaction. The β decay rate will be modified due to the presence of crystal anapole moment. If this modification could be detected, the hypothesis for the anapole moment and its coupling to weak interaction will be verified for the first time; if this modification could not be detected by this method, an upper limit of up to 10^{-6} for the coupling of anapole moment to weak process should be given. This experiment will give direct verification to Zel'dovich's assumption.

References

1. Zel'dovich, Ya.B. (1958) Electromagnetic interaction with parity violation, *Sov. Phys. JETP*, **6**, 1184–1186.
2. Radescu, B.B. (1986) On the electromagnetic properties of Majorana fermions, *Phys, Rev. D*, **32**, 1266–1268.
3. Dubovik, V.M. and Tugushev, V.V. (1990) Toroid moments in electrodynamics and solid state physics, *Phys. Rep.*, **187**, 145–202.
4. Fruchart, D. Bertaut, E.F. and Madar, R. (1971) Structure Magnetique et Rotation de Spin de Mn_3NiN, *Solid State Comm.*, **V9**, 1793–1797.
5. Rekalo, M.P. (1978) Scattering of polarized leptons by hadrons and the anapole moment of leptons and quarks, *Soy. J. Nucl. Phys.*, **28**(6), 852–853.

23

Three-dimensional reconstruction of electron momentum densities and occupation number densities of Cu and CuAl alloys

G. DÖRING, K. HÖPPNER, A. KAPROLAT and W. SCHÜLKE

University of Dortmund, Institute of Physics, Otto Hahn Straße 4, D-44221 Dortmund, Germany

1. Introduction

Fermi surfaces belong to the most frequently investigated properties of solids and lots of experimental methods have been developed for this purpose. These are, for example, the de Haas–van Alphén effect, the cyclotron resonance method, positron annihilation and the Compton scattering spectroscopy. One problem of all methods based on electron transport is the necessity of high purity and low temperature of the sample [1], which is difficult to obtain especially for disordered substitutional alloys as some of these materials show a phase transition at low temperatures and the use of such methods is limited to very dilute alloys because of impurity scattering. The angular correlation of positron annihilation radiation (ACAR) [2] and the Compton scattering spectroscopy are not subjected to such restrictions and provide the possibility to obtain not only Fermi surface parameters but after an appropriate reconstruction procedure the full three-dimensional electron momentum space density and the occupation number density. Admittedly the ACAR method has the disadvantage that the influence of the positron cannot be neglected and therefore one can only get information about the so-called two-photon momentum density $\rho^{2\gamma}$. For this reason it is primarily the Compton scattering spectroscopy which gives access to information about the Fermi surfaces of disordered substitutional alloys.

In this chapter we will have a closer look at the methods of the reconstruction of the momentum densities and the occupation number densities for the case of CuAl alloys. An analogous reconstruction was successfully performed for LiMg alloys by Stutz *et al.* in 1995 [3]. It was found that the shape of the Fermi surface changed and its included volume grew with Mg concentration. Finally the Fermi surface came into contact with the boundary of the first Brillouin zone in the [110] direction. Similar changes of the shape and the included volume of the Fermi surface can be expected for CuAl [4], although the higher atomic number of Cu compared to that of Li leads to problems with the reconstruction, which will be examined.

Section 2 describes the experimental determination of Compton profiles of Cu and $Cu_{0.953}Al_{0.047}$ in some detail. Section 3 describes the data evaluation and Section 4 the method of the reconstruction. Section 5 presents the results and finally Section 6 concludes.

Paul G. Mezey and Beverly E. Robertson (eds.), Electron, Spin and Momentum Densities and Chemical Reactivity, 303–312

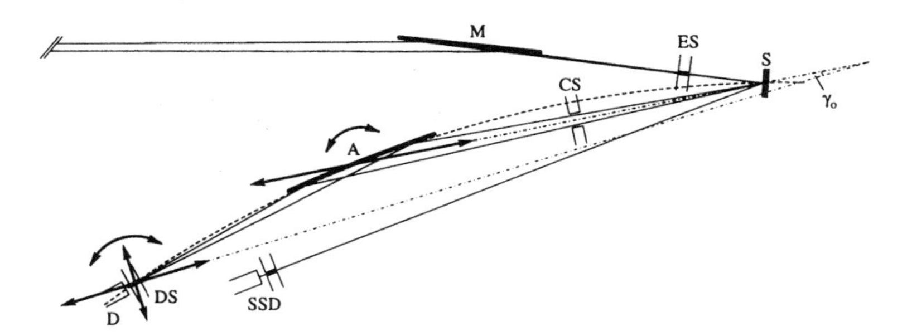

Figure 1. Compton spectrometer at ESRF ID15b [5].

2. Compton profile measurements on Cu and $Cu_{0.953}Al_{0.047}$

The Compton profile measurements on Cu and $Cu_{0.953}Al_{0.047}$ were performed at ID15b of the ESRF. Figure 1 shows the setup of the scanning-type Compton spectrometer used. It consists of a Si (311) monochromator (M), a Ge (440) analyzer (A) and a NaI detector (D). The signal of an additional Ge solid state detector (SSD) was used for normalization. ES, CS and DS denote the entrance slit, the collimator slit and the detector slit, respectively. For each sample 10 different directions were measured with approximately 1.5–2 $\times$ 10^7 total counts per direction. The incident energy was 57.68 keV for the Cu and 55.95 keV for the $Cu_{0.953}Al_{0.047}$ measurement.

Unfortunately a problem with the measurements of Cu Compton profiles cropped up which was caused by an unwanted (551) reflection of the analyzing crystal, which hit the detector when the incident angle at the analyzer became greater than approximately 7.7°. This caused an increase of the measured intensity. Therefore this region of the Compton profile could not be used at the expense of statistics. The problem was solved before starting the measurements on $Cu_{0.953}Al_{0.047}$ by using a differently orientated Ge (440) analyzing crystal. Figure 2 shows a typical raw Compton profile of Cu in the [100] direction.

3. Data evaluation

Only the valence Compton profiles are needed for the reconstruction of the momentum density and the occupation number density. So one has to subtract an appropriate core Compton profile. Furthermore the contribution of the multiple scattered photons to the measured spectra has to be taken into account (for example by a Monte Carlo simulation [6]). Additionally one has to take heed of the fact that the efficiency of the spectrometer is energy dependent, so the data must be corrected for energy dependent effects which are the absorption in the sample and in the air along the beam path, the vertical acceptance of the spectrometer and the reflectivity of the analyzing crystal. The relativistic derivation of the relationship between the Compton cross section and the Compton profile leads to a further correction factor [7]. Finally a background subtraction and a normalization of the valence profiles to the number of valence

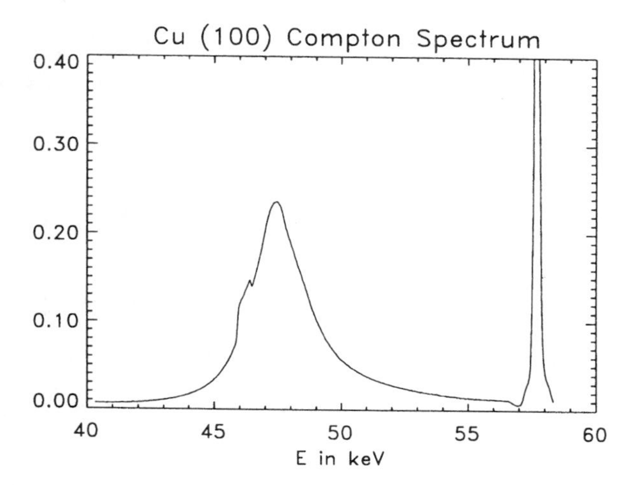

Figure 2. Measured Compton profile of Cu in [100] direction.

electrons has to be performed. In the process one faces the problem that the last two steps cannot be separated, as the normalization constant depends on the background, which can only be fitted to the outer regions of the Compton profile $|p_z| > p_0$ when this constant is known. Therefore these two steps have been performed simultaneously by an iterative process. For $|p_z| > p_0$ the background and the normalization constant are given by

$$B(p_z) = I(p_z) - \frac{1}{CK(p_z)}\left[J_{\rm core}(p_z) + M(p_z)\right] \tag{1}$$

and

$$C = \frac{\bar{Z}_{\rm val} + \int_{-p_0}^{+p_0} J_{\rm core}(p_z)\,\mathrm{d}p_z + \int_{-p_0}^{+p_0} M(p_z)\,\mathrm{d}p_z}{\int_{-p_0}^{+p_0} K(p_z)\left[I(p_z) - B(P_z)\right]\mathrm{d}p_z}, \tag{2}$$

where

$$\begin{aligned} J(p_z) &= J_{\rm val}(p_z) + J_{\rm core}(p_z) \\ &= J_{\rm core}(p_z) \quad \text{for } |p_z| > p_0, \end{aligned} \tag{3}$$

I is the measured intensity, J the Compton profile, M the multiple scattering contribution, K the energy dependent correction factor, B the background, C the normalization constant and $\bar{Z}_{\rm val}$ the mean number of valence electrons. Figure 3 shows a valence Compton profile of Cu obtained by this procedure.

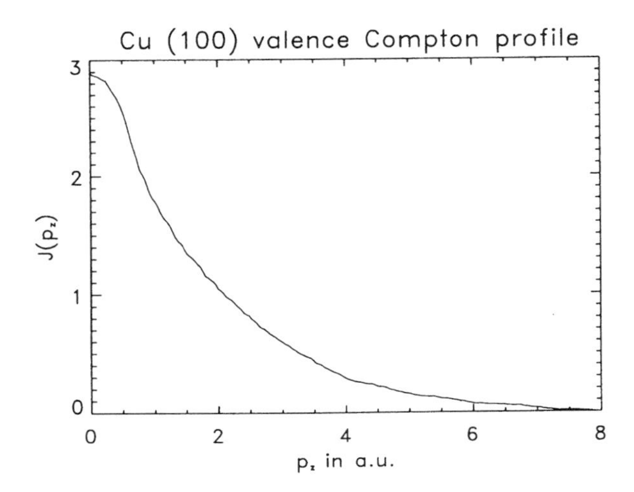

Figure 3. Valence profile of Cu in [100] direction.

4. Reconstruction of the momentum densities and the occupation number densities

Using the valence profiles of the 10 measured directions per sample it is now possible to reconstruct as a first step the full three-dimensional momentum space density. According to the Fourier Bessel method [8] one starts with the calculation of the Fourier transform of the Compton profiles which is the reciprocal form factor $B(z)$ in the direction of the scattering vector $\mathbf{q}$. The full $B(\mathbf{r})$ function is then expanded in terms of cubic lattice harmonics up to the 12th order, which is to take into account the first 6 terms in the series expansion. These expansion coefficients can be determined by a least square fit to the 10 experimental $B(z)$ curves. Then the inverse Fourier transform of the expanded $B(\mathbf{r})$ function corresponds to a series expansion of the momentum density, whose coefficients can be calculated from the coefficients of the $B(\mathbf{r})$ expansion.

For the reconstruction of the occupation number density $n(\mathbf{k})$ in the repeated zone scheme one uses the reciprocal form factor at lattice translation vectors $\mathbf{R}$, as $n(\mathbf{k})$ can be written as [9]

$$n(\mathbf{k}) = \sum_{\mathbf{R}} B(\mathbf{R}) \mathrm{e}^{-\mathrm{i}\mathbf{kR}}. \tag{4}$$

The Fermi surface can be determined using

$$\int_{n(\mathbf{k}) \geq c} \mathrm{d}^3 k = \frac{4}{3} \pi \left(k_{\mathrm{F}}^0\right)^3, \tag{5}$$

where $\left(k_{\mathrm{F}}^0\right)^3$ is the Fermi radius of the free electron gas sphere with a corresponding number of valence electrons. The Fermi surface is then given by $n(\mathbf{k}) = c$.

The main problem in using this method is the statistical error of the Compton profiles which calculates to

$$\sigma(B) = \frac{B(0)}{\sqrt{N_{\text{tot}}}}, \tag{6}$$

where N_{tot} is the total number of counts per spectrum. According to the definition of the reciprocal form factor $B(\mathbf{r})$ as the Fourier transform of the momentum density $\rho(\mathbf{p}).B(0)$ is simply the integration of $\rho(\mathbf{p})$ over the whole momentum space which results in the total number of electrons Z.

After calculating the Fourier transform of the Compton profiles one observes that the amplitude of its oscillations becomes smaller than this statistical error when $|\mathbf{r}|$ is greater than 15 a.u. and therefore the $B(\mathbf{r})$ function cannot be used for $|\mathbf{r}| > 15$ a.u. On the other hand if one wants to get results for Cu with a similar statistical error compared to the results of the Li reconstruction the number of counts needed is given by

$$N_{\text{Cu}} = \left(\frac{Z_{\text{Cu}}}{Z_{\text{Li}}}\right)^2 N_{\text{Li}}. \tag{7}$$

In the case of the LiMg momentum density and occupation number density reconstruction of Stutz *et al.*, who collected 6×10^6 counts for Li and 6×10^7 counts for LiMg, this would mean that 6×10^8–6×10^9 counts per spectrum were required, which hardly can be accomplished in a reasonable amount of time even at modern synchrotron radiation sources.

5. Results

The following figures show the results of the reconstructions using the described methods. Figures 4 and 6 show the momentum density anisotropy of Cu and $Cu_{0.953}Al_{0.047}$ respectively in the $(1\bar{1}0)$ plane. The anisotropy is obtained by neglecting the first, isotropic term of the series expansion of $\rho(\mathbf{p})$ in cubic lattice harmonics.

One can clearly see the large positive anisotropy in the [111] direction near the boundary of the first Brillouin zone (BZB). It is caused by the [111] high momentum component, which produces a continuous distribution of the momentum density across the BZB, as the Fermi surface has contact with the BZB in this direction. In the other directions, especially in [100], calculations show a steep decrease of the momentum density at the Fermi momentum and therefore a negative deviation from the spherical mean value.

Furthermore the absolute values of the anisotropies show an increase with the Al concentration. Admittedly it has to be noticed that the errors as shown in Figures 5 and 6 are large in the high symmetric directions [10], especially in the [100] direction. Another source of error is the truncation of the $B(\mathbf{r})$ function mentioned above, which causes oscillations (Figure 7).

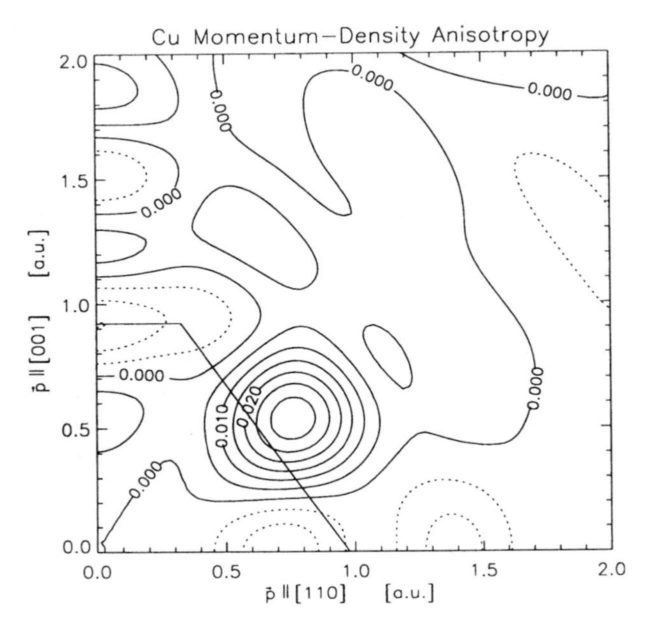

Figure 4. Momentum density anisotropy of Cu; the solid line marks the boundary of the first Brillouin zone; solid and dashed contour lines mark positive and negative anisotropies, respectively.

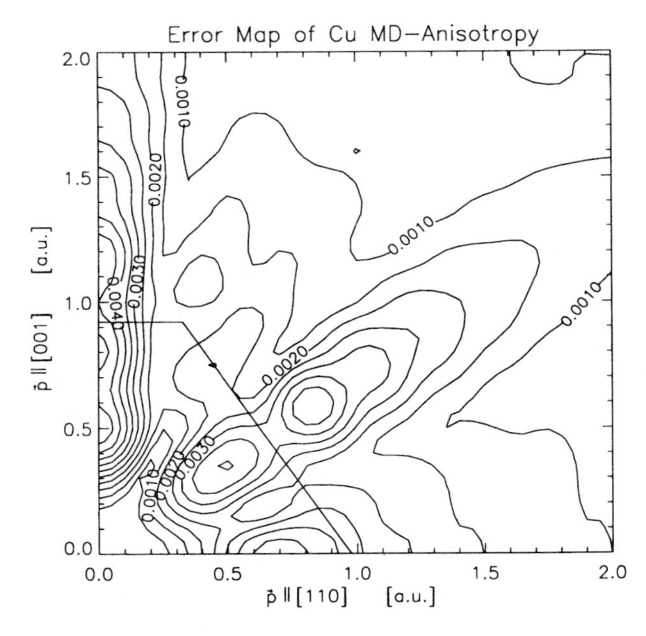

Figure 5. Error Map of Figure 4.

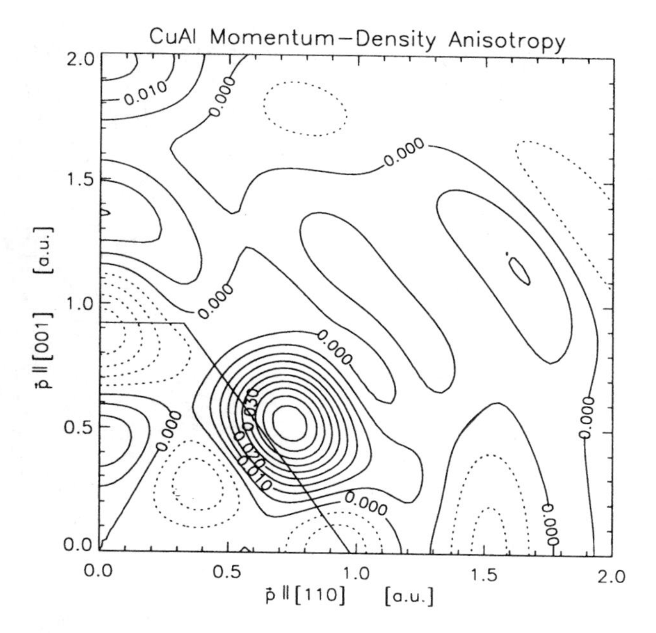

Figure 6. Momentum density anisotropy of $Cu_{0.953}Al_{0.047}$; the solid line marks the boundary of the first BZB; solid and dashed contour lines mark positive and negative anisotropies, respectively.

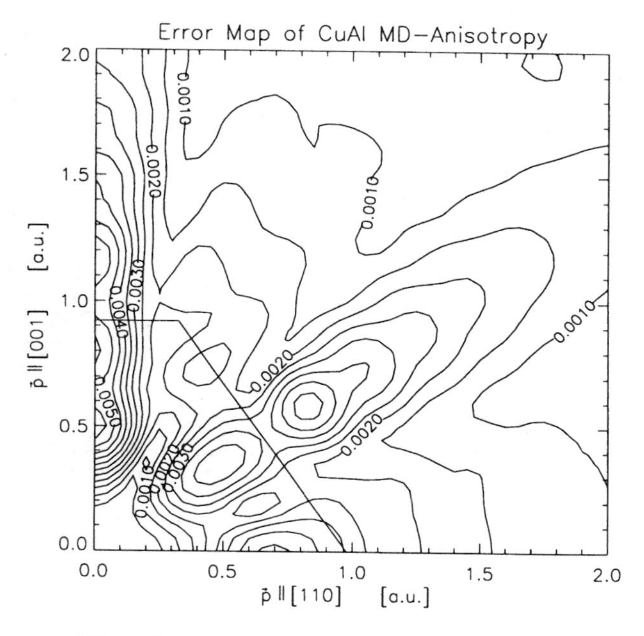

Figure 7. Error map of Figure 6.

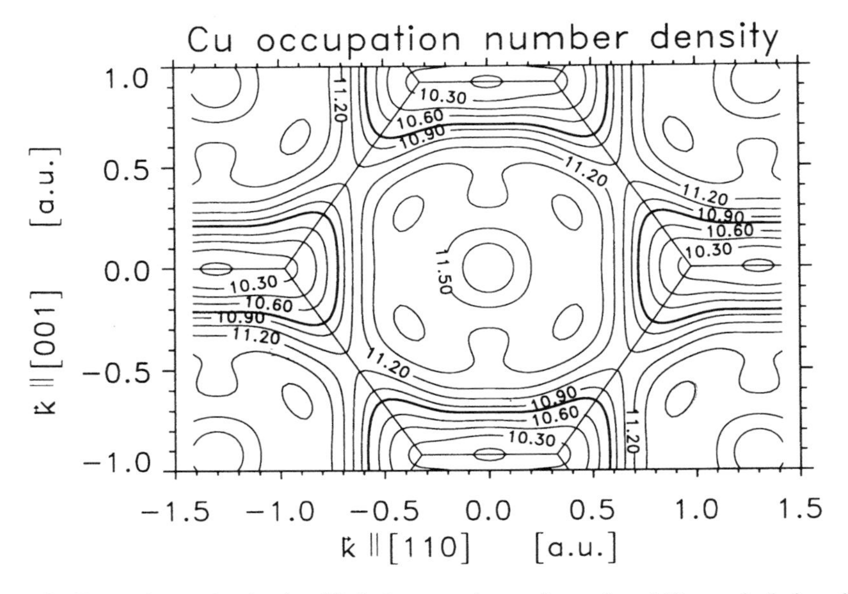

Figure 8. Occupation number density of Cu in the repeated zone scheme; the solid line marks the boundary of the first BZB; the bold contour line marks the Fermi surface.

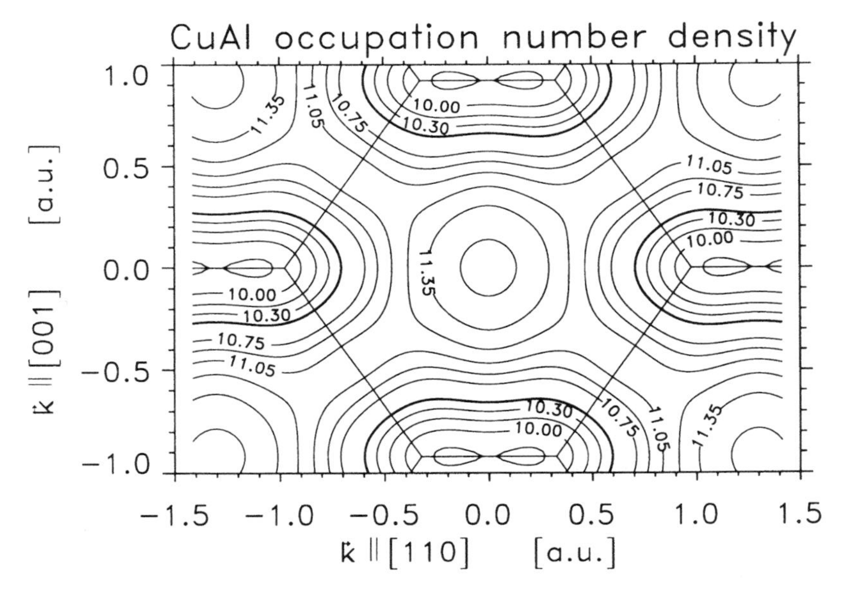

Figure 9. Occupation number density of $Cu_{0.953}Al_{0.047}$ in the repeated zone scheme; the solid line marks the boundary of the first BZB; the bold contour line marks the Fermi surface.

Figures 8 and 9 show the occupation number densities of Cu and $Cu_{0.953}Al_{0.047}$ again in the $(1\bar{1}0)$ plane. The bold line marks the crossing of the Fermi surface as determined according to Equation (5) with this plane. The well-known 'neck' structure in the $\langle 111 \rangle$ directions can be seen clearly as well as an increase of the neck radius with Al concentration. This increase is about 30% which is in agreement with calculations [4, 11] and ACAR experiments [12, 13]. Admittedly the reconstruction suffers from the early truncation of the series expansion of $n(\mathbf{k})$ according to Equation (4).

6. Conclusions

The reconstruction of the electron momentum densities and the occupation number functions of Cu and $Cu_{0.953}Al_{0.047}$ could not produce results on an equal profound base as those based on the results of Li and LiMg reconstructions. This would need approximately 100 times the number of counts per spectrum which was not achieved.

Nevertheless features of the [111] high momentum component were found in the form of anisotropies of the momentum density near the boundary of the first BZB in the [111] and the [100] directions. These anisotropies increase with Al concentration. Measurements on $Cu_{0.1}Al_{0.9}$ will show if this effect persists at higher Al concentration.

The Fermi surfaces exhibit the well-known neck structure in the $\langle 111 \rangle$ directions. Their radii also increase with Al concentration.

On the other hand both the momentum densities and the occupation number functions were influenced by the early truncation of the $B(\mathbf{r})$ function due to inadequate statistics. The effect of this influence has to be studied, which could best be done by a reconstruction using calculated Compton profiles.

Acknowledgements

We would like to thank the beamline scientists of ID15 of the ESRF. T. Buslaps, V. Honkimäki, A. Shukla and P. Suortti for their valuable help with the measurements and F. Maniawski and J. Kwiatkowska for their help with the sample preparation. We are also grateful to J. Felsteiner for sending us his Monte-Carlo simulation program for the multiple scattering correction.

References

1. Coleridge, P.T., Templeton, I.M., and Vasek, P., (1984) *J. Phys. F*, **14**, 2963.
2. Mijnarends, P.E., (1979) *Electron momentum densities in metals and alloys*. In *Positrons in Solids*, (Ed.) Hautojärvi, P., Springer.
3. Stutz, G., Wohlert, F., Gabriel, K.J., Kaprolat,A., Schülke, W., Shiotani, H., Sakurai, Y., Tanaka, Y., Ito, M., and Kawata, H., Sagamore XI, Extended Abstracts, Brest 1994, p. 75.
4. Raychaudhuri, M., and Chatterjee, S., (1982) *Phys. Stat. Sol. (b)*, **112**, 105.
5. Honkimäki, V., private communication.
6. Felsteiner, J., Pattison, P., and Cooper, M., (1974) *Philos. Mag.*, **30**, 537.
7. Ribberfors, R., (1975) *Phys. Rev. B*, **12**, 3136.
8. Hansen, H., (1980) *Reconstruction of the electron momentum distribution from a set of experimental Compton profiles*, Hahn Meitner Institute (Berlin), Report HMI B 342.

9. Schülke, W., (1977) *Phys. Stat. Sol. (b)*, **82**, 299.
10. Schülke, W., Stutz, G., Wohlert, F., and Kaprolat, A., (1996) *Phys. Rev. B*, **54**, 14381.
11. Bansil, A., Ehrenreich, H., Schwarz, L., Watson, and R.E., (1974) *Phys. Rev. B*, **9**, 445.
12. Murray, B.W., and McGervey, J.D., (1970) *Phys. Rev. Lett.*, **24**, 9.
13. Thompson, A., Murray, B.W., and Berko, S., (1971) *Phys. Lett.*, **37**, 461.

24

X-ray and neutron studies of *cis*-enol systems at liquid helium temperatures

GEORG K.H. MADSEN and CLAIRE WILSON

Department of Chemistry, University of Århus, DK-8000 Århus C, Denmark

1. Introduction

Gilli *et al.* [1] have proposed the resonance assisted hydrogen bond (RAHB) model to account for the very short O–H · · · O and N–H · · · O distances observed in conjugated systems containing hydrogen bonds. Their model for a RAHB system in a *cis*-enol fragment is illustrated by the scheme shown in Figure 1, which suggests that the resonance introduces partial charges with the appropriate signs to strengthen the hydrogen bond. The energy of the system will consequently be lowered as the positive hydrogen nucleus moves towards the negative keto oxygen atom. Thus Gilli's RAHB model can be perceived as a feedback mechanism which maintains zero partial charge on the opposite oxygens. The increase in polarization which is due to resonance is neutralized by a shift in the proton position in the hydrogen bond.

The scheme in Figure 2 illustrates a possible alternative explanation for the observation that bond lengths in *cis*-enol systems are intermediate between single and double bonds. If the molecules have statistically disordered enol systems, the hydrogen atoms of the hydrogen bond will be distributed over two positions in the crystal structure. Indeed this was the case for the C polymorph of naphtazarin above 110 K; at this temperature there is a second-order transition to a state with an ordered enol hydrogen [2].

A low-temperature study of structure of benzoylacetone was undertaken as a test of these models. The crystal structure of benzoylacetone has previously been determined by neutron diffraction [3]. It was found that the enolic hydrogen has a very large displacement amplitude between the two oxygens. In diffraction studies of such systems (particularly X-ray diffraction) it can be difficult to locate the hydrogen position in a hydrogen bond sufficiently accurately. Especially for room-temperature diffraction studies the observed atomic displacement of the hydrogen atom around its average position is usually so substantial that it is impossible to judge whether the hydrogen

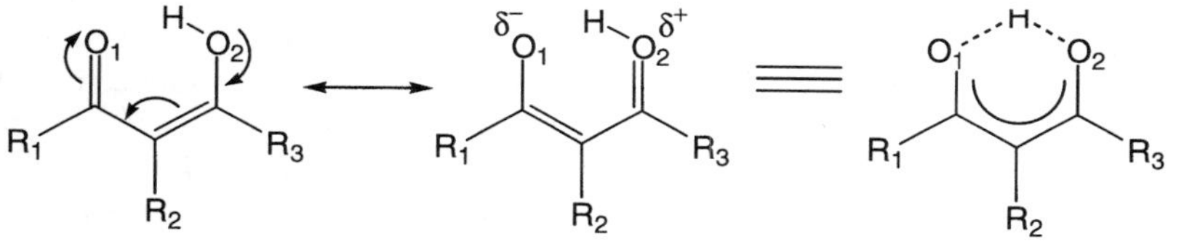

Figure 1. The resonance assisted hydrogen bonding model.

Paul G. Mezey and Beverly E. Robertson (eds.), Electron, Spin and Momentum Densities and Chemical Reactivity, 313–321

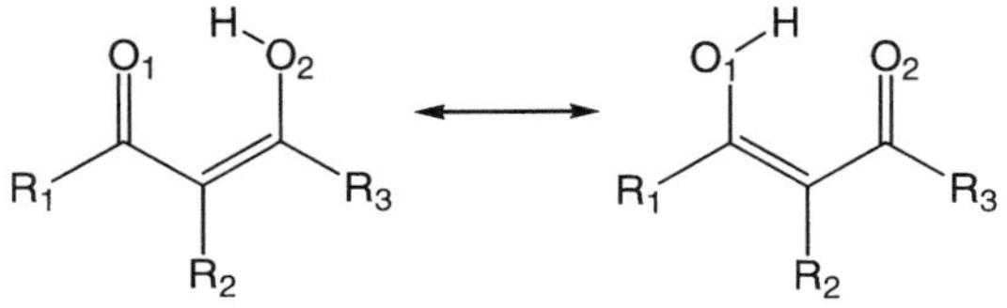

Figure 2. The two molecules of the disordered model.

atom is distributed over two positions (statistically or residing in a double minimum) or is moving in a shallow potential well. If the thermal energy of the hydrogen atom is sufficiently high, it is conceivable that a double minimum potential may be disguised as dynamic disorder. Thus it is clearly desirable to carry out diffraction studies of hydrogen bonding at the lowest possible temperatures [4].

Benzoylacetone was studied using both X-ray and neutron diffraction. Four X-ray data sets at four different temperatures were collected, namely at room temperature, 160, 20 and 8 K. Furthermore a neutron data set was collected at 20 K.

Nitromalonamide ($C_3H_5N_3O_4$, $R_1 = R_3 = NH_2$, $R_2 = NO_2$) was chosen as a further example of a very short intramolecular keto–enol O–H · · · O hydrogen bond. It has one of the shortest known O–H · · · O distances at 2.38 Å[5]. Both low-temperature X-ray (10 K) and neutron data sets (15 K) have been collected to examine whether our results from benzoylacetone are of a general nature.

2. The structure and electron density of benzoylacetone

Based on an analysis of the positional and thermal parameters determined in the 20 K neutron study, the benzoylacetone structure was concluded to be ordered (Figure 3). Despite the low temperature the enol hydrogen was observed to have a large atomic displacement parallel to the O–O interatomic vector. This was interpreted as a hydrogen vibrating in a low barrier potential well that is characteristic of strong hydrogen bonding [6].

This implies that replacement of the enol hydrogen in benzoylacetone by deuterium might give a double minimum well. Very large atomic displacement parameters were also found for the methyl hydrogens. This could naturally lead one to speculate whether a coupling between the enol hydrogen and the methyl hydrogen is present. As a deuterium atom in the keto–enol group would be expected to be localized on one oxygen or statistically distributed between two positions, one would furthermore expect it to have a smaller vibrational amplitude. A possible coupling between the vibration of the enol hydrogen and the rotational vibration of the methyl group hydrogens could mean that deuteration of the short hydrogen bond would result in a decrease in the large vibrational amplitudes of the methyl hydrogens. These interesting ideas will have to be tested by low-temperature neutron diffraction.

The charge density study of benzoylacetone [8] revealed that the Laplacian at the bond critical points between the enol hydrogen and the oxygens has a negative value. This means that the bonds between that hydrogen and both the oxygens have covalent character. Furthermore the populations of the spherical valence parts of the multipole

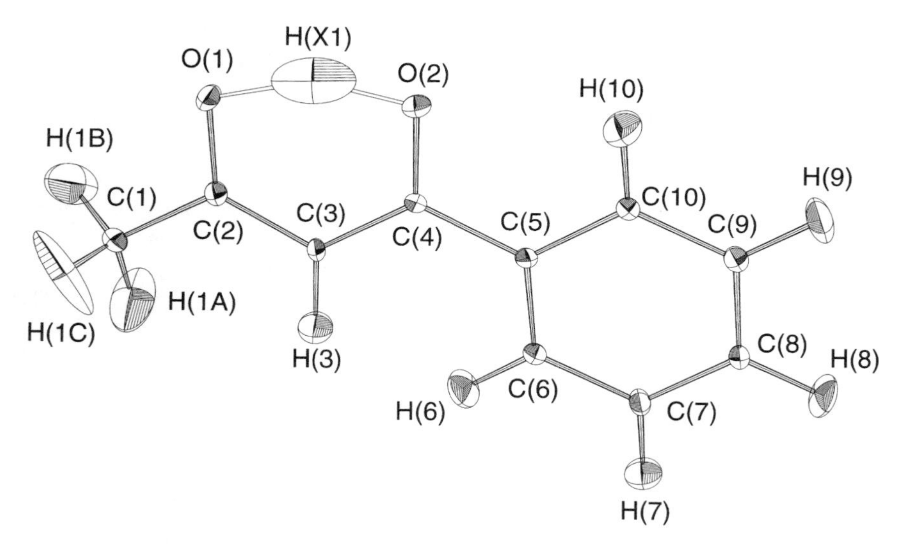

Figure 3. ORTEP [7] drawing of the benzoylacetone molecule showing 50% probability ellipsoids.

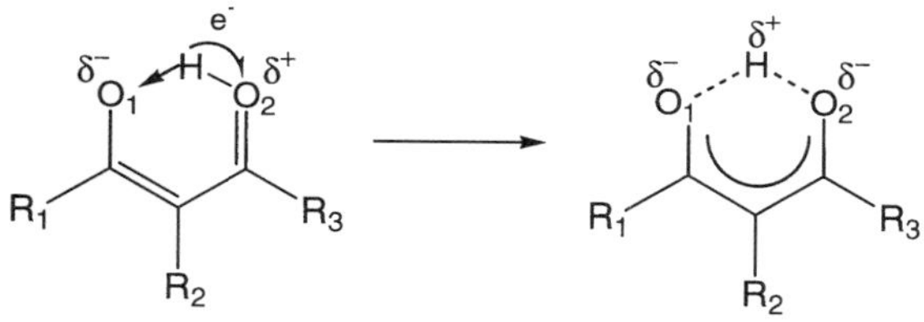

Figure 4. The modified RAHB model.

model revealed large formal charges on the enol oxygens and on the enol hydrogen, as shown in Figure 4. We have therefore suggested that the resonance assisted hydrogen bonding model must be modified slightly to give the more adequate scheme shown below; a mechanism which drives the charges in the ring toward symmetry, rather than driving toward zero partial charges.

Detailed documentations of the structure and experimental electron density and of the thermal behavior of benzoylacetone at different temperatures have been submitted for publication [8, 9]. Therefore the following account will concentrate on giving a preliminary account of the structure and the electron density of nitromalonamide, and comparisons between the two studies.

3. Experimental X-ray diffraction study of nitromalonamide

Nitromalonamide was synthesized according to Hantzsch [10]. Single crystals were grown by evaporation from a methoxy-ethanol solution. The crystal used for data collection was glued to a few carbon fibres stuck on a copper wire for better thermal contact and was fitted on a cold station of a type 202 DISPLEX closed-cycle helium refrigerator mounted on a type 512 HUBER four-circle diffractometer. All accessible data $2\theta \leq 90^\circ$ were collected in one quadrant. Symmetry equivalent reflections were collected out to $2\theta \leq 70^\circ$ in two quadrants, selected so that one octant was measured three times in all. Further experimental details are given in Table 1.

For the multipolar modeling [11, 12] of the X-ray diffraction data the program XD [13] was used. The atomic density contributions are parametrized into a core term, ρ_{core}, a spherical valence term, ρ_{valence}, and a set of multipolar functions:

$$\rho_{\text{atom}}(\mathbf{r}) = P_{\text{core}}\rho_{\text{core}}(\mathbf{r}) + \mathrm{P}_{\text{valence}}\kappa^3\rho_{\text{valence}}(\mathbf{r}) + \sum_{l=0}^{l_{\max}} \kappa_l' R_l(\kappa_l'\mathbf{r}) \sum_{m=0}^{l} P_{lm\pm}\mathrm{d}_{lm\pm}(\theta, \varphi).$$

Table 1. Experimental details.

Crystal Data	X-ray Study
Chemical formula	$C_3N_3O_4H_5$
Chemical formula weight	147.0902
Space group	Orthorhombic P $2_12_12_1$
a(Å)	4.862(1)
b(Å)	4.980(1)
c(Å)	21.938(5)
Z	4
Radiation	MoKα
Wavelength (Å)	0.7114
$\mu(\text{mm}^{-1})$	0.170
Temperature (K)	10(1)
Crystal morphology	Colorless crystal bounded by ±[100] 0.525 mm, ±[010] 0.220 mm, ±[001] 0.0075 mm
Data collection	
Diffractometer	Type 512 HUBER
Scan method	ω–2θ
Transmission factors	0.954 – 0.987
No. of measured reflections	8300
No. of unique reflections	3762
R_I	1.96%
Range h, k, l	$h = -9, 9; k = -8, 9; l = -42, 34$
Refinement on	F^2
$R(F)$, $R(F^2)$ (%)	2.20, 2.52
$wR(F)$, $wR(F^2)$ (%)	3.05, 5.03
GoF	1.168
N_{obs}	3507, $I > 2\sigma(I)$
N_{par}	310
$N_{\text{obs/par}}$	11.31
Weighting scheme	$\sigma(F^2)$
$(\Delta/\sigma)_{\max}$	0.16

On the carbons, nitrogens and oxygens expansions up to octapole level were introduced, whereas the expansions were limited to quadrupole level for the hydrogen atoms. All atoms were given a κ expansion/contraction parameter for the spherical monopole term, and all atoms except the hydrogens were given κ' parameters to expand or contract the non-spherical poles. The κ and κ' values on O(1) and O(3), on N(1) and N(3), on C(1) and C(3) and on O(21) and O(22) were constrained to be equal.

Neutron diffraction study of nitromalonamide

The 15 K neutron diffraction data were collected using monochromatic thermal neutrons with a wavelength of 0.955 Å on D19, a four-circle neutron diffractometer on the H11 beam at the Institut Laue-Langevin, Grenoble. The diffractometer is equipped with a $64 \times 4^{\circ}$ position sensitive detector and data were collected in normal-beam Weissenberg geometry. The three-dimensional count distribution around each reciprocal-lattice point was corrected for background and reduced to an integrated intensity I by a method that minimizes the relative standard deviation, $\sigma(I)/I$[14]. The neutron diffraction data are presently being processed and refined. The atomic parameters determined in that study will be introduced into the refinement of the X-ray data for an X–N refinement.

4. The structure of nitromalonamide

The discussions of the structure and the electron density are based on the structure found by a full multipole refinement of the X-ray data with the hydrogen positions fixed at the neutron values and the hydrogen thermal parameters fixed at scaled neutron values (Figure 5).[1] The interatomic distances and intramolecular bond angles are given in Table 2.

The nitromalonamide molecule is almost completely planar. The largest intramolecular torsion angle being -3.3° between C(1)–O(1)–H(X)–O(3). In contrast to the structure of benzoylacetone, the crystal structure of nitromalonamide is characterized by hydrogen bonded networks. The main feature of the hydrogen bonded networks are the ribbons of, almost co-planar, hydrogen bonded molecules related by pure translation along the a- and b-axes. The ribbons stack in layers along the c-axis, each layer consisting of parallel ribbons running in the orthogonal direction with respect to the ribbons in the neighboring layer.

The most noticeable thing about the structure of nitromalonamide is the lack of symmetry in the keto–enol ring system inspite of the short intramolecular hydrogen bond. According to the RAHB model there should be correlation between a short O–O distance and the degree of symmetry of the ring. However, the enol hydrogen is asymmetrically placed and the two C–O bonds are of markedly different length. In nitromalonamide the C–OH bond is significantly longer than the C=O bond.

[1] The neutron data are still being processed. The neutron values reported here have been taken from the best refinement obtained till now.

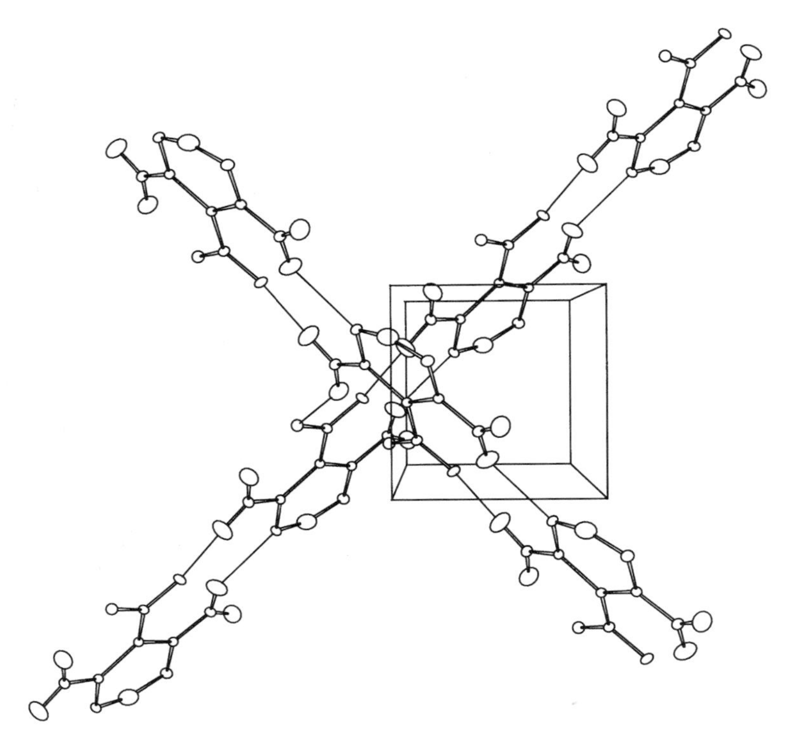

Figure 5. ORTEP [7] drawing of hydrogen bonded network showing two ribbons in neighboring layers. The structure is viewed along the *c*-axis with the *a*-axis vertical and the *b*-axis horizontal.

An apparently important limitation of Gilli *et al.*'s [1] study is the omission of all structures that were involved in intermolecular hydrogen bonding from the crystal structure correlations used to validate their RAHB model. The asymmetry of the *cis*-enol ring of nitromalonamide indicates that intermolecular hydrogen bonds can pertubate the resonance in the ring, and it seems clear that one must understand how intermolecular interactions effect the intramolecular hydrogen bonding before the full story of RAHB can been told. It has also become obvious that the O–O distance that has been used to classify the strength of hydrogen bonding systems [6], is not a fully adequate characterization.

Electron density of nitromalonamide

Nitromalonamide crystallizes in the non-centrosymmetric space group $P2_12_12_1$. This potentially adds extra uncertainty to the multipole refinement because of more ambiguity in phase assignments of the reflections, than for a centrosymmetric space group. Haouzi *et al.* [15] have shown how refinement of multipole populations in non-centrosymmetric space groups can lead to unreasonable results. This is due to odd order multipoles that are invariant to the space group symmetry and therefore mostly

Table 2. Interatomic distances and selected bonding angles.

	Interatomic distances/ Å		Interatomic distances/ Å		Angle/ °
O(1)–O(3)	2.3916(7)	O(1)–H(1)	2.3925(4)	C(1)–O(1)–H(X)	104.6(1)
O(1)–H(X)	1.1373(5)	O(1)–H(4)^{1}	2.2861(5)	C(3)–O(3)–H(X)	103.7(1)
O(3)–H(X)	1.3035(5)	O(22)–H(1)^{1}	1.9294(6)	O(1)–H(X)–O(3)	156.9(1)
C(1)–O(1)	1.3030(6)	O(22)–H(4)	1.8559(6)	C(2)–C(1)–O(1)	118.2(1)
C(3)–O(3)	1.2867(7)			C(2)–C(3)–O(3)	118.0(1)
C(1)–C(2)	1.4556(7)	O(3)–H(3)	2.4521(5)	C(1)–C(2)–C(3)	118.6(1)
C(2)–C(3)	1.4634(7)	O(3)–H(3)^{3}	2.0673(5)	C(1)–O(1)–H(1)	56.6(1)
C(1)–N(1)	1.3201(8)			C(1)–O(1)–H(4)^{1}	131.4(1)
N(1)–H(1)	1.0134(6)	O(21)–H(2)	1.9002(5)	C(3)–O(3)–H(3)	55.4(1)
N(1)–H(2)	1.0108(5)	O(21)–H(2)^{4}	2.1293(5)	C(3)–O(3)–H(3)^{3}	119.7(1)
C(3)–N(3)	1.3260(8)			N(2)–O(21)–H(2)	110.6(1)
N(3)–H(3)	1.0072(5)			N(2)–O(21)–H(2)^{4}	107.9(1)
N(3)–H(4)	1.0056(6)			N(2)–O(22)–H(4)	110.0(1)
C(2)–N(2)	1.3923(7)			O(1)–H(1)–N(1)	69.0(1)
N(2)–O(21)	1.2510(7)			O(22)–H(1)1–N(1)^{1}	157.7(1)
N(2)–O(22)	1.2566(7)			O(21)–H(2)–N(1)	123.1(1)
				O(21)–H(2)–N(1)^{4}	141.3(1)
				O(3)–H(3)–N(3)	66.5(1)
				O(3)–H(3)3–N(3)^{3}	142.1(1)
				O(1)–H(4)–N(3)	136.9(1)
				O(22)–H(4)1–N(3)^{1}	126.6(1)

Symmetry operators $^{1}x, y, z$, translation -1 1 0; $^{2}1/2 - x, -y, 1/2 + z$; $^{3} - x, 1/2 + y, 1/2 - z$, translation 2 0 1; $^{4}1/2 + x, 1/2 - y, -z$, translation 0 -1 1.

influence the phase of the calculated structure factors. Refining the populations of these poles can lead to a singular least-squares normal-equations matrix. In the present space group only the O2− multipole is invariant under the space group symmetry. An ill-determined linear combination of poles would result in large standard deviations on a group of population parameters and large correlations between these. This has not been observed, and furthermore none of the O2− poles refine to significant values. Therefore we do not believe that the uncertainty in the phase assignments has seriously compromised the accuracy of the electron density. An experimental deformation density map for nitromalonamide is shown in Figure 6.

The overall electronic features of the bonding that were found in the study of benzoylacetone have been fully confirmed by the study of nitromalonamide. Table 3 lists the valence shell populations of the pseudoatoms in nitromalonamide. As was the case with benzoylacetone there is a large positive formal charge on the hydrogen involved in the strong intramolecular hydrogen bond, and negative charge on the enol oxygens. The study of nitromalonamide thus confirms that the polarization of the O–H–O bond is an inherent part of RAHB. The charges on the carbons are larger than in the *cis*-enol ring of benzoylacetone. This is caused by the fact that the ring substituents, a nitro and two amino groups, are more electronegative/electropositive respectively than the ring substituents in benzoylacetone. The increased polarization

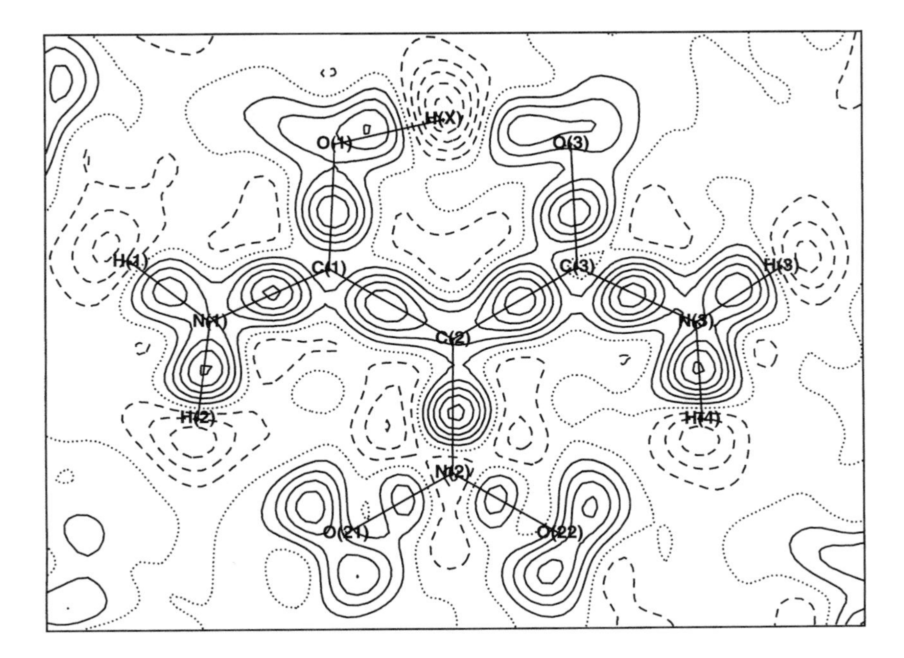

Figure 6. Experimental deformation map for nitromalonamide. The contour interval is $0.1\,e/Å^3$. The dotted line is the zero contour. Solid lines are positive contours, broken lines are negative contours; $\sin\theta/\lambda < 0.7$. The plane shown is the one spanned by the C(1)–O(1) and the C(1)–C(3) vectors.

Table 3. Spherical valence populations of the nitromalonamide multipole model.

Atom	$P^{valence}$	Atom	$P^{valence}$
O (1)	6.3477	N (1)	4.9209
O (3)	6.3653	N (2)	4.7690
O (21)	6.3756	N (3)	5.1075
O (22)	6.4092	H (1)	0.8269
C (1)	3.8779	H (2)	0.7498
C (2)	4.1262	H (3)	0.8332
C (3)	3.8938	H (4)	0.7711
		H (X)	0.6243

of the ring due to the substituents is probably the reason why O· · ·O is so extremely short.

A topological analysis of the total static density has been carried out. The analysis is not complete, and will not be discussed in any great detail in the present context. It is worth mentioning however that similar results as found for benzoylacetone were obtained. The values of ρ_b and $\nabla^2\rho_b$ at the O(1)– H(X) and O(3)– H(X) bond critical points consistently have high values of ρ_b compared to the 'normal' hydrogen bonds and negative Laplacians, thus indicating that both the O–H bonds have a covalent nature.

5. Conclusion

An important lesson learned from the studies of naphtazarin [2], benzoylacetone [8] and nitromalonamide has been that the detailed structure of these types of compounds can only be reliably determined by introducing results of low-temperature neutron diffraction studies in the analysis of the low-temperature X-ray diffraction data. Furthermore it has been found that information about the bonding of the enol hydrogen can be extracted from the thermal parameters of the enol hydrogen. This underlines the importance of the neutron diffraction study in these cases.

The combination of the neutron structural model with a multipole model of the X-ray data measured at a matching temperature, has enabled us to obtain detailed information about the electron density distribution. This has revealed new information about the bonding in *cis*-enol systems.

References

1. Gilli, G., Bellucci, F., Ferretti, V. and Bertolasi, V. (1989) *J. Am. Chem. Soc.*, **111**, 1023–1028.
2. Herbstein, F.H., Kapon, M., Reisner, G.M., Lehmannn, M.S., Kress, R.B., Wilson, R.B., Shiau, W.I., Duesler, E.N., Paul, I.C. and Curtin, D.Y. (1985) *Proc. R. Soc. Lond.*, **399**, 295–319.
3. Jones, R.D.G. (1976) *Acta Cryst. Sect. B*, **32**, 2133–2136.
4. Larsen, F.K. (1995) *Acta Cryst. Sect. B.*, **51**, 468–482.
5. Simonsen, O. and Thorup, N. (1979) *Acta Cryst. Sect. B.*, **35**, 432–435.
6. Hibbert, F. and Emsley, J. (1990) *Adv. Phys. Org. Chem.*, **26**, 255–379.
7. Johnson, C.K., Program ORTEPII. Report ORNL-5138, Oak Ridge National Laboratory, Oak Ridge, TN 1976.
8. Madsen, G.K.H., Iversen, B.B., Larsen, F.K., Kapon, M., Reisner, G.M. and Herbstein, F.H. (1998) *J. Am. Chem. Soc.*, **120**, 10040–10045.
9. Herbstein, F.H., Kapon, M., Reisner, G.M., Madsen, G.H.K., Iversen, B.B. and Larsen, F.K (1999) The crystal structure of benzoylacetone (1-Phenyl-1,3- butadione) in the temperature range 300 to 8 K (in press).
10. Hantzsch, A. (1907) *Chem. Ber.*, **40**, 1523–1532.
11. Stewart, R.F. (1973) *J. Chem. Phys.*, **58**, 1668–1676.
12. Hansen, N.K. and Coppens, P. (1978) *Acta Cryst. Sect. A*, **34**, 909–921.
13. Koritzansky, T., Howard, S., Mallison, P.R., Su, Z., Richter, T. and Hansen, N.K. (1995) XD, a computer program package for multipole refinement and analysis of charge densities from diffraction data, Institute for Crystallography, Berlin.
14. Wilkinson, C., Khamis, H.W., Stansfield R.F.D. and McIntyre, G.J. (1988) J. *Appl. Cryst.*, **21**, 471–478.
15. El Haouzi, A., Hansen, N.K., Le Hénaff, C. and Protas, J. (1996) *Acta Cryst. Sect. A*, **52**, 291–301.

25

Effect of pressure on the Compton scattering of metallic Li

GENDO OOMI[1], FUMINORI HONDA[1], TOMOKO KAGAYAMA[1], FUMITAKE ITOH[2], HIROSHI SAKURAI[2], HIROSHI KAWATA[3] and OSAMU SHIMOMURA[3]

[1]*Department of Mechanical Engineering and Materials Science, Kumamoto University, Kumamoto 860, Japan*
[2]*Department of Electrical Engineering, Gunma University, Kiryu, Gunma 376, Japan*
[3]*Photon Factory, National Laboratory for High Energy Physics, Tsukuba, Ibaraki 305, Japan*

1. Introduction

It is well known that the energy profiles of Compton scattered X-rays in solids provide a lot of important information about the electronic structures [1]. The application of the Compton scattering method to high pressure has attracted a lot of attention since the extremely intense X-rays was obtained from a synchrotron radiation (SR) source. Lithium with three electrons per atom (one conduction electron and two core electrons) is the most elementary metal available for both theoretical and experimental studies. Until now there have been a lot of works not only at ambient pressure but also at high pressure because its electronic state is approximated by free electron model (FEM) [2, 3]. In the present work we report the result of the measurement of the Compton profile of Li at high pressure and pressure dependence of the Fermi momentum by using SR.

2. Experimental methods

The high pressure was generated by using Bridgman type sintered diamond anvils having a face of 3 mm in diameter. A beryllium disk (0.5 mm in thickness) was used as a gasket because it is transparent to X-rays. Before using the gasket, it was annealed at 500° C for 3 h to get good ducitility. Compton scattering measurements were carried out by using monochromatized 59.34 keV X-ray beam from a multipole wiggler [4]. The angle to observe the Compton scattered X-rays was 90°. The scattered photons were detected by means of a solid state detector. Polycrystalline Li was used as a sample because the contribution from the core electrons to the Compton profile is small. The Li sample, which is easily oxidized in the air, was placed carefully in a small hole (0.5 mm in diameter) drilled in the center of the Be gasket without any pressure medium.

Paul G. Mezey and Beverly E. Robertson (eds.), Electron, Spin and Momentum Densities and Chemical Reactivity, 323–326

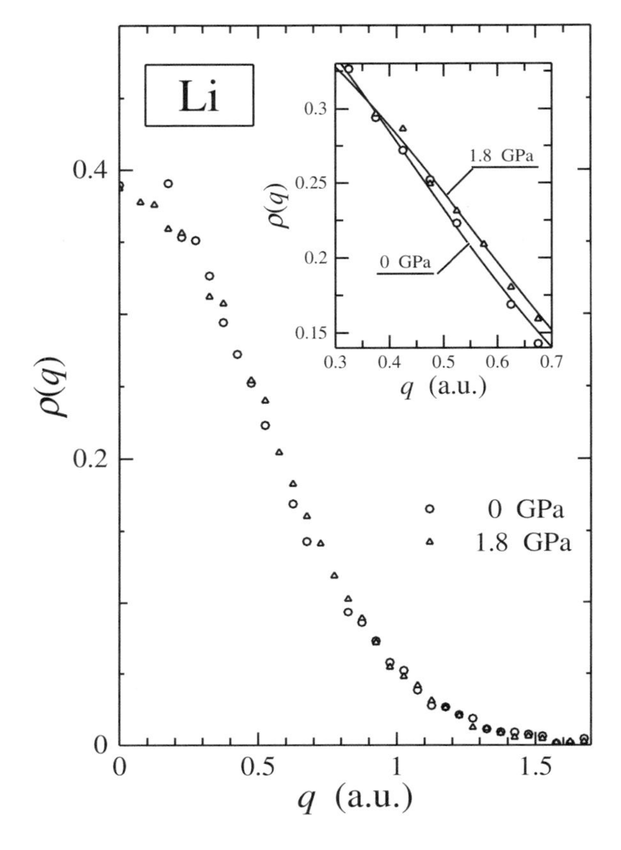

Figure 1. $\rho(q)$ of Li as a function of q at 0 and 1.8 GPa. The inset is the data in an extended scale. The solid lines show least squares fitting.

3. Results and discussion

The scattered photon intensity, $I_s(E, P)$, from Li at a pressure P was estimated in the following way where E is the photon energy. First the X-ray intensity at ambient pressure without the Li sample is observed (i.e.the background), $I_0(E)$. The background $I_0(E)$ is assumed to be independent of pressure. Second the intensity including the Li sample is observed, $I(E, P)$. It is assumed that $I_s(E, P)$ has the following form,

$$I_s(E, P) = I(E, P) - \alpha(P) I_0(E), \tag{1}$$

where α is an adjustable parameter depending on the pressure. The process to extract the Compton profile $J(q)$ from $I_s(E, P)$ has been described previously [5]. We analyze the observed, $J(q)$ on the basis of FEM. In the framework of the FEM,

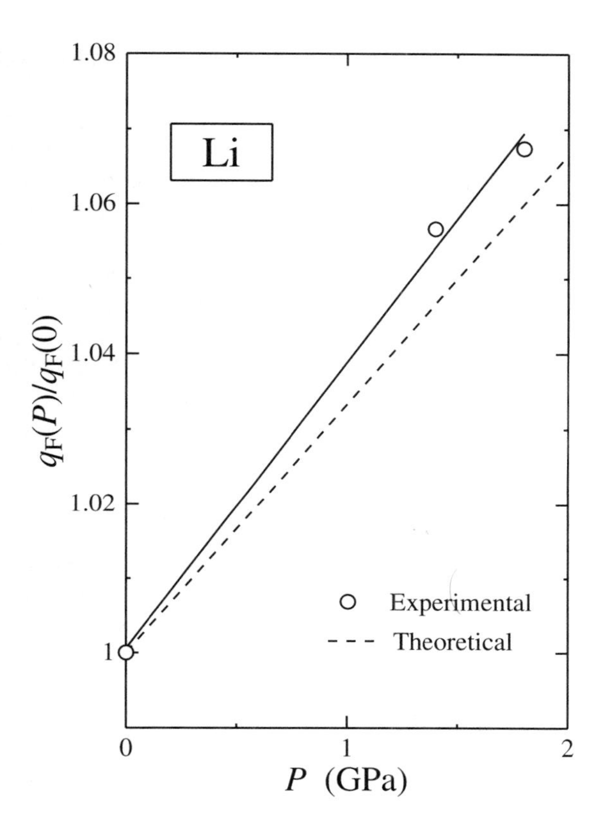

Figure 2. Relative change of q_F with pressure. The dotted line shows the result of FEM.

$J(q)$ can be described as

$$J(q) = \frac{3z}{4q_F^3}(q_F^2 - q^2) \tag{2}$$

Here z is the number of valence electrons per atom and q_F is the Fermi momentum given by

$$q_F = \left(\frac{3\pi^2 N}{V}\right)^{1/3}, \tag{3}$$

where V is the volume and N is the number of electrons. By differentiating Equation (2) with respect to q, we obtain

$$\rho(q) = -\frac{1}{q}\frac{\partial J(q)}{\partial q} = +\frac{3z}{2q_F^3} = \text{const.} \tag{4}$$

This result means that $\rho(q)$is constant in the range $-q_F \leq q \leq q_F$. At finite temperature, however, $\rho(q)$ has a finite width of $k_B T$ at q_F due to the Fermi distribution function. Figure 1 shows the $\rho(q)$ as a function of ρ both at 0 and 1.8 GPa. The inset indicates $\rho(q)$ near q_F in the extended scale. It is found that $\rho(q)$ at 1.8 GPa shifts towards right hand side compared with that at 0 GPa, which means that q_F at 1.8 GPa is larger than that at 0 GPa. q_F was determined as the value of q at the full width at half maximum. From this result, we obtained q_F as a function of pressure. Figure 2 shows the values of q_F as a function of pressure. q_F is found to increase with pressure having a coefficient, $q_F^{-1}\partial q_F/\partial P = 3.8 \times 10^{-2}\,\mathrm{GPa}^{-1}$. On the other hand by using Equation (3), we estimated the pressure change of q_F as $q_F^{-1}\partial q_F/\partial P = (1/3)\kappa$, where κ is the compressibility of Li. By using $\kappa = 0.1\,\mathrm{GPa}^{-1}$ for Li, the pressure coefficient is estimated to be $3.3 \times 10^{-2}\,\mathrm{GPa}^{-1}$, which is shown by dotted line in Figure 2. This result indicates that the pressure dependence estimated from FEM is in good agreement with that obtained in the present work.

References

1. Williams, B., (1997), Compton Scattering, New York, McGraw-Hill.
2. Phillips, W.C. and Weiss, R.J. (1968), *Phys.Rev.*, B, **171**, 790.
3. Oomi, G., Mohammed, M.A.K. and Woods, S.B., (1987), *Solid State Commun.*, **62**, 141.
4. Yamamoto, S., Kawata, H.,Kitamura, H., Ando, M., Sakai, N. and Shiotani, N. (1989), *Phys. Rev. Lett.*, **62**, 2672.
5. Itoh, F., Honda, T. and Suzuki, K. (1979) *J. Phys. Soc. Jpn.*, **47**, 122.

26

Magnetic Compton profile in uranium chalcogenide compounds UX (X = Se, Te)

F. ITOH[1], H. HASHIMOTO[1], H. SAKURAI[1], A. OCHIAI[2], H. AOKI[3] and T. SUZUKI[3]

[1] *Department of Electronic Engineering, Gunma University, Kiryu, Gunma 376-8515, Japan*
[2] *Department of Materials Science and Technology, Niigta University, Igarashi 8050, Niigata 950–2181, Japan*
[3] *Department of Physics, Tohoku University, Aoba, Aramaki, Sendai 980-8579, Japan*

1. Introduction

Magnetic Compton scattering is a well-established technique to investigate the momentum distribution of electrons with unpaired spins in ferro- and ferrimagnetic materials, using circularly polarized X-rays. There are some features inherent to the magnetic Compton scattering technique under the impulse approximation [1]. One is that the magnetic Compton profile (MCP) sees only 'spin' contribution to the magnetic moment, i.e. no orbital contribution can be reflected in MCP. Therefore, if the total magnetization is measured by some independent technique, one can separately obtain spin and orbital contributions to the magnetic moment by combining MCP with the traditional magnetization measurement [2]. The second is that the momentum distributions of different groups (3d, 4f, 5f, conduction-like electrons, etc.) have different characteristic MCPs, therefore one can deduce site-selective magnetic information even in alloys and compounds.

Uranium monochalcogenide compounds UX (X = S, Se, Te) undergo the ferromagnetic phase transition at the temperature $T_c = 180$, 174, and 104 K, respectively. The magnetic moment of UX increases with increasing atomic number of chalcogenide element. This is believed to come from the degree of localization of 5f electrons because the U–U separation inferred from the structural data increases from US to UTe. However, the magneto-optical properties do not show monotonic correspondence against the atomic number of chalcogenide element; i.e. Kerr rotation angle of US, USe and UTe are 2.6°, 3.3° and 3.1°, respectively [3], which suggest non-simple scheme of spin–orbit interaction between U and chalcogenide element. Therefore, it would be interesting to study the spin and orbital contribution of UX compounds separately by magnetic Compton scattering.

In this paper, we report MCP of USe and UTe which have been carried out at AR-NE1 station of KEK, Japan, and try to separate the spin and orbital contributions of magnetic moments by combining MCP with the magnetization measurement. Furthermore, we discuss the degree of localization of 5f electrons of these samples by decomposing the MCP into localized component and itinerant component.

Paul G. Mezey and Beverly E. Robertson (eds.), Electron, Spin and Momentum Densities and Chemical Reactivity, 327–331

2. Experiment

MCP experiments were performed at AR-NE1 station of KEK (National Laboratory of High Energy Physics), Japan, using circularly polarized X-rays with the incident X- ray energy of 60 keV emitted from the elliptical multipole wiggler. Figures 1 and 2 show MCPs of USe and UTe, which have been measured at 150 and 80 K, respectively.

Both MCPs are found to be negative, which shows the spin moments of these samples are aligned anti-parallel to the magnetic field applied as is already reported previously [4]. The MCP, $J_{\mathrm{mag}}(p_z)$, is decomposed into two components; one is high momentum component and the other is low momentum component. The former is well fitted in the range greater than 2 a.u. by a model profile $J^{5\mathrm{f}}_{\mathrm{mag}}(p_z)$ which is calculated from 5f Hartree–Fock wave function of uranium atom. This shows that the localized spin moment comes from 5f electrons of uranium atom. The latter is defined as the difference, $J_{\mathrm{mag}}(p_z) - J^{5\mathrm{f}}_{\mathrm{mag}}(p_z)$, and the low momentum component is found to be close to the profile expected from 6d Hartree–Fock wave function of uranium atom. This fact suggests that the narrow component, i.e. the diffused component, mainly consists of 6d electrons hybridized with s and p electrons. The absolute values of spin moment, $\mu_{\mathrm{S}}(5\mathrm{f})$, and diffused moment, $\mu_{\mathrm{S}}(\mathrm{diff})$, are calculated from the area under the curve, and are shown in Table 1 together with the neutron results in the parentheses for comparison.

The orbital contribution μ_{L} has been deduced by combining spin moments μ_{S} (all) by the MCP with the magnetization measurement M by an equation $\mu_{\mathrm{L}} = \mu_{\mathrm{S}}(\mathrm{all}) - M$.

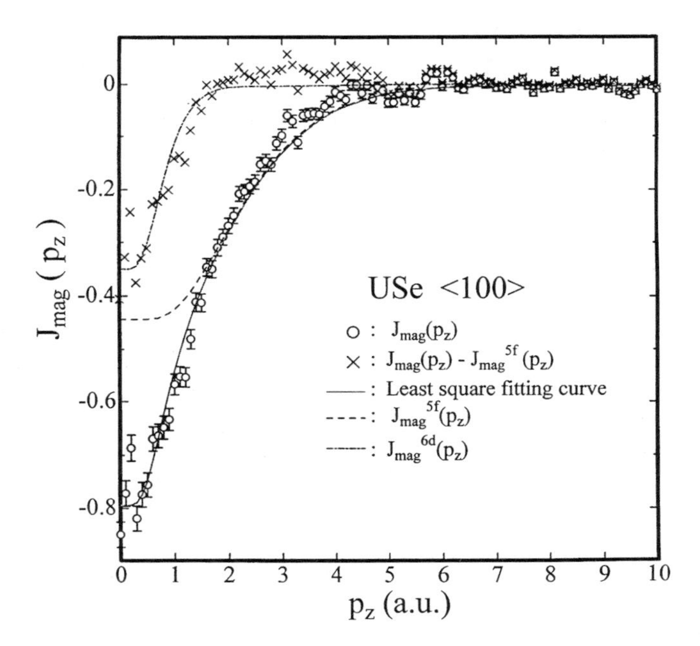

Figure 1. Magnetic Compton profile of USe at 150 K.

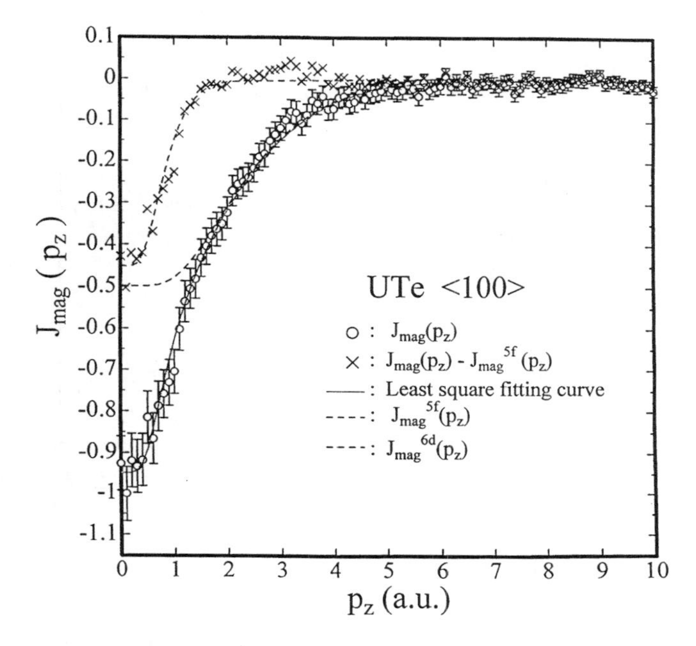

Figure 2. Magnetic Compton profile of UTe at 80 K.

Table 1. Separation of spin and orbital moments from magnetic scattering and magnetization measurement in USe and UTe compounds (in bohr magneton unit).

	M	μ_s(all)	μ_L (5f)	μ_s (5f)	μ_s (diff)	μ_{total} (5f)
USe	1.79	−1.78	3.57	−1.39	−0.39(−0.18)	2.18(2.0)
UTe	1.87	−1.70	3.57	−1.28	−0.42(−0.34)	2.29(2.25)

These are summarized together with previous results [4] in Table 1. It is seen from Table 1 that the total moment of each compound is controlled by the orbital moment. Although the magnetic moment increases with going down from X = Se to Te, the absolute value of spin moment decreases while that of orbital moment changes little. These results imply that some orbital quenching happens to occur due to f–d hybridization between U atoms and/or f–p hybridization between U atom and chalcogenide atom.

Figure 3 shows difference of MCP between USe and UTe,

$$\Delta J_{\text{mag}}(p_z) = J_{\text{mag}}(p_z; \text{USe}) - J_{\text{mag}}(p_z; \text{UTe}), \tag{1}$$

where each Compton profile is normalized to a same area. Two model Compton profiles are compared with experimental. One is the difference (a) when no hybridization effects are taken into account (dashed curve) and the other is the difference (b) when Hartree–Fock wave functions of 4s, 4p electrons from Se, and 5s, 5p electrons from

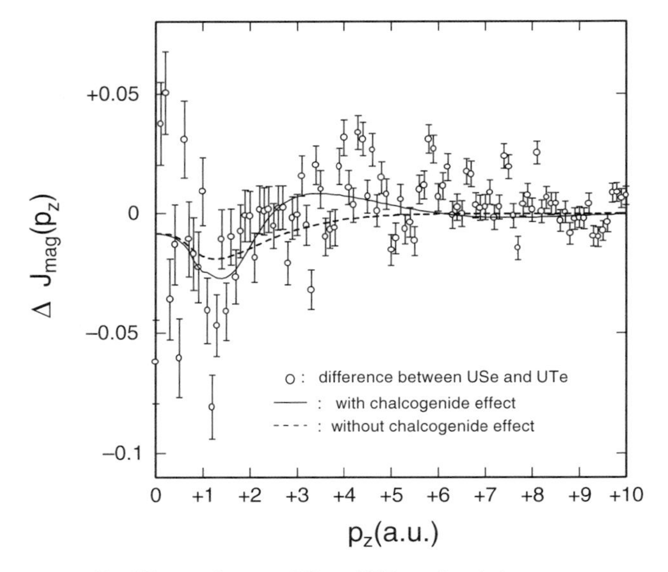

Figure 3. Compton profile difference between USe and UTe, $\Delta J_{\mathrm{mag}}(p_z)$.

Te are considered to participate in the hybridization bond in addition to 5f and 6d electrons from uranium element. As shown in Figure 3, there is better agreement between experiment and the model calculation (b). This suggests that there exist some hybridization effects between U and chalcogenide atom in UX compounds. Measurements of circular magnetic X-ray dichroism have been performed for UTe sample and magnetic polarization on $M_{4,5}$-absorption edges (d–f transition) of uranium atom and $L_{2,3}$-edges of Te (p–d transition) have been successfully observed [5]. The observation of magnetic dichroism at L-edges of Te atom strongly supports the existence of hybridization between 4p electrons of Te and 5f electrons of U in UTe compound.

3. Conclusion

The MCP measurements have been performed on USe and UTe monochalcogenide compounds at 150 and 80 K, respectively. The results are summarized as follows:

1. The spin moments of both USe and UTe are aligned anti-parallel to the magnetic field. That is to say, the whole magnetization is dominated by the orbital moments of these compounds.
2. The spin moments were decomposed into localized 5f component, $\mu_S(5f)$, and diffused components, $\mu_S(\mathrm{diff})$. Combining magnetization measurement with this decomposition, the orbital contribution , $\mu_L(5f)$, has been deduced .
3. The increase of the total magnetic moment from USe to UTe is the result of the decrease of spin moment from 5f electrons due to the stronger hybridization effects between U atom and Te atom.

References

1. Sakai, N., (1996) *J. Appl. Crys.*, **29**, 81.
2. Cooper, M.J., Zukowski, E. Collins, S.P., Timms, D.N., Itoh, F. and Sakurai, H. (1992) *J. Phy.: Condens. Matter*, **4**, L399.
3. Brooks, M.S.S. Johanson, B. and Skriver, H.L., *Handbook on the Physics and Chemistry of the Actinides*, Freeman, A.J. and Lander, G.H., (Eds.), North-Holland, Amsterdam, 1984, Vol. 1, p. 153.
4. Sakurai, H., Hashimoto, H., Ochiai, A., Suzuki, T., Ito, M. and Itoh, F., (1995) *J. Phys: Condens. Matter*, **7**, L599.
5. Hashimote, H., (1997) Doctor theses, Gunma University.

Understanding Chemical Reactivity

1. Z. Slanina: *Contemporary Theory of Chemical Isomerism.* 1986 ISBN 90-277-1707-9
2. G. Náray-Szabó, P.R. Surján, J.G. Angyán: *Applied Quantum Chemistry.* 1987 ISBN 90-277-1901-2
3. V.I. Minkin, L.P. Olekhnovich and Yu. A. Zhdanov: *Molecular Design of Tautomeric Compounds.* 1988 ISBN 90-277-2478-4
4. E.S. Kryachko and E.V. Ludeña: *Energy Density Functional Theory of Many-Electron Systems.* 1990 ISBN 0-7923-0641-4
5. P.G. Mezey (ed.): *New Developments in Molecular Chirality.* 1991 ISBN 0-7923-1021-7
6. F. Ruette (ed.): *Quantum Chemistry Approaches to Chemisorption and Heterogeneous Catalysis.* 1992 ISBN 0-7923-1543-X
7. J.D. Simon (ed.): *Ultrafast Dynamics of Chemical Systems.* 1994 ISBN 0-7923-2489-7
8. R. Tycko (ed.): *Nuclear Magnetic Resonance Probes of Molecular Dynamics.* 1994 ISBN 0-7923-2795-0
9. D. Bonchev and O. Mekenyan (eds.): *Graph Theoretical Approaches to Chemical Reactivity.* 1994 ISBN 0-7923-2837-X
10. R. Kapral and K. Showalter (eds.): *Chemical Waves and Patterns.* 1995 ISBN 0-7923-2837-X
11. P. Talkner and P. Hänggi (eds.): *New Trends in Kramers' Reaction Rate Theory.* 1995 ISBN 0-7923-2940-6
12. D. Ellis (ed.): *Density Functional Theory of Molecules, Clusters, and Solids.* 1995 ISBN 0-7923-3083-8
13. S.R. Langhoff (ed.): *Quantum Mechanical Electronic Structure Calculations with Chemical Accuracy.* 1995 ISBN 0-7923-3264-4
14. R. Carbó (ed.): *Molecular Similarity and Reactivity: From Quantum Chemical to Phenomenological Approaches.* 1995 ISBN 0-7923-3309-8
15. B.S. Freiser (ed.): *Organometallic Ion Chemistry.* 1996 ISBN 0-7923-3478-7
16. D. Heidrich (ed.): *The Reaction Path in Chemistry: Current Approaches and Perspectives.* 1995 ISBN 0-7923-3589-9
17. O. Tapia and J. Bertrán (eds.): *Solvent Effects and Chemical Reactivity.* 1996 ISBN 0-7923-3995-9
18. J.S. Shiner (ed.): *Entropy and Entropy Generation.* Fundamentals and Applications. 1996 ISBN 0-7923-4128-7
19. G. Náray-Szabó and A. Warshel (eds.): *Computational Approaches to Biochemical Reactivity.* 1997 ISBN 0-7923-4512-6
20. C. Sándorfy (ed.): *The Role of Rydberg States in Spectroscopy and Photochemistry.* Low and High Rydberg States. 1999 ISBN 0-7923-5533-4
21. P.G. Mezey and B.E. Robertson (eds.): *Electron, Spin and Momentum Densities and Chemical Reactivity.* 2000 ISBN 0-7923-6085-0

Kluwer Academic Publishers – Dordrecht / Boston / London